From Pole to Pole

Series Editors

Roland Kallenborn
Guido di Prisco
David Walton
Susan Barr

For further volumes:
http://www.springer.com/series/7132

Guido di Prisco · Cinzia Verde
Editors

Adaptation and Evolution in Marine Environments, Volume 1

The Impacts of Global Change on Biodiversity

Springer

Dr. Guido di Prisco
Institute of Protein Biochemistry
CNR,Via Pietro Castellino 111
80131 Napoli
Italy

Dr. Cinzia Verde
Institute of Protein Biochemistry
CNR,Via Pietro Castellino 111
80131 Napoli
Italy

ISSN 2193-7338
e-ISSN 2193-7346
ISBN 978-3-642-43102-9
ISBN 978-3-642-27352-0 (eBook)
DOI 10.1007/978-3-642-27352-0
Springer Heidelberg New York Dordrecht London

Printed on acid-free paper

Springer is part of Springer Science+Business Media (www.springer.com)

Preface

The Series "From Pole to Pole: Polar Environmental Research during the International Polar Year 2007–2009" (Springer Verlag; Series Editors: Roland Kallenborn, University of Life Sciences, Norway; Guido di Prisco, National Research Council, Italy; David Walton, British Antarctic Survey, UK; Susan Barr, Directorate for Cultural Heritage, Norway) was conceived to report achievements of environmental research during the 4th International Polar Year (IPY) 2007–2009.

The major aim of this series is to provide up-dated science-based information on IPY research results and perspectives in all disciplines. It is intended as a starting point to gather information on specific environmental research topics within IPY activities, which would be difficult or even impossible to collect in another way. The volumes will provide scientific general information on the concepts, findings and scientific motivation on the various IPY projects and research activities, thereby directing the interested reader to further information dealing with the scientific aspects of the research. The scientific value of the series will grow in the years to come, because the volumes will also be available in e-book format, and a continuous up-date on references and information sources is expected for several years, supported by the Series Editors and the Publisher.

Marine Biology will provide two Volumes under the general title "Adaptation and Evolution in Marine Environments—The Impacts of Global Change on Biodiversity".

The present volume (Volume 1) will address two themes:

1. Biodiversity and the Environment.
2. Response to Stress—Adaptations.

The authors who have contributed to Volume 1 describe the concept, aim, and first findings of the respective IPY projects, providing information, equipped with exhaustive reference lists and relevant web pages, on results and research perspectives feeding into the framework of IPY 2007–2009.

We would like to express our gratitude to the authors and referees of the papers collected together in this volume.

Volume 2 will address different themes, and will be published in the next months.

Guido di Prisco
Cinzia Verde

Letter from the Editorial Team

The first two International Polar Years both failed to coordinate and distribute their assembled data adequately and to ensure its proper analysis, resulting in a less than satisfactory legacy from what had been considerable international efforts. Recognising this, the Third International Polar Year (International Geophysical Year) made extensive plans to ensure its contributions would be both accessible and used, establishing the World Data Centres as a major new initiative. In the early preparatory stages of the latest International Polar Year (IPY 2007–2009) the importance of providing for the legacy of this demanding international research effort was made clear, with priority being given to planning for well-organised dissemination and coordinated publication of the results, data evaluations and scientific findings. It was with this in mind that we proposed our publication project (IPY Project No. 79) in the form of the book series 'From Pole to Pole: Environmental Research within the International Polar Year 2007–2009' with over 50,000 scientists involved in a myriad of projects, there was an obvious need for a guide to the principal findings and the key papers within environmental science fields.

The 'From Pole to Pole' book series is intended to serve as a comprehensive publication framework for the documentation of environmental research activities performed during the IPY period. The book series is not intended to be a typical collection of original scientific project publications/chapters in the form of standard monographs. It is rather a bibliographic, science-based information source and a starting point for interested scientists and the public to access condensed information on specific environmental research topics within the IPY activities. The volumes will provide scientifically sound general information on the concepts, findings and scientific motivation of the various relevant research activities and will direct the interested reader to more detailed scientific papers, web-based information and other publications which will provide the detailed data and their analyses. The compilation of citations and references within the book volumes will be an important component for the assessment of progress in each area, and the scientific significance and value will grow as the series develops.

The volumes will also be available in e-book format, which will allow continuous up-dating of references and information sources (including Internet pages and databases) by the editorial team on an annual basis, thus keeping the works topical as a living reference source.

The forthcoming volumes (11 volumes are currently planned) will cover an extensive spectrum of environmental research including adaptation and evolution, geomonitoring, geology, cryospheric processes, polar biodiversity, polar climates, the Arctic and Southern oceans, as well as pollution and atmospheric monitoring. It is expected that this documentation will provide a comprehensive picture of most of the environmental research performed within the IPY framework.

At the first official stock-taking of the IPY during the Oslo Science Conference (OSC, June 2010)—where the findings and the implications of the research were initially evaluated—it became very clear that the IPY endeavour as a whole had proved to be an unprecedented success for polar research. IPY efforts have contributed to a new and comprehensive understanding of global environmental processes in the cryo-, hydro-, bio-, geo-, atmos- and anthroposphere, from both a social and natural scientific perspective. What was also clear at this largest-ever polar science meeting was that it would require continued efforts to make sure that the results of the IPY research would be easily available and properly documented for future research and evaluation processes.

This book series aims to make an important contribution to that documentation process. The editorial team is not only looking forward to assisting in the development of those volumes already planned, but also invites colleagues and experts to propose other topics not yet covered as potential volumes in the series 'From Pole to Pole: Environmental Research within the International Polar Year 2007–2009'.

With this first volume on the history of the International Polar Years (edited by Susan Barr and Cornelia Lüdecke), our concept has finally begun to be realised. We congratulate the volume editors wholeheartedly for an excellent historical overview of the scientific work and implication of IPY research activities during the past, and look forward to working with the volume editors and Springer Verlag to publish the remaining titles in this new series.

Aas, Norway — Roland Kallenborn
Naples, Italy — Guido di Prisco
Cambridge, UK — David Walton
Oslo, Norway — Susan Barr

Editorial Introduction

Over the past 130 years, there have been four occasions for scientists of all disciplines from around the world to join forces in concerted, international cooperative activities for explorations and investigations in the polar regions. Each of these occasions was labelled as "International Polar Year" (IPY). Every IPY produced advancement in geographical exploration and scientific knowledge, extending the understanding of many phenomena that influence global systems in the planet and paving the way to political agreements among governments.

The first volume of this Series, "The History of the International Polar Years (IPYs)", edited by Susan Barr and Cornelia Lüdecke, is an excellent historical overview of the scientific work and implications of IPY research activity in the past. We felt, however, that a short historical summary of the previous IPYs would be helpful to the reader to place IPY 2007–2009 into context, therefore we begin this Introduction with such a summary, before describing the scientific frame and the common concepts throughout the IPY research presented herewith and attempting to identify historical and international contexts.

The idea of organising International Polar Years was inspired by Karl Weyprecht, Austrian explorer and naval officer. He was a scientist and co-commander of the Austro-Hungarian Polar Expedition of 1872–1874. Weyprecht's polar experience convinced him that solutions to the fundamental problems of meteorology and geophysics were likely to be found in the proximities of the poles, and that such investigations could not be tackled by a single nation but needed coordinated international efforts. His belief set a legacy for the future IPYs. Unfortunately, Weyprecht had no chance to see the acceptance of his concepts, because he died before the occurrence of IPY.

In 1879 the International Polar Commission was established, and it was agreed that an IPY (the first) would be held in 1882–1883, to coincide with a transit of Venus across the face of the sun (December 6th, 1882). Weyprecht's pioneering concepts set an important precedent for international science cooperation. The decision to cooperate rather than compete, and to focus on scientific endeavours rather than acquisition of territory, was an audacious approach that left a lasting example for future ventures. It saw the participation of 12 countries:

Austria–Hungary, Canada, Denmark, Finland, France, Germany, The Netherlands, Norway, Russia, Sweden, UK, USA. Fifteen expeditions took place, 13 to the Arctic, 2 to the Antarctic. The Jan Mayen Island, Alaska, Greenland, Tierra del Fuego, South Georgia, Novaja Semlja, Spitzbergen were amongst the expedition targets. Fourteen meteorological stations were set around the North Pole, and the observations (meteorology, geomagnetism, auroral phenomena, ocean currents, tides, structure and motion of ice and atmospheric electricity) were integrated by over 40 meteorological observatories around the world. The enormous amount of information formed the basis of our knowledge of the Earth's magnetic field and climate. The first IPY was mostly focussed on the past and present circumpolar Arctic environment.

The need to take advantage of the new inventions and discoveries to advance geophysical knowledge led the International Meteorological Organisation in 1927–1928 to pave the way for a second IPY, aimed at investigating global implications of the newly discovered "Jet Stream". It proposed magnetic, auroral and meteorological observations at a network of stations in the polar regions that would advance general knowledge of terrestrial magnetism, marine and aerial navigation, wireless telegraphy and weather forecasting. Thus, 50 years after the first IPY, the second IPY was organised in 1932–1933. Forty-six nations participated, and brought progress to transport by air, sea and land (advent of the airplane, motorised sea and land transport), meteorology, magnetism, atmospheric science, and "mapping" of ionospheric phenomena that advanced radio science and technology. Arctic research was supported by establishing 40 permanent stations, many of which are still active today. The second Byrd Antarctic expedition (USA) established a winter meteorological station, the first research station inland from the coast, 125 miles south of Little America Station on the Ross Ice Shelf. A world data centre was created and coordinated by the International Meteorological Organisation.

The first two IPYs suffered enormous logistic difficulties. Together with harsh privations, survival was a main concern, and tragic episodes did occur. These factors of course strongly limited the amount of time devoted to science.

In the 1950s, rocketry, radar and seismography were among the issues that inspired a third IPY. The International Council of Scientific Unions broadened the issue from polar studies in the upper atmosphere to geophysical research, renaming this effort "International Geophysical Year" (IGY). Sixty-seven countries and more than 70 national institutions participated in this cooperative venture. We can consider IGY, which took place from July 1957 to December 1958, as the third IPY, occurring 75 years after the first IPY and 25 years after the second IPY. It envisaged the peaceful use of newly developed (for military purposes) technologies, with the aim to achieve research advances, particularly in the upper atmosphere. Revision of many notions about the Earth's geophysics are due to IGY's discoveries and synoptic observations. For instance, the theory of continental drift was confirmed, enabling us to understand the formation of continents and oceans. The Van Allen Radiation Belt encircling the Earth was discovered by a US satellite. In fact, IGY saw the dawn of the space age with the launch of the first satellites (1957: Sputnik I,

USSR; 1958: Explorer I, USA). The first estimates of the size of Antarctica's ice sheet was obtained by traversing the continent for the first time. IGY led to strong development of research in many disciplines. The scientific, institutional, and political legacies of IGY lasted for decades, providing countless science achievements, and continue to the present. A notable political result founded on IGY was the ratification of the Antarctic Treaty in 1961, which established that the Antarctic continent would be dedicated to peaceful research.

During each of the three IPYs, scientists from all over the world came together to organise intensive scientific and exploration programmes in the polar regions, generating important advances in scientific and geographical knowledge. From laying the foundations of our understanding of nature's global systems to launching the modern space age, IPYs set the stage for many international scientific collaborations as well as a long-standing political accord.

Half a century after IGY, in 2007 the fourth IPY began (www.ipy.org). It was sponsored by the International Council for Science (ICSU) and the World Meteorological Organisation (WMO). The International Planning Group established within ICSU was co-chaired by Robin Bell and Chris Rapley. The Director of the IPY International Programme Office was David Carlson (see Volume 1 of the Series). IPY 2007–2009 has been the largest ever international programme of scientific research in the Arctic and Antarctic regions, building upon the long legacy, established in the previous IPYs, of international cooperation, scientific achievement and societal benefits.

The importance and complexity of the fourth IPY deserves adequate description of the outcomes flowing from such a vast international initiative. Two important fora have been organised to meet this target. One of them, the "IPY Oslo Science Conference", took place in 2010 in Norway. The second one, "IPY—From Knowledge to Action", will privilege the collective dissemination of the first scientific results and perspectives; this rendezvous is scheduled in 2012 in Montreal (Canada).

The Series of Springer Verlag books "From Pole to Pole: Polar Environmental Research during the International Polar Year 2007–2009" is a strong contribution for libraries of institutions of the whole world. It is ideally complementing the many articles on IPY research which are increasingly being published in scientific journals.

Marine Biology is providing its own contribution, summarising the achievements of this science area in two Volumes on "Adaptation and Evolution in Marine Environments—The Impacts of Global Change on Biodiversity". The present volume is the first one. The chapters describe investigations that pertain to IPY projects. All of them will undergo further developments in the decades to come. Field work, logistic challenges, and potential follow-up activities are included in each chapter, identifying linkages with investigations described in other chapters.

The ideas and concepts of each chapter are briefly outlined below.

The volume opens with an introductory overview describing the origin and future developments of the international, multi- and cross-disciplinary programme "Evolution and Biodiversity in the Antarctic—The Response of Life to Change"

(EBA). Launched by the Scientific Committee for Antarctic Research (SCAR) in 2004, EBA assembles almost one hundred teams and covers most of Antarctic biological research in the marine, terrestrial and freshwater realms.

SCAR (www.scar.org) is the major organisation coordinating research in the Antarctic and Southern Ocean (SO) region. With SCAR's support, EBA facilitates interdisciplinary interaction for the integrated approach required for unravelling the role of the polar environments to modulate the Earth system. By feeding information, EBA enhances SCAR's ability to address key issues raised within the Antarctic Treaty System. SCAR provides an opportunity to inform non-biological disciplines of the ultimate necessity of the programme, namely to contribute to understand the impact of Climate Change on Antarctic ecosystems. Most SCAR nations participate in EBA, that acts as a major route for capacity building in new SCAR members and those with reduced logistic and financial resources, and contributes to a wide variety of international programmes. EBA includes sub-Antarctic islands, inland to remote nunataks as well as northward to the Magallanes Strait, stretches across the SO down to the deep ocean as well as the shelves, and links with northern polar studies. The objectives are to understand the evolution and diversity of life in the Antarctic, to determine how these have influenced the properties and dynamics of present Antarctic and SO ecosystems, and to make predictions on how organisms and communities will respond to current and future environmental change.

The chapter describes the EBA Work Packages, each focussed on a specific area of science. A major marine focus has occurred during IPY. Antarctica is conventionally described as having very limited terrestrial biodiversity. This exists in the form of isolated "islands" of terrestrial habitat surrounded by inhospitable ocean or ice. These fragmentary habitats provide an ideal "evolutionary laboratory", allowing questions to be addressed on both relatively short (e.g. isolation of populations during the Pleistocene) and long (e.g. post-Gondwanan) evolutionary scales. This part will be illustrated in the Series volume addressing Terrestrial Biology.

An explicit aspect of EBA is to compare and where possible integrate results from the marine, terrestrial, and limnetic environments. The programme is interdisciplinary in that it brings together a wide range of biological disciplines to tackle a series of sharply focussed questions. It utilises state-of-the-art enabling technologies in molecular biology, ecophysiology, microbiology, taxonomy and organismal biology. It liaises with the relevant physical, geological and historical disciplines to ensure regular interaction and use of the most recent data and insights in interpreting the biological results. It involves fieldwork and laboratory work, both in the Antarctic and in home institutions. It requires extensive international collaboration. Exploration of some areas requires new technologies, for example benthic landers, remotely operated vehicles (ROVs) for the deep sea, autonomous underwater vehicles (AUVs) for work beneath ice shelves.

The timing of IPY has overlapped with that of EBA; the EBA and IPY activities were conceived in parallel, and the IPY Initial Outline Science Plan (April 2004) indicated the ability of EBA to provide a significant contribution to IPY. The research and projects are all under the umbrella of EBA, and cross-linkages will continue in the future. By undertaking a focussed initiative elucidating the spatial distribution of marine and terrestrial diversity, EBA is leaving a legacy of biodiversity information and the tools with which to explore it, which is a hallmark of an IPY programme. In fact, EBA has contributed substantially to IPY.

Following the EBA introductory chapter, the contents of this volume address two main themes, which are strictly connected and complementary to each other, namely Theme 1: Biodiversity and the Environment, and Theme 2: Response to Stress—Adaptations. Before listing the first group of chapters grouped in Theme 1 (Biodiversity and the Environment), a comment is pertinent. EBA has been a Lead Project of IPY, liaising with SCAR programmes in other fields. Some SCAR projects are integral parts of EBA, and have been described or mentioned in many of the Chapters. The most important ones deserve a brief description.

The first programme is the "Census of Antarctic Marine Life" (CAML; www.caml.aq), which was performed in 2004–2010 under the auspices of the "Census of Marine Life" (CoML). Polar regions experience greater rates of climate change than elsewhere on the planet. The faunas are uniquely adapted to their extreme environments, and may be vulnerable to shifts in climate. There is an urgent need to establish the state of these communities, and in particular their diversity, if we are to understand the impact of climate change. Current knowledge of Antarctica's marine biodiversity is patchy. We know more about the surface of the moon than we do about the sea floor. Almost nothing is known about the mesopelagic, bathy/abysso-pelagic and benthic fauna of the slopes and deep-sea abyssal plains, nor about the tiny organisms (bacteria, archaea, protists, viruses, nanoplankton) in the sea and other habitats, nor about the faunas associated with hydrothermal vents, cold seeps and seamounts. CAML was a 5 year project that, during IPY 2007–2009, focussed on the ice-bound oceans of Antarctica. The coincidence with IPY made CAML a once-in-a-lifetime opportunity to conduct a comprehensive study of the evolution and biology of a vast and fascinating region of the Earth. It comprised the part of CoML that deals with the SO. Its objective was to study the evolution of life in Antarctic waters to determine how this had influenced the diversity of the present biota, and to use these observations to predict how it might respond to future change. The project integrated knowledge across all regions, biomes, habitats and fields of study to strengthen our knowledge of ecosystem dynamics in this high-latitude, frozen ocean system. Only through a multi-scale level of investigation will a better understanding of the diversity and status of Antarctica's marine life be obtained. CAML's aims included: (1) undertaking a species inventory of the Antarctic slopes and abyssal plains; (2) undertaking an inventory of benthic fauna under disintegrating ice shelves; (3) undertaking an inventory of plankton, nekton and sea-ice associated biota at all levels of biological organisation from viruses to vertebrates; (4) assessing critical habitats for Antarctic top predators; (5) developing a coordinated network of interoperable databases for all Antarctic biodiversity data. It employed modern genomic techniques and contributed to the project Barcode of Life. It interacted with the Arctic Ocean Diversity project (ArcOD), drawing comparisons between the Arctic ocean and the SO. Reference to earlier "Discovery" voyages permitted assessment of faunal changes occurring over the past 60–70 years. CAML revealed many species new to science. Sampling sites will be revisited in the future for further comparisons. It is establishing a comprehensive Antarctic marine database. The essential element of CAML was its international structure, involving utilisation of ships of many nations. Young researchers had the opportunity to participate, at sea and in subsequent data analysis. Beginning in 2005, the Alfred P. Sloan Foundation (New York, USA) funded coordination activities for 5 years, in order to cover the IPY time frame. The importance of CAML has been referred to, and/or clearly appears in Chaps. 1, 2, 4–6, 8, 9, 11.

The second programme is the series of cruises named ICEFISH: "International Collaborative Expedition to collect and study Fish Indigenous to Sub-Antarctic Habitats". In a world experiencing global climate changes, loss of biodiversity and overfishing, the biotas of the Antarctic and the sub-Antarctic offer compelling natural laboratories for understanding the evolutionary impact of these processes. Since IGY (1957–1958), fish biologists from the Antarctic-Treaty nations have made impressive progress in the knowledge of the Antarctic ichthyofauna. However, research integration into the broader marine context has been limited, largely due to lack of access to sub-Antarctic fishes. The latter, in particular those of the dominant suborder Notothenioidei, are critical for a complete understanding of the evolution, population dynamics, eco-physiology and eco-biochemistry of their Antarctic relatives. The ICEFISH programme was designed to fill these critical gaps in our knowledge. Many of the authors of Chaps. 1, 4, 6, 8, 11 realised the importance of the initiative, and took advantage of the possibility of working on a large number of commonly unavailable species. Before the initiation of IPY, ICEFISH-2004 (www.icefish.neu.edu) was the first comprehensive international survey of the sub-Antarctic marine habitat of the South Atlantic sector, onboard the icebreaker R/V Nathaniel B. Palmer; fishing at Burwood Banks, Falkland Islands/Islas Malvinas, Shag Rock, South Georgia, South Sandwich Islands, Bouvetøya, and Tristan Da Cunha, at depths ranging from tide pools to the abyss (5,400 m). The aims were: (1) systematics and evolutionary studies to relate sub-Antarctic notothenioids to their Antarctic relatives through morphological, molecular and cytological analyses; (2) life-history strategies and population dynamics to characterise the composition, distribution, habitat preferences and diets of the sub-Antarctic species, and larval recruitment; (3) physiological, biochemical and molecular-biology studies of organ and tissue systems to analyse the evolutionary basis of the adaptations of high-Antarctic notothenioids relative to their ancestral stock; (4) genomic resources of sub-Antarctic notothenioids.

Because notothenioids occupy high trophic niches, they constitute an important sentinel taxon for monitoring the impact of climate change on loss of biodiversity and on community dynamics, and of depletion caused by marine fisheries in the SO. ICEFISH contributes to better understanding the effect of this impact by adding the essential contribution provided by the knowledge of the sub-Antarctic within the SO scenario. It also contributes to development of a baseline understanding of these ecosystems, one against which future changes in species distribution and survival may be evaluated judiciously. Sampling provided voucher specimens of sub-Antarctic fishes, deposited in museum collections around the world, as well as genomic resources for polar marine biologists.

The outputs of ICEFISH comprise training PhD students, media coverage, publications, congress proceedings, input to databases (e.g. Genebank), to CCAMLR (Convention on the Conservation of Antarctic Marine Living Resources), ANDEEP-SYSTCO (see below), interactions with other SCAR programmes: for example, it addresses a specific part (marine ichthyofauna) of the SCAR programmes EBA and CAML. ICEFISH is also important to undertake comparative studies on the biogeography, evolution, and adaptation of fishes thriving in the Antarctic—sub-Antarctic latitudinal gradient, and in the Arctic (being thus relevant to TEAM-Fish, see below, and Chap. 3). As an intermediate geographical system between the polar extremes, the sub-Antarctic and its marine fish fauna will provide vital information pertinent to a global synthesis of the characteristics of marine ecosystems.

The third programme (Chap. 2) is "ANtarctic benthic DEEP-sea biodiversity: colonisation history and recent community patterns—SYSTem COupling" (ANDEEP-SYSTCO), crucial for EBA and CAML (and CoML). It requires a far greater effort than what can be achieved by any single nation. It builds on international and interdisciplinary investigations to add an innovative aspect to polar biological research by involving scientists from atmospheric science, climatology, hydrography, planktology, physical oceanography, geophysics, geology, sedimentology, bathymetry, etc. to shed light on atmospheric-pelagic-benthic coupling processes, using innovative technology, e.g. modern satellites, very fine-meshed plankton samplers, novel sea-bed landers, ROVs, plankton suctors, etc., and to train a new generation of polar scientists. Important issues to address are: measurements of atmospheric parameters such as aerosols, ozone, reflectivity, UV irradiance, or volcanic activity (SO_2 will inform about the particle load of the atmosphere, and the magnitude of light penetration); influence of atmospheric processes on plankton in the water column, of the biogeochemistry of surface water on primary productivity, biomass and diversity of nanoplankton, vertical changes in the plankton community to abyssal depths; biology of abyssal key species; role of the bottom-nepheloid layer for recruitment (larvae) of benthic animals; influence of quantity and quality of food sinking through the water column on abyssal life; functional morphology and physiology of abyssal animals (measurements of ^{15}N and ^{13}C to estimate the trophic position of dominant pelagic and benthic animals, to determine carbon flow to the consumers); effects of sedimentology, biogeochemistry, and pore water on benthic life (palaeontology); sedimentation rates and processes over time (bathymetric mapping).

Chapter 2 also describes deep-sea isopod biodiversity in the SO. The SO deep sea does not bear any barriers isolating its fauna from adjacent deep-sea basins. Isolation between shelf and deep-sea faunas is reflected in the habitat faunal composition. The SO deep-sea fauna is the least studied, and its species richness, patterns of distribution, endemism and interesting faunal characteristics are outlined in this chapter, with the first attempts to explain the driving forces of the patterns, including coupling processes between the pelagic and benthic realms. In the framework of EBA-IPY, ANDEEP-SYSTCO conducted the first comprehensive survey of megafaunal, macrofaunal and meiofaunal deep-water communities in the Atlantic sector of the

SO. The programme addresses the processes responsible for the differences in biodiversity, investigating the ecology of dominant abyssal species, examining the functioning of abyssal communities, and trying to understand atmospheric-pelagic-benthic coupling processes and gain initial insights into the trophic structure of the SO deep sea. Based on current climate change, potential future scenarios are hypothesised, and the importance of biodiversity studies is emphasised for the establishment of a robust benchmark against which future faunal changes can be measured.

Turning to the Arctic, the history, current status and prospects of the EBA-IPY TUNU Programme (current acronym, TEAM-Fish: "TUNU-Programme: Euro-Arctic Marine Fishes—Diversity and Adaptation"), an ongoing international scientific effort that addresses the diversity of species, populations and communities in Arctic marine fishes across the Euro-Arctic region, is outlined in Chap. 3. TUNU (East Greenland in modern Greenlandic language), organised and managed by the University of Tromsø, studies the climate and the marine fish fauna of the North-East Greenland Fjord Systems. The diversity and distribution of fish species in NE Greenland is practically unknown. The warming trends reported for Arctic waters and—in particular—NE Greenland fjords make studies of the fish fauna and its response to climatic changes an unprecedented challenge for Arctic marine ecology. This is the scientific background of this multi-year programme, conducted with the ice-strengthened R/V Jan Mayen as operational base. Genetic and demographic structuring, trophic relationships and physiological adaptations are viewed on a broad evolutionary time scale and in the context of novel climate and human stressors. Baseline transects for long-term monitoring cover hundreds of stations in NE Greenland. A growing TUNU Collection (Bergen Museum) forms an exceptional reference for detailed taxonomic and phylogenetic studies of Euro-Arctic fishes. TUNU includes PhD students and scientists from the EU, USA, and Russia. The following three main goals make up the scientific framework of TUNU: (1) to conduct a zoogeographical mapping and quantify the marine fish fauna at selected sites along the NE Greenland coast, between 77°N (Danmarkshavn) and 70°N (Scoresby Sund), and from the innermost part of the fjords to the continental slope; (2) to gather basic hydrographical data—e.g. depth profiles of temperature, salinity, and density—at the sites; (3) to revisit and repeat investigations at key sites to obtain long-term data on possible interannual changes in fish-composition and hydrographical regimes. TUNU is interdisciplinary and multidisciplinary, in the sense that the distribution and diversity of fishes is closely linked to genetic (molecular genetics, cytogenetics), physiological (blood chemistry, metabolism) and hydrographical studies. TUNU investigates fjord systems in NE Greenland that are pristine from a scientific point of view. The sea-ice cover has been significantly reduced in the area during the last three decades and this makes NE Greenland an excellent site to study effects of a changing marine environment on the marine fauna. Many Arctic fish species are physiologically adapted to live within a very narrow thermal zone (from −1.8 to +1°C), and even slight increases in temperature (and concomitant reductions in salinity) are deemed to have profound effects on their composition (diversity) and spatial distribution.

The role of sea ice on the life history of *Pleuragramma antarcticum* is analysed in Chap. 4. The life of marine plants and animals is influenced by sea ice, which in polar fish has driven the evolution of biological responses that allows them to avoid freezing. The discovery that *P. antarcticum*, a fish playing a pivotal role in the coastal system, uses seasonal sea ice as nursery ground, stimulated research to understand its life cycle and the relationship with sea ice. Evidence from IPY activities highlights that such relationship is a major feature in its early life history and reproduction, calling for future work on predictions about impacts that changes in the sea-ice dynamics may have on the coastal Antarctic ecosystem.

Factors such as gene flow, mutation, genetic drift and selection affect the evolution of biodiversity (Chap. 5). The structure of the fish populations of the SO should be homogenised by the Antarctic Circumpolar Current. Some species do indeed show evidence for strong connectivity, with genotypes being shared across the full range. However, species-specific life-history traits influence the patterns of most taxa such that distinct populations are identified. In some cases, fishing and climate change impact the genetic structure in a measurable way. Hence, management measures are recommended. Quota systems have been implemented for some time, and marine protected areas are progressively being identified.

Molecular phylogeny has changed systematics, allowing better taxonomy in molecular trees. IPY has stimulated coordinated access to the Antarctic for large population samplings. Although not the only criterion, barcoding is modifying the flow chart of taxonomy, leaving routine identification to the barcode tool. Barcodes consist first in sequencing a gene of reference (for most vertebrates the mitochondrial gene encoding cytochrome oxidase I) for well identified specimens deposited in collections. Barcoding of Antarctic fishes is rapidly increasing. Identifications of notothenioid eggs and larvae are increasingly reliable. Taxonomy of notothenioids is therefore alive and well; the names of some genera should disappear to render their stem genus monophyletic. Taxonomy is facing deep changes in its interactions between traditional morphology-based taxonomic skills and DNA sequence-based approaches through "integrative taxonomy". Although dominant, notothenioids are not the only taxonomic component of the Antarctic fish diversity; they account for 90% of the biomass of the shelf and are the most studied Antarctic fishes, but there are also liparids and zoarcids. Gene amplification, sequencing and computing power are driving the rise of phylogenetics, providing criteria to know whether a monophyletic group of species is reliable. Each dataset has a source of potential errors with regard to species interrelationships; recovering the same clade from independent data is a strong indication of reliability. The phylogeny of notothenioids is now clear at the level of genera and at the interspecific level, except for the family Nototheniidae. All this is discussed in Chap. 6, together with the description of the 8 families of the suborder and their species interrelationships. Other important evolutionary issues (species flocks, dating of notothenioids, their origin) are also discussed.

Theme 2 (Response to Stress—Adaptations) assembles the second group of chapters.

Within IPY, the recognition of the role of the polar habitats in climate changes, that has awakened great interest in the evolutionary biology of polar organisms, is duly taken under consideration. The latter are exposed to strong environmental constraints, and it is important to understand how they have adapted to cope with these challenges and to what extent their refined adaptations may be upset by current climate changes. As in the previous theme, all contributions, although tackling a range of different organisms, geographical sites, adaptations (from molecules to organisms and communities), are clearly complementary.

In conjunction with evolutionary cold adaptation, the concept of "disaptation" is discussed in Chap. 7. Endemic Channichthyidae (icefishes) live permanently at or near freezing and are a paradigm of disaptation among adult vertebrates, because of their loss of hemoglobin and, in some species, myoglobin (see also Chap. 11). Therefore ice fishes, as natural "knockouts", permit to analyse the epigenetic compensatory mechanisms and their multilevel integration in this original phenotype. The functional significance of the cardio circulatory compensations (hypervolemia, near-zero hematocrit, low blood viscosity, large-bore capillaries, increased vascular diameter, cardiomegaly and large cardiac output, high blood flow with low systemic pressure and resistance) to face the challenge of hypoxia induced by the loss of respiratory pigments, are highlighted in this chapter. The phenotypic plasticity/vulnerability of this exceptional fish family may open new scenarios in environmental and evolutionary physiology.

A major question in polar biology is to understand whether vertebrate and invertebrate organisms have the genetic opportunity to adapt, and/or the physiological plasticity to tolerate new climate conditions. Identifying the key adaptations allowing survival is a complex task, as recalled in Chap. 8. In fact, "adaptation" itself is a slippery concept, but we will leave the debate on this provocative statement to other occasions. Undoubtedly one key adaptation is the fish ability to avoid freezing in seawater, constantly below the freezing point of the body fluids, by year-round biosynthesis of antifreeze compounds capable to block the growth of ice microcrystals inside the organism. This is particularly important for fish living in shallow water or around ice sheets and shelves where supercooled water generates showers of small ice crystals. Other adaptations are less clear-cut. The reduced enzyme activity at lower temperatures is a general problem, but presumably there are several ways to mitigate this effect. Cold denaturation is also likely to be problematic; again, experience suggests there are many ways to provide solutions. Going back to the introductory general question, rapid warming is under way in the Peninsula region, and is expected to happen soon in the rest of the continent. In this scenario, will Antarctic fish be able to cope with the current changes and survive? Although this is another question that is difficult to answer, there is some experimental evidence suggesting that a mere increase in temperature is in itself survivable. But any temperature change will be accompanied by important changes in ecological niches, that are likely to become the dominant factor. The latter changes may alter the habitat and affect interactions with other species. Presumably, their consequences are essentially imponderable.

There are very strong perspectives to witness important progress in developing new knowledge of Antarctic biology at the genomic level over the next decades. The rapid technological advances in genomic technologies have begun to be applied to Antarctic invertebrates (Chap. 9), but also to a wide array of marine organisms, including fish, urchins, starfish, nemerteans, amphypods, ciliates, krill, etc. These advances are providing further opportunities for advancing across several biological fields, as molecular biology is gradually being integrated as a standard tool in a wide array of applications, e.g. understanding basic adaptations, responses (e.g antioxidants) to change in sea temperature and acidity (e.g. effect on shells and exogenous skeleton), understanding gene flow between populations and evolution. These integrated approaches have supported a dramatic increase in efforts to identify novel genes and proteins, essential for life in this extreme environment. As in many of the chapters, the importance of "omics" is highlighted as a more and more essential tool to support studies needed to also determine whether life in Antarctica requires shared common survival strategies. Adaptations provide extra problems when faced with warming or acidification of the oceans, especially for slow physiological rates, slow growth and long developmental periods. The SO may be one of the first environments to see large-scale effects of climate change on the organisms living there, and more tools are now available to understand how its unique fauna will cope.

Undoubtedly, the SO and Antarctic continent pose extreme challenges for organisms, from bacteria to vertebrates, to survive there, under low temperature and high oxygen content, namely the main environmental stressors over million of years. Therefore, globins involved in oxygen storage and delivery (as well as in other putative functions) are bound to have an important physiological role. The last two chapters analyse the role of hemoprotein in the adaptive evolution of polar microorganisms and fish.

Since the sequencing of the human genome, completed in 2003, the speed of advance in DNA sequencing has increased dramatically, allowing genomic analysis of all polar life forms, especially microbes. In the past, the diversity of microorganisms inhabiting cold environments was mainly investigated in terms of distribution with no attention to their functional role in some important processes. Currently the "omics" methodologies offer new tools to investigate new biochemical pathways and to understand the evolutionary principles of adaptation and tolerance/resistance to extreme conditions. In Chap. 10, some examples on how cold temperatures may affect the physiology of microorganisms are reported. The focus of the chapter is on the molecular mechanisms revealed by recent biochemical and genetic studies that are shedding light on microbial adaptations to cold. A short overview on the biological function of oxygen-binding proteins in bacteria and their potential role in radical scavenging and cellular metabolism is provided, together with the awareness of the central role of oxygen and oxidative/nitrosative stress in regulating adaptive responses at cellular and molecular levels.

The evolutionary adaptations of the fish cardiovascular system have been vital for their survival (see also Chap. 7), and are just beginning to be understood at the molecular level. Following fragmentation of the supercontinent Gondwana, the biogeography scenario is a suitable background to the phyletically basal families of the dominant suborder Notothenioidei. These fish offer invaluable opportunities for investigating cold adaptation. In the Antarctic, the availability of phylogenetically related taxa in a wide range of latitudes has led comparative physiologists and biochemists to address the molecular bases of environmentally driven phenotypic gain and, conversely, loss of function. The function of hemoproteins has taught important lessons (Chap. 11). In the process of cold adaptation, the evolutionary trend of Antarctic and sub-Antarctic notothenioids (see also ICEFISH) has produced unique specialisations, including modification of hematological characteristics, e.g. decreased amounts and multiplicity of hemoglobin's. The genomic implications of the lack of hemoglobin in Channichthyidae, the notothenioid crown group, and the hemoglobin structure, function and phylogeny are analysed in detail. This work includes comparisons with the Arctic ichthyofauna. Such an integrative approach can provide answers to the question of how Antarctic and Arctic fish will respond, and whether they will be able to adjust, to ongoing Global Warming, already in full action in the polar regions. Adaptation and phylogeny (see also Chap. 6), in particular of the oxygen-transport system in fish, seem to be based on evolutionary changes involving levels of biological organisation higher than the hemoglobin structure.

In conclusion, the above notes underscore the importance of EBA, which acts as an umbrella to all cooperative and cross-disciplinary research efforts outlined in this volume. EBA is providing a long-term legacy to IPY 2007–2009, in particular for evolutionary and biodiversity information. It envisages links with the Arctic. EBA will end in 2013, and the EBA community is actively engaged in planning future developments. This large programme has direct relevance to Global Change, because it addresses the impacts of the latter on biodiversity, adaptations and community dynamics, providing useful information that can be extended also to temperate latitudes.

Guido di Prisco
Cinzia Verde

Contents

Contributors

Daniela Amelio Department of Pharmaco-Biology, University of Calabria, Arcavacata di Rende, 87030 Cosenza, Italy, e-mail: daniela.amelio@unical.it

Angelika Brandt Zoological Museum, Martin-Luther-King-Platz 3, 20146 Hamburg, Germany, e-mail: abrandt@uni-hamburg.de

Jørgen S. Christiansen Department of Arctic and Marine Biology, Fisheries and Economics, University of Tromsø, NO-9037 Tromsø, Norway, e-mail: jorgen.s.christiansen@uit.no

Melody S. Clark British Antarctic Survey, Natural Environment Research Council, High Cross, Madingley Road, Cambridge CB3 0ET, UK, e-mail: mscl@bas.ac.uk

Peter Convey British Antarctic Survey, Natural Environment Research Council, High Cross, Madingley Road, Cambridge CB3 0ET, UK, e-mail: pcon@bas.ac.uk

Guido di Prisco Institute of Protein Biochemistry, CNR, Via Pietro Castellino 111, 80131 Napoli, Italy, e-mail: g.diprisco@ibp.cnr.it

Filippo Garofalo Department of Pharmaco-Biology, University of Calabria, Arcavacata di Rende, 87030 Cosenza, Italy, e-mail: filigaro@libero.it

Laura Ghigliotti DipTeRis, University of Genoa, 16132 Genoa, Italy, e-mail: laura.ghigliotti@unige.it

Daniela Giordano Institute of Protein Biochemistry, CNR, Via Pietro Castellino 111, 80131 Napoli, Italy, e-mail: d.giordano@ibp.cnr.it

Philippe Koubbi Laboratoire d'Océanographie de Villefranche, Université Paris 06, UMR 7093, 06230 Villefranche-sur-mer, France, e-mail: koubbi@obs-vlfr.fr

Guillaume Lecointre Département Systématique et Evolution, Muséum National d'Histoire Naturelle, 57 rue Cuvier, 75231 Paris cedex 05, France, e-mail: lecointr@mnhn.fr

Craig Marshall Department of Biochemistry and Genetics Otago, University of Otago, PO Box 56, Dunedin, New Zealand, e-mail: craig.marshall@otago.ac.nz

Lloyd S. Peck British Antarctic Survey, Natural Environment Research Council, High Cross, Madingley Road, Cambridge CB3 0ET, UK, e-mail: lspe@bas.ac.uk

Daniela Pellegrino Departments of Cellular Biology, University of Calabria, Arcavacata di Rende, 87030 Cosenza, Italy, e-mail: danielapellegrino@unical.it

Eva Pisano DipTeRis, University of Genoa, 16132 Genoa, Italy, e-mail: pisano@unige.it

Jennifer Rock Department of Zoology, University of Otago, Dunedin 9054, New Zealand, e-mail: jennifer.rock@otago.ac.nz

Roberta Russo Institute of Protein Biochemistry, CNR, Via Pietro Castellino 111, 80131 Napoli, Italy, e-mail: r.russo@ibp.cnr.it

Bruno Tota Departments of Cellular Biology, University of Calabria, Arcavacata di Rende, 87030 Cosenza, Italy, e-mail: tota@unical.it

Marino Vacchi ISPRA, Rome and MNA, University of Genoa, 16132 Genoa, Italy, e-mail: m.vacchi@unige.it

Anton P. Van de Putte Belgian Biodiversity Platform, Royal Belgian Institute for Natural Sciences, Vautierstraat 27, 1000 Brussels, Belgium, e-mail: anton.vandeputte@bio.kuleuven.be

Cinzia Verde Institute of Protein Biochemistry, CNR, Via Pietro Castellino 111, 80131 Napoli, Italy, e-mail: c.verde@ibp.cnr.it

Filip A. M. Volckaert Laboratory of Animal Diversity and Systematics, Katholieke Universiteit Leuven, Charles Deberiotstraat 32, 3000 Leuven, Belgium, e-mail: filip.volckaert@bio.kuleuven.be

Part I
Introductory Overview

Chapter 1
The Origin of the SCAR Programme "Evolution and Biodiversity in the Antarctic"

Guido di Prisco and Peter Convey

Nothing in biology makes sense except in the light of evolution
Theodosius Dobzhansky (1973)

1.1 Evolution

Evolutionary biology covers almost all aspects of biology: "Evolution is the major unifying principle of biology, and evidence of evolutionary processes pervades all levels of biological organisation from molecules to ecosystems" (Eastman 2000).

Evolution is best studied where it can be seen without interference from confounding factors. The development of evolutionary theory was driven by observations in isolated island systems, such as the Galápagos Islands and the Indo-West Pacific archipelago. Antarctica is very rarely included among notable evolutionary sites, with the focus rather being on islands and locations such as Hawaii, Australia, Madagascar, the East African Great Lakes and Lake Baikal. Perhaps the reason that Antarctica has been under-appreciated in an evolutionary context is that much research carried out there has emphasised aspects of extreme biology rather than unifying principles of evolutionary biology. This approach has yielded valuable information about molecular and organ system function under extreme environmental stresses such as low temperature and desiccation, but has not focussed attention on aspects of biology that the Antarctic biota shares with biotas elsewhere. However, recent research on the Antarctic fauna and flora has provided powerful insights into evolutionary processes, as well as raising the visibility of Antarctic evolutionary biology.

G. di Prisco (✉)
Institute of Protein Biochemistry, National Research Council,
Naples, Italy
e-mail: g.diprisco@ibp.cnr.it

P. Convey
British Antarctic Survey, Cambridge, UK
e-mail: pcon@bas.ac.uk

G. di Prisco and C. Verde (eds.), *Adaptation and Evolution in Marine Environments, Volume 1*, From Pole to Pole, DOI: 10.1007/978-3-642-27352-0_1,

The exploration of remote habitats such as those of Antarctica has proved to be far more valuable than merely providing documentation of the existence of unusual faunas and floras. Discoveries made at the Galápagos archipelago stimulated Charles Darwin to begin erecting the framework for evolutionary thought. Darwin's voyage in HMS *Beagle* (1831–1836), and subsequent research by numerous scientists, has made the Galápagos the premier evolutionary site in the world. Darwin and HMS *Beagle* also visited the Falkland Islands and Tierra del Fuego and, at about the same time (1839–1843), James Clark Ross with HMS *Erebus* and *Terror*, and other explorers (Wilkes and Dumont d'Urville), were exploring the high latitudes of Antarctic coastal waters. During these voyages, many species were discovered, including an endemic tribe of four seals, six species of penguins and about a dozen species of fish of the dominant suborder Notothenioidei. However, a comprehensive picture of the fauna did not emerge. Antarctica remained unappreciated as a continental island with highly endemic marine and terrestrial biotas, and still has not gained wide recognition as a centre of evolution.

The isolation of such an extreme environment makes the Antarctic an important natural laboratory for evolutionary work. Its relatively well known tectonic and climatic history provides the context for evolution. Over geological time the combination of isolation and climate change has led to a biota rich in endemic taxa (Clarke and Johnston 2003; Pugh and Convey 2008; Griffiths et al. 2009; Convey et al. 2012). The apparently simple ecosystems on land contrast with diverse marine benthic systems on the continental shelves and in the Southern Ocean deep-sea (Gutt and Starmans 1998; Chown et al. 2000; Brandt et al. 2004a, b). It is important to seek an understanding of the reasons behind these striking differences in one of Earth's major biomes.

1.2 The "Polar Amplification"

Compared to temperate and tropical latitudes, contemporary climate warming is having stronger impacts in the Arctic, where sea-ice is predicted to disappear in summer in a few decades (Walsh and Timlin 2003; Johannessen and Miles 2000; Overpeck et al. 2006; Moline et al. 2008). Antarctica, which has a decisive role in driving the world's climatic and oceanic circulations, has not been spared. Contrasting trends are present in different parts of this vast region, but the western side of the Antarctic Peninsula is currently undergoing one of the fastest rates of warming on the planet (Turner et al. 2009a; Convey et al. 2009). It has recently been realised that the formation of the anthropogenic ozone hole in the latter decades of the twentieth century has been responsible for shielding the bulk of the Antarctic continent from the impacts of global warming (Turner et al. 2009b); as the damage to the global ozone layer is predicted to recover over the next century, considerable warming and linked environmental changes are predicted to impact much of the continent. Any large reduction of annual sea-ice causes displacement of key invertebrate and fish species of the trophic web, whose reproductive processes, closely associated to sea-ice, are upset (Moline et al. 2008). In both polar environments, changes have complex and interacting effects, and an impact on any level in a food web can propagate through to affect other taxa. For example, the progressive disappearance of sea-ice algae in the Arctic may eventually accelerate the extinction of the polar bear.

The predicted "polar amplification" of global anthropogenic warming is supported by evidence of acceleration of glacier retreat, sea-ice thinning and permafrost degradation. The vulnerability of species to recent and past climate change raises the possibility that human influence may cause a major extinction event in the near future, and detailed analyses have led to the alarming conclusion that many species could be driven to extinction over the next 50 years (Hoffmann and Sgrò 2011). Initiatives and debates are under way for the urgent development of new policies aimed at reducing and mitigating impacts of human activity. Global warming prompts scientists and governments to consider the risk of extinction of species inhabiting environments influenced by ice. Such an event may be averted only by resorting to concerted, multidisciplinary, international programmes aimed at understanding life processes, evolution and adaptations in the polar regions, and finally protecting polar life.

1.3 Biodiversity

The largest challenge facing mankind is the management of the Earth System to ensure a sustainable future. It is essential to understand the functioning of the Earth System in the context of natural and anthropogenic changes. Human health is closely connected to the health of the planet, hence to changes in the environment. The concept of a "sustainable world" is strongly linked to that of "biodiversity". Biological diversity, as defined by the Convention on Biological Diversity, encompasses 'the variability among living organisms from all sources including, inter alia, terrestrial, marine and other aquatic ecosystems and the ecological complexes of which they are part; this includes diversity within species, between species, and of ecosystems' (Anonymous 2009). All life becomes possible because biodiversity makes the planet what it is. Biodiversity is a complex concept depending on factors such as diversity of genes, individuals, species, habitats, as well as—above all—their interconnections and relationships (Pimm 2009).

The diversity of life can be regarded as one of the main outcomes of the evolutionary process. The large differences in genetic and taxonomic diversity among Antarctic habitats and broad systematic groups provide a unique opportunity to understand the mechanisms responsible for speciation and extinction, and diversity at the molecular, functional, organismal and population levels.

Contemporary climate changes will influence the biodiversity of marine, terrestrial and limnetic habitats alike. Every aspect of an organism's biology, from cellular biochemistry to food web and habitat, can be affected, and all organisms are vulnerable. The world is undergoing environmental changes at an unprecedented rate, due to the combined effects of natural variability and human activity. Global warming appears to be the master driver of changes (Rosenzweig et al. 2008).

Anthropogenic contributions, such as the accelerated input of greenhouse gases into the atmosphere, increased levels of harmful short-wavelength UV-B radiation due to ozone depletion, marine overexploitation and largely uncontrolled land-use change, are crucial in driving environmental change and/or greatly increasing its speed. Awareness of this is steadily spreading within the global population, despite attempts driven by commercial and political interests to disseminate scepticism. Present patterns of biodiversity and distribution are a consequence of factors and processes working on physiological, ecological and evolutionary timescales (Walther et al. 2002; Pörtner and Knust 2007; Pörtner and Farrell 2008; Peck 2011), which can be modified and driven by environmental changes. Even though warming is currently of greater magnitude in the air than in the sea, many typically stenothermal marine polar species may be particularly vulnerable to even small changes, as only a small window of temperature can be tolerated (Pörtner et al. 2007). The pressure to investigate biodiversity at high latitudes is increasing, and understanding the impact of past, current and predicted change on biodiversity and the consequences for ecosystem adaptation and function is a major research theme today.

1.4 EBA: Describe the Past, Understand the Present, Predict the Future

Evolution and Biodiversity in the Antarctic—The Response of Life to Change (EBA) is a multidisciplinary programme aimed at understanding life processes, evolution and adaptations in Antarctic marine and terrestrial environments. In 2004 the Scientific Committee on Antarctic Research (SCAR) launched this vast venture, involving over 600 scientists from virtually all SCAR member countries. Co-chaired by di Prisco (IBP-CNR, Italy) and Convey (British Antarctic Survey, United Kingdom), the programme commenced in 2006 and is planned to end in 2013. EBA inherited the tasks and legacy of previous successful SCAR marine and terrestrial biological initiatives, such as Biological Investigation of Marine Antarctic Systems and Stocks (BIOMASS), Biological Investigations of Terrestrial Antarctic Systems (BIOTAS), Ecology of the Antarctic Sea-Ice Zone (EASIZ), Evolution of Antarctic Organisms (EVOLANTA) and Regional Sensitivity to Climate Change in Antarctic Terrestrial Ecosystems (RiSCC). EBA was itself created by SCAR's decision to merge EVOLANTA and RiSCC in 2006, before their planned end. To illustrate these foundations of the SCAR programme EBA, we provide here a brief overview of some of these predecessor programmes.

1.4.1 Ecology of the Antarctic Sea-Ice Zone

The EASIZ programme ran from 1994 to 2004, and involved over 150 scientists from more than 17 countries. The object of the programme was to increase understanding of the structure and dynamics of the Antarctic coastal and shelf marine ecosystems. These are amongst the most complex and productive marine ecosystems both in the region and globally, and are considered to be the most susceptible to global and regional environmental variation. Because at that time water-column studies were served by existing international programmes, e.g. Southern Ocean-Joint Global Ocean Flux Study (SO-JGOFS), Southern Ocean-Global Ocean Ecosystem Dynamics Programme (SO-GLOBEC), and national oceanographic programmes, EASIZ concentrated on an integrated study of the water-column and benthos, and bentho-pelagic coupling. Particular attention was paid to those features that make the biology of this ice-dominated ecosystem so distinctive, and to understanding seasonal, inter-annual, and long term changes.

Work in the EASIZ programme led to the overturning of some previous paradigms (for example that the Antarctic marine system is species-poor) and replaced these with a revised picture linking the assemblage structure and population dynamics to the glacial-marine setting. The legacy of EASIZ is a wide-ranging reassessment of the diversity, history and ecology of the Antarctic benthos, its coupling to water-column processes, and a fundamental revision of our view of physiological adaptation to temperature in polar marine organisms (e.g. Arntz and Clarke 2002; Clarke et al. 2005, 2006; Clarke and Arntz 2006).

1.4.2 Evolution of Antarctic Organisms

The SCAR Workshop "Evolutionary Biology of Antarctic Organisms" in Curitiba, Brazil, 1999 (Rodhouse et al. 2000) and the subsequent meeting at Down House, Kent, United Kingdom (2000) identified three key areas of research:

1. *What is the relationship between gene flow in Antarctic species and circulation patterns in the atmosphere and ocean?*
2. *What are the effects of global climate changes and variability on evolutionary processes?*
3. *Do evolutionary processes in the Antarctic, now and in the past, differ from those in other parts of the world, including the Arctic?*

The notable lack of focus on the evolutionary biology of Antarctic organisms (an issue later taken up by EBA) was at the origin of this programme, which was approved by SCAR in 2000 and had an originally planned lifespan of 8 years. The broad aim was to provide a framework for research to improve understanding of the evolutionary history and biology of the unique Antarctic biota, and to integrate this with developing knowledge of the climatic and tectonic context within which this evolution has occurred and continues to occur. The key scientific questions concerned evolutionary response to global change, as well as involving Antarctic/Arctic comparisons. Important aspects included adaptive radiation and evolutionary history, gene flow, adaptation, life cycles, macro-evolution at the inter-species level, micro-evolution at the intra-species level, and biodiversity. The programme was planned at a time when there was an explosive development of molecular methods, driven by research on the human genome, which revolutionised biology and provided the tools to explore the function of individual genes. Molecular biology now has the potential to revolutionise the fields of evolutionary biology and ecology.

EVOLANTA was a largely marine programme partially running in parallel with EASIZ, but some terrestrial and limnetic studies were included, and research also had some aspects in common with the terrestrially-focused RiSCC programme (see below). The proceedings of the final EVOLANTA symposium (Eastman et al. 2004) included contributions on molecular phylogeny, genome dynamics, and protein structure and function, as well as on abundance, species diversity, and distribution patterns (Huiskes 2007). EVOLANTA stimulated research on evolution by providing a framework for studies across the range of evolutionary timescales, habitats and species diversity, embracing the possibility of laboratory and field-based investigations of natural and anthropogenic evolutionary trends. It facilitated links with SCAR and non-SCAR programmes that interface on evolutionary biology.

1.4.3 Regional Sensitivity to Climate Change in Antarctic Terrestrial and Limnetic Ecosystems

The creation of RiSCC was originally motivated by the Northern Hemisphere International Tundra Experiment (ITEX). The foundation of the RiSCC programme was to study changes and patterns in species diversity and organismal performance in terrestrial and limnetic environments around Antarctica, based around the premise that latitude and altitude could act as proxies when testing predictions of future climate change. RiSCC established an 'Antarctic Environmental Gradient' extending over 40° of latitude from sub-Antarctic Marion Island (47°S) to the continental Antarctic Transantarctic Mountains (87°S), encompassing the range of climatic zones present in the entire region. The primary goals of RiSCC were to (1) understand the likely response of Antarctic biotas to changing climates, and (2) contribute to the development of theory applying to interactions between climate change, indigenous and introduced species, and ecosystem functioning. RiSCC developed a strong international science network and acted as a catalyst for over 200 scientific publications, involving scientists from over 18 nations, and the coordination of several successful international collaborative expeditions.

The RiSCC programme ran for just over 5 years (2000–2005), before being absorbed into EBA. It terminated with the production of the volume 'Trends in Antarctic Terrestrial and Limnetic Ecosystems, Antarctica as a Global Indicator' (Bergstrom et al. 2006). This volume comprises contributions on biodiversity, colonisation processes, and biogeography, including chapters on genetic studies of the origin, diversity and evolution of Antarctic terrestrial organisms, all set in the context of regional and global trends of environmental change (Huiskes 2007).

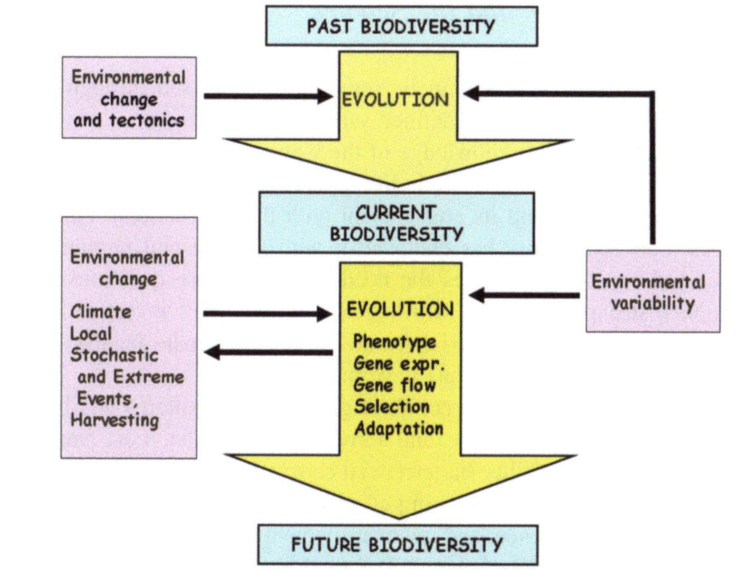

Fig. 1.1 Flow chart of evolution and biodiversity in the Antarctic (EBA)

1.5 The Role and Legacy of EBA

1.5.1 Historical Background

SCAR underwent a major structural change in 2002. The previously existing SCAR Working Groups were combined into three Standing Scientific Groups, in Life Sciences (SSG-LS), Physical Sciences (SSG-PS) and Geosciences (SSG-GS). A small number of Scientific Research Programmes (SRPs) was envisaged, with each SSG coordinating one major one. These SCAR SRPs effectively provide umbrellas or scientific fora in which Antarctic scientists, whose research is funded by their various national research agencies and mechanisms, can meet and communicate in order to more effectively coordinate and plan their research, and identify and achieve future and pressing research priorities.

EBA, illustrated in Fig. 1.1, began taking shape in 2004–2005. It is based on important aspects of the philosophy of EASIZ, and the two major then-existing predecessor programmes, EVOLANTA and RiSCC, were merged into it (Summerhayes 2011). Following considerable discussion, the SCAR Delegates approved EBA as the flagship SRP of the SSG-LS. As well as acting as an umbrella and coordinating body for international scientific collaboration and interaction, EBA assists in providing scientific advice to the Antarctic Treaty parties and the Commission for the Conservation of Antarctic Marine Living Resources (CCAMLR).

1.5.2 Work Packages and the Science

EBA's marine, terrestrial and limnetic research is addressed through five Work Packages (Table 1.1), each with defined sub-themes.

EBA aims at improving knowledge of the ways in which evolution has driven and is driving life, and of how life is likely to change. It is the only SCAR SRP having this major goal, and its engagement with the international Antarctic biological research community has endowed it with the relevant human resources, expertise, and capability. Besides the need to tackle these questions in all biological fields, strong integration with parallel SCAR SRPs investigating climate and tectonic history are an essential feature, because intimate feedbacks between the living and abiotic environments have modulated both.

Among the ecological factors controlling distribution patterns and biodiversity of the modern Antarctic marine and terrestrial biota, the most important are temperature, water availability (in terrestrial ecosystems), ice cover, oxygen, light, UV-B and wind. EBA is investigating all of these aspects, providing information on the resistance, resilience and recovery potential of cold-adapted Antarctic communities under current changes. The evolution of Antarctic organisms has taken millions of years, as the continent gradually cooled, to build cold adaptations suitable to allow survival and reproduction. Hence, the critical examination of the ability of cold-adapted organisms and communities to cope with changes is likely to provide a major contribution to the understanding of evolutionary processes that are of relevance to all life on Earth, including those at lower latitudes of more immediate importance to human populations.

As it separated from the other southern continents, Antarctica played a key role in altering ocean circulation and forcing the global climate toward cooling and glaciation. The Antarctic continental shelf offers several striking examples of faunal change and radiation, with fish and some invertebrate groups among the best studied. During the past 40 million years there has been a nearly complete replacement of the fish fauna on the Antarctic shelf, and a diverse, cosmopolitan temperate fauna from the late Eocene has been succeeded by the highly endemic, cold-adapted modern fauna. The present fish fauna of the continental shelves is dominated by the radiation of a single group of perciform fishes, the Notothenioids. This radiation is a rare example of a marine species flock (Eastman and Clarke 1998).

Antarctica is a continental-scale island about twice the size of Australia, with the dominant fauna inhabiting the surrounding sub-zero marine waters rather than the ice-covered landmass. As isolating conditions developed over the past 40 million years, the marine and terrestrial faunas became adapted to new shelf and terrestrial habitats and their ranges became highly circumscribed. Rates of species endemism reach 97% or even 100% in the case of some marine and terrestrial groups, though more typically being in the range of 30–50% (Pugh and Convey 2008; Griffiths et al. 2009; Convey et al. 2009, in press). Evolutionary biologists are drawn to these isolated habitats because of their unusual faunas. In this sense, the waters of the Antarctic shelf are comparable to, but less well known than,

Table 1.1 EBA work packages and sub-themes

WP1: Evolutionary history of Antarctic organisms	Vicariance and radiations: When did the key radiations of Antarctic taxa take place?
	Impact of glaciation on land (habitat modification/loss and timing and extent of isolation), and at sea (evolutionary links between continental shelf and slope or deep-sea species)
	Phylogeography: geographical structure and relationships in the Antarctic biome
	Evolutionary history of Antarctic microorganisms (both prokaryotic and eukaryotic)
WP2: Evolutionary adaptation to the Antarctic environment	Limits to organism performance: adaptation to the environment constraining physiol performance
	Physiological and genomic adaptations that allow organisms to survive: the extent to which these are special to the Antarctic or simply variants of general adaptations exhibited elsewhere
	Ability of Antarctic organisms to cope with daily, seasonal and longer-term environmental changes
	Behavioural and morphofunctional adaptations
	Adaptation and plasticity (genotype and phenotype)
WP3: Patterns of gene flow and consequences for population dynamics: isolation as a driving force	Population structure and dynamics in the context of evolutionary biology
	Natural and anthropogenic dispersal processes: immigration/emigration of organisms, intra-Antarctic dispersal; the role of advective/transport processes in gene flow and population structure
	Genetic structure: differences among and between Antarctic and non-Antarctic populations
	The extent to which populations of Antarctic organisms exist as metapopulations
WP4: Patterns and diversity of organisms, ecosystems and habitats in the Antarctic, and controlling processes	Spatial and temporal variations in diversity: variation of diversity at different spatial scales within the Antarctic and within defined time frames
	Response to latitudinal and environmental gradients: local, regional and global
	History of key evolutionary radiations
	Unknown areas: patterns of diversity and biotic composition of unexplored but important areas (e.g. deep sea, inland nunataks, subglacial lakes)
WP5: Impact of past, current and predicted future environmental change on biodiversity and ecosystem function	Interactions between introduced and indigenous species in selected environments undergoing change
	Effect of abiotic change on biota
	Modelling interactions between change and organism responses in order to predict biotic change
	Impact of biological feedback on climate

classic evolutionary sites mentioned above. Research on the Antarctic fauna will provide insights into micro-evolutionary and macro-evolutionary processes in a variety of habitats ranging from subglacial lakes to polar deserts and ice-covered shelf waters. The Antarctic fauna and flora also occupy a variety of extremes in the spectrum of habitats and environmental gradients where life can exist on Earth (Peck et al. 2006)—they provide a glimpse of the wide scope of adaptation and evolution in habitats once thought to be incompatible with life.

For many years a major role of genetics in population biology has been to clarify how populations, and species themselves, are structured. If two or more populations of a species become geographically separated, they are likely to diverge genetically through processes including random genetic drift, mutation and natural selection. *Gene flow* is the movement of genetic information among populations within a species, and exerts a major influence on the rate and pattern of evolution. The key roles that oceanic and atmospheric circulation patterns in the Antarctic have played will explain colonisation processes and population structure in relation to the physical environment.

For many organisms, levels of gene flow anticipated on the basis of life history characteristics are not always achieved. In the Southern Ocean it might appear there are few barriers to dispersal of planktonic or other marine animals, other than that presented by the Antarctic Circumpolar Current and in particular the Antarctic Polar Front (Clarke et al. 2005). However, it has been hypothesised that great depths (>4000 m) between continental shelves and/or oceanographic phenomena, such as confluences and gyres, may restrict or preclude dispersal of genes between geographically separated populations. Such depths probably constitute insuperable barriers for the adults, but pelagic larvae may or may not be able to cross, depending on length of larval life, distance to be crossed, current direction and oceanic barriers. The importance of the latter is becoming increasingly apparent (Convey et al. in press), although few explicit studies yet exist. One of the few studies of pelagic larval dispersal in the Antarctic (Allcock et al. 1997) shows that an oceanic barrier between Shag Rocks and South Georgia prevents transport of octopus larvae between these two geographically close areas. Spatial patterns of gene flow dictated by oceanic or atmospheric features will play a central role in defining population structure in many Antarctic species. Planktonic animals, e.g. salps, ctenophores and krill, are likely to only move by selecting currents at different depths, but this would suggest that, over an extended timescale, genetic mixing cannot be avoided. However, this is a broad generalisation which it is already clear does not apply to all potentially suitable species, and much more work is still required to clarify the importance of isolation and other factors on gene flow (e.g. see Hoffmann et al. 2011a, b). Work by Zane et al. (1998) demonstrated a genetic break between South Georgia and Weddell Sea krill populations, underscoring the potential importance of oceanic barriers to gene flow. Additional data are needed to understand the extent to which the oceanography of the Antarctic Ocean, coupled with the biology of individual species, may affect the genetic isolation of planktonic organisms. In an analogous fashion, it is becoming clear that long term isolation and barriers to gene flow have been important factors that have defined and continue to define the evolution of Antarctic terrestrial diversity (Convey and Stevens 2007; Convey et al. 2008). Most recently, strong indications are becoming apparent that this even applies to patterns of Antarctic microbial diversity, potentially overturning a long-standing hypothesis of microbial global ubiquity through the lack of apparent limitation to microbial dispersal (Vyverman et al. 2010).

Antarctic life has experienced cycles of global environmental change, driven by periodic glaciations. Regional-scale and short-term climatic variations seem more frequent and intense in recent years. All organisms, both terrestrial and marine, are susceptible to environmental changes, but small or non-motile organisms are particularly vulnerable. Our current understanding is that many species are susceptible to current climate change, with those of the marine environment being particularly vulnerable (see below). Organisms are thermally adapted for low temperature growth and have evolved physiological and biochemical mechanisms to cope with the cold. Stenothermal organisms intolerant to temperature changes may be influenced adversely by a 1–2° change in ambient temperature. Others have adapted to survive over a wide thermal range and may fit into phase-changing niches, with cycles of freeze-thaw over seasonal and geological time (Morley et al. 2009; Barnes et al. 2010; Somero 2010).

Over geological time Antarctic environmental conditions and available habitats have changed dramatically. The fossil record, stretching back over 500 million years, provides a broad outline of evolutionary history of the continent and its biota. The first signs of temperate biotas, marine and terrestrial, are present in the middle Devonian (c. 375 Mya). Some elements of the living marine biotas can be traced back to the Early Cretaceous period (130 Mya). During the Cretaceous, the highly seasonal, then warm, high-CO_2 climate and unstable landscape of high-latitude Gondwana may have been the centre of origin for gymnosperm and angiosperm taxa that subsequently spread northward, providing sources for temperate Southern Hemisphere floras. The earliest cold climate marine faunas are thought to date from the latest Eocene–Oligocene (c. 35 Mya). Conditions on land fluctuated greatly between cold and warm during the Tertiary (c. 65–2 Mya), and terrestrial biotas changed accordingly. Habitats now supporting terrestrial faunas and floras have been continuously available for lengths of time ranging from several million to only a few 1,000 years (Convey et al. 2008), while contemporary glacial retreat continues to expose new areas of ice-free ground for colonisation (Cook et al. 2005). The retreat of formerly tidewater glaciers to form beach-heads, and the loss of large areas of floating ice shelves, exposes new intertidal, subtidal and shelf area either to first colonisation or to the displacement of previous sub-ice-shelf communities, and links previously isolated areas of terrestrial habitat.

Many Antarctic habitats are characterised by environmental extremes which may be either constant or fluctuate over a variety of time scales (Peck et al. 2006). An understanding of the molecular, physiological and behavioural mechanisms by which Antarctic organisms are adapted to survive, grow and reproduce under such conditions will provide fundamental insight into evolution and the biological basis for adaptation in the polar environment, as well as differences and similarities among non-polar organisms. The *life cycle* of an organism is an evolved response to its environment, integrating a wide range of genomic, biochemical, physiological and ecological processes and factors. Understanding life cycles of organisms thriving under extreme conditions thus provides insight into which features of the environment have been important in the evolution of the Antarctic biota.

In a bipolar perspective, as far as species diversity is concerned, adaptive radiations of fish, isopods and amphipods are known from the Antarctic but not the Arctic. Physiological similarities and differences, especially related to cold adaptation, could be considered as aspects of both organismal and molecular diversity. Variation in the rates of micro-evolution in organisms from the two poles is worthy of study. Morphism may be a common evolutionary response in low-diversity boreal lakes and polar shelf habitats, and may entail genetic components. Molecular biology has shown that genes coding the antifreeze proteins of northern cods and southern notothenioids, with identical function, have different evolutionary histories (Cheng et al. 1997; Cheng and Chen 1999).

Modelling, with its long history of fundamental contribution to evolutionary biology, will be necessary to the achievement of EBA's aims. In the Antarctic, the importance of modelling is at least twofold. Firstly, in the application of existing models (for example for studies on life history evolution) to the particular conditions in the Antarctic. Secondly, there is the role of physical models and their integration with biological knowledge. EBA linked-programmes such as ICED (Integrating Climate and Ecosystem Dynamics in the Southern Ocean; www.iced.ac.uk; Murphy et al. 2008) provide a good example of the way in which this integrative and cross-disciplinary approach is being taken forward. Modelling is anticipated to contribute to studies on the evolution of life history, population genetics, speciation/extinction processes and oceanic and atmospheric circulation and other physical processes (Gutt et al. 2012). Other important tools are *workshops and symposia*, essential for the delivery of this programme. A *web site* (www.eba.aq) and a *Newsletter* are key to the coordination of EBA and the dissemination of information across its wide array of international participants.

Because of the fundamental importance of evolution in biology, EBA has links, and benefits from synergy, with *several other programmes and bodies* in the wider context, including:

- Commission for the Conservation of Antarctic Marine Living Organisms (CCAMLR; www.ccamlr.org)
- Antarctic Benthic Deep-Sea Biodiversity—System Coupling (ANDEEP-SYSTCO; www.andeep-systco.com)
- Microbiological and Ecological Responses to Global Environmental Changes in Polar Regions (MERGE)
- Subglacial Antarctic Lake Environments (SALE; www.sale.scar.org)
- International Collaborative Expedition to Collect and Study Fish Indigenous to Sub-Antarctic Habitats (ICEFISH; www.icefish.neu.edu)
- Latitudinal Gradient Programme (LGP, Victoria Land; www.lgp.aq)
- Long Term Ecological Research (LTER) Programme

- Census of Marine Life (CoML; www.coml.org), and Census of Antarctic Marine Life (CAML; www.caml.aq)
- Marine Fishes of North East Greenland (TUNU-MAFIG)

1.5.3 How Will EBA Develop?

EBA is a wide overarching umbrella, endorses many large international polar programmes, including acting as a focus for several shorter-term IPY programmes, and acts as a forum for discussion and a coordinating body for a large part of the Antarctic biological research community. As it approaches the end of its planned lifespan in 2013, it is now appropriate for EBA to help catalyse the development of future and more focussed SCAR science activities. Synergy is a long-lasting legacy, and we will continue fostering multi-national and cross-disciplinary initiatives by (1) encouraging integration across all biological disciplines; (2) upgrading synergy with the wider physical sciences community, including modelling, palaeoscience, geophysics, glaciology, oceanography and climatology, in order to establish firmer links between evolution and tectonics, climate evolution and glacial processes; (3) maintaining links with bodies such as the Intergovernmental Panel on Climate Change (IPCC), United Nations Environment Programme (UNEP) and the Committee for Environmental Protection (CEP) of the Antarctic Treaty System.

The EBA and wider biological community is currently developing the next generation of SCAR Life Sciences SRPs. In the most recent SCAR Meeting (Buenos Aires, Argentina, 2010), SCAR Delegates approved proposals to lead Antarctic biology towards two new SRPs focussed on distinct but complementary aspects of Polar Biology, each encompassing marine, freshwater and terrestrial organisms and ecosystems. These proposals now go through a 2-year planning and approval cycle, with the aim of being in place as EBA itself comes to its formal end.

1. *Antarctic Thresholds - Ecosystem Resilience and Adaptations (AnT-ERA)*— This proposal is strongly characterised by the application of 'omics' approaches, and will examine biological processes (intra-/inter-cellular → organism → community → ecosystem) to define their tolerance limits and determine resistance and resilience to change. It will address a range of key questions, including:

   - What are the genetic underpinnings to the life history, organism plasticity and physiological adaptations of polar organisms?
   - How do species traits impact community interactions and stability? Will invasive species have catastrophic impacts on these community interactions, and thus on ecosystem processes?
   - What are the likely consequences of a changing environment for ecosystem functioning and services?

2. *State of the Antarctic Ecosystem* (*AntEco*)—Biodiversity dictates how ecosystems function and underpins the life-support system of our planet. The aims of this programme are to focus on patterns of biodiversity within the Antarctic, sub-Antarctic and Southern Ocean regions, and to provide the scientific foundation for biodiversity conservation and management. Objectives: we propose to explain what biodiversity is there from the level of molecules (genetic diversity) through species to ecosystem diversity, how it got there, why it is there, what threatens it, and also provide recommendations for its management and conservation. Details include: modelling approaches to understand distribution, abundance and diversity of Antarctic biota, identification of patterns of biodiversity, biodiversity hotspots and areas with unique biodiversity, protection of unique biodiversity areas, assessing microevolutionary processes and the capacity for adaptation to climatic change impacts, identifying vulnerable groups.

Acknowledgements EBA is greater than the sum of its parts, and we firstly gratefully acknowledge the enthusiastic participation and contributions of the vast community of involved scientists. We specifically wish to gratefully acknowledge the late Edith Fanta, who (besides being active in CCAMLR, SCAR and IPY) gave outstanding, efficient and enthusiast contributions first to launching EVOLANTA and then EBA. EBA research also falls in the framework of the project CAREX (Coordination Action for Research Activities on Life in Extreme Environments), European Commission FP7 call ENV.2007.2.2.1.6. We are grateful to Prof. Andy Clarke for his helpful and constructive comments on the manuscript.

References

Allcock AL, Brierley S, Thorpe JP, Rodhouse PG (1997) Restricted gene flow and evolutionary divergence between geographically separated populations of the Antarctic octopus *Pareledone turqueti*. Mar Biol 129:97–102

Anonymous (2009) Convention on biological diversity. http://www.cbd.int/

Arntz WE, Clarke A (eds) (2002) Ecological studies in the Antarctic sea ice zone. In: Results of EASIZ midterm symposium. Springer, Berlin, p 277

Barnes DKA, Peck LS, Morley S (2010) Acute temperature sensitivity of Antarctic invertebrates determines colonisation potential, biogeography and resilience to environmental change. Global Change Biol 16:3164–3169

Bergstrom DM, Convey P, Huiskes AHL (eds) (2006) Trends in Antarctic terrestrial and limnetic ecosystems: Antarctica as a global indicator. Springer, Dordrecht

Brandt A, Brökeland W, Brix S, Malyutina M (2004a) Diversity of southern ocean deep-sea isopoda (Crustacea, Malacostraca)—a comparison with shelf data. Deep-Sea Res II Special ANDEEP 51(14–16):1753–1768

Brandt A, De Broyer C, Gooday AJ, Hilbig B, Thomson MRA (2004b) Introduction to ANDEEP (ANtarctic benthic DEEP-sea biodiversity: colonization history and recent community patterns) a tribute to Howard L. Sanders. Deep-Sea Res II Special ANDEEP 51(14–16): 1457–1465

Chen L, DeVries AL, Cheng C-HC (1997) Evolution of antifreeze glycoprotein gene from a trypsinogen gene in Antarctic notothenioid fish. Proc Natl Acad Sci U S A 94:3811–3816

Cheng C-HC, Chen L (1999) Evolution of an antifreeze glycoprotein. Nature 401:443–444

Chown SL, Gaston KJ, Gremmen NJM (2000) Including the Antarctic: insights for ecologists everywhere. In: Davison W, Howard-Williams C, Broady P (eds) Antarctic ecosystems: models for wider ecological understanding. New Zealand Natural Sciences, Christchurch, pp 1–15

Clarke A, Arntz WE (2006) An introduction to EASIZ (ecology of the Antarctic sea ice zone): an integrated programme of water column, benthos and bentho-pelagic coupling in the coastal environment of Antarctica. Deep Sea Res Part II 53:803–814

Clarke A, Barnes DKA, Hodgson DA (2005) How isolated is Antarctica? Trends Ecol Evol 20:1–3

Clarke A, Johnston NM (2003) Antarctic marine benthic diversity. Oceanogr Mar Biol Ann Rev 41:47–114

Clarke A, Arntz WE, Smith CR (eds) (2006) EASIZ: Ecology of the Antarctic sea ice zone. Deep-Sea Res Part II, 53: 803–1140

Convey P, Barnes DKA, Griffiths H, Grant S, Linse K, Thomas DN (2012) Chapter 15: Biogeography and regional classifications of Antarctica. In: Rogers AD, Johnston NM, Murphy E, Clarke A (eds) Antarctica: an extreme environment in a changing world. Blackwell, Oxford (in press)

Convey P, Stevens MI (2007) Antarctic biodiversity. Science 317:1877–1878

Convey P, Gibson J, Hillenbrand C-D, Hodgson DA, Pugh PJA, Smellie JL, Stevens MI (2008) Antarctic terrestrial life–challenging the history of the frozen continent? Biol Rev 83:103–117

Convey P, Stevens MI, Hodgson DA, Smellie JL, Hillenbrand C-D, Barnes DKA, Clarke A, Pugh PJA, Linse K, Cary SC (2009) Exploring biological constraints on the glacial history of Antarctica. Quatern Sci Rev 28:3035–3048

Cook A, Fox A, Vaughan D, Ferrigno J (2005) Retreating glacier fronts on the Antarctic Peninsula over the past half-century. Science 308:541–544

Dobzhansky T (1973) Nothing in biology makes sense except in the light of evolution. Am Biol Teach 35:125–129

Eastman JT (2000) Antarctic notothenioid fishes as subjects for research in evolutionary biology. Antarctic Sci 12:276–287

Eastman JT, Clarke A (1998) A comparison of adaptive radiations of Antarctic fish with those of non-antarctic fish. In: di Prisco G, Pisano E, Clarke A (eds) Fishes of Antarctica: a biological overview. Springer, Berlin, pp 3–26

Eastman JT, Gutt J, di Prisco G (eds) (2004) Adaptive evolution of Antarctic marine organisms. Antarctic Sci 16: 1–89

Gutt J, Starmans A (1998) Structure and biodiversity of megabenthos in the Weddell and Lazarev seas (Antarctic): ecological role of physical parameters and biological interactions. Polar Biol 20:229–247

Gutt J, Zurell D, Thomas J, Bracegirdle TJ, Cheung W, Clark MS, Convey P, Danis B, David B, De Broyer C, di Prisco G, Griffiths H, Laffont R, Peck LS, Pierrat B, Riddle MJ, Saucede T, Turner J, Verde C, Wang Z, Grimm V (2012) Correlative and dynamic species distribution modelling for ecological predictions in the Antarctic: a cross-disciplinary concept. Polar Res (in press)

Griffiths HJ, Barnes DKA, Linse K (2009) Towards a generalized biogeography of the Southern Ocean benthos. J Biogeogr 36:162–177

Hoffmann AA, Sgrò CM (2011) Climate change and evolutionary adaptation. Nature 470: 479–485

Hoffmann JI, Clarke A, Linse K, Peck LS (2011a) Effects of brooding and broadcasting reproductive modes on the population genetic structure of two Antarctic gastropod molluscs. Polar Biol 158:287–296

Hoffmann JI, Peck LS, Linse K, Clarke A (2011b) Strong population genetic structure in a broadcast-spawning Antarctic marine invertebrate. J Heredity 102:55–66

Huiskes A (2007) Evolution and biodiversity in the Antarctic: the response of life to change. Antarctic Sci 19:279–281

Johannessen OM, Miles MW (2000) Arctic sea ice and climate change–will the ice disappear in this century? Sci Prog 83:209–222

Moline MA, Karnovsky NJ, Brown Z, Divoky GJ, Frazer TK, Jacoby CA, Torres JJ, Fraser WR (2008) High latitude changes in ice dynamics and their impact on polar marine ecosystems. Ann NY Acad Sci 1134:267–319

Morley SA, Hirse T, Pörtner H-O, Peck LS (2009) Geographical variation in thermal tolerance within southern ocean marine ectotherms. Comp Biochem Physiol A 153:154–161

Murphy EJ, Cavanagh RD, Johnston NM, Reid K, Hofmann EE (eds) (2008) Integrating climate and ecosystem dynamics in the southern ocean (ICED): a circumpolar ecosystem programme. Science plan and implementation strategy. GLOBEC Report No. 26/IMBER Report No 2, GLOBEC International project office, Plymouth

Overpeck JT, Otto-Bliesner BL, Miller GH, Muhs DR, Alley RB, Kiehl JT (2006) Paleoclimatic evidence for future ice-sheet instability and rapid sea-level rise. Science 311:1747–1750

Peck LS (2011) Organisms and responses to environmental change. Mar Gen 4:237–243

Peck LS, Convey P, Barnes DKA (2006) Environmental constraints on life histories in Antarctic ecosystems: tempos, timings and predictability. Biol Rev 81:75–109

Pimm SL (2009) Climate disruption and biodiversity. Curr Biol 19:R595–R601

Pörtner H-O, Farrell AP (2008) Physiology and climate change. Science 322:690–692

Pörtner H-O, Knust R (2007) Climate change affects marine fishes through the oxygen limitation of thermal tolerance. Science 315:95–97

Pörtner H-O, Peck L, Somero GN (2007) Thermal limits and adaptation in marine Antarctic ectotherms: an integrative view. Philos Trans R Soc B: Biol Sci 362:2233–2258

Pugh PJA, Convey P (2008) Surviving out in the cold: Antarctic endemic invertebrates and their refugia. J Biogeog 35:2176–2186

Rodhouse PG, Fanta E, di Prisco G, Hureau J-C (eds) (2000) Evolutionary biology of Antarctic organisms. Antarctic Sci 12:257–393

Rosenzweig C, Karoly D, Vicarelli M, Neofotis P, Wu Q, Casassa G, Menzel A, Root TL, Estrella N, Seguin B, Tryjanowski P, Liu C, Rawlins S, Imeson A (2008) Attributing physical and biological impacts to anthropogenic climate change. Nature 453:353–357

Somero GN (2010) The physiology of climate change: how potentials for acclimatization and genetic adaptation will determine "winners" and "losers". J Exp Biol 213:912–920

Summerhayes CP (2011) A history of SCAR, 2004–2010. SCAR Occasional Publication, SCAR, Cambridge. ISBN 978-0-948277-26-9

Turner J, Bindschadler R, Convey P, di Prisco G, Fahrbach E, Gutt J, Hodgson DA, Mayewski PA, Summerhayes CP (eds) (2009a) Antarctic Climate Change and the Environment. Scientific Committee for Antarctic Research, Cambridge, UK. xi + 526. http://www.scar.org

Turner J, Comiso JC, Marshall GJ, Lachlan-Cope TA, Bracegirdle TJ, Maksym T, Meredith MP, Wang Z, Orr A (2009b) Non-annular atmospheric circulation change induced by stratospheric ozone depletion and its role in the recent increase of Antarctic sea ice extent. Geophys Res Lett 36:L08502. doi:10.1029/2009GL037524

Vyverman W, Verleyen E, Wilmotte A, Hodgson DA, Willem A, Peeters K, Van de Vijver B, De Wever A, Leliaert F, Sabbe K (2010) Evidence for widespread endemism among Antarcticmicro-organisms. Polar Sci 4:103–113

Walsh JE, Timlin MS (2003) Northern hemisphere sea-ice simulations by global climate models. Polar Res 22:75–82

Walther G-R, Post E, Convey P, Parmesan C, Menzel M, Beebee TJC, Fromentin J-M, Hoegh-Guldberg O, Bairlein F (2002) Ecological responses to recent climate change. Nature 416:389–395

Zane L, Ostellari L, Maccatrozzo L, Bargelloni L, Battaglia B, Patarnello T (1998) Molecular evidence for genetic subdivision of Antarctic krill (*Euphausia superba* DANA) populations. Proc R Soc London: Biol Sci 265:2387–2391

Part II
Theme 1: Biodiversity and the Environment

Chapter 2
Southern Ocean Deep-Sea Isopod Biodiversity Research: From Census to Ecosystem Functioning

Angelika Brandt

2.1 Introduction

The isolation of Antarctica makes this continent a perfect evolutionary laboratory for studies of marine biodiversity and biogeography. Attempts to describe and explain patterns of species diversity have become a major goal in biological research since the pioneering deep-sea investigations in the early 60s of the last century (Sanders et al. 1965; Sanders and Hessler 1969). Analyses of large-scale (global) patterns of deep-sea biodiversity include latitudinal gradients, apparent decreases in species richness among a number of taxa from the equator towards the poles (Poore and Wilson 1993; Rex et al. 1993, 1997). The Atlantic sector of the Southern Ocean (SO) does not seem to follow this pattern (Brandt et al. 2007b) although the deep seafloor of the SO remains the least studied even though it is the largest single benthic habitat (Clarke and Johnston 2003).

The Southern Ocean benthic marine shelf flora and fauna has been impacted over millions of years by plate tectonics and the resulting changes in the global climate. Tertiary palaeogeologic and -oceanographic changes (e.g. Brown et al. 2006; Zachos et al. 2008) successively generated the psychrosphere and the cold Antarctic Deep Water which is linked to the surface water in polar areas. These key events of cooling and glaciation followed by warmer periods shaped the evolution of many SO marine species. Following evolutionary extinctions some benthic marine invertebrate taxa radiated, others are remnants from the progressive retraction of cosmopolitan taxa established during the Jurassic and Cretaceous when Antarctica was still under greenhouse conditions, some disjunct distribution patterns resulted from vicariance events following the disintegration of

A. Brandt (✉)
Zoological Museum, Martin-Luther-King-Platz 3,
20146 Hamburg, Germany
e-mail: abrandt@uni-hamburg.de

G. di Prisco and C. Verde (eds.), *Adaptation and Evolution in Marine Environments, Volume 1*, From Pole to Pole, DOI: 10.1007/978-3-642-27352-0_2,

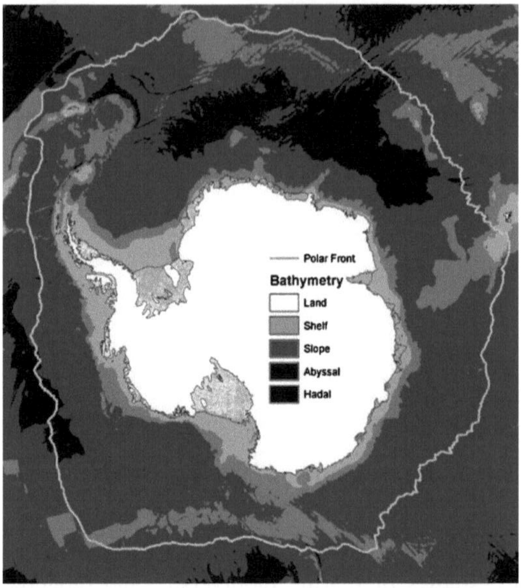

Fig. 2.1 Bathymetry of the southern ocean, showing shelf, slope, abyssal and hadal regions (Kaiser and Barnes 2008)

the supercontinent Gondwana. Thus, the present Southern Ocean fauna has changed in biodiversity and composition over geological time scales and species have actively migrated in and out of the SO (depending on dispersal capabilities) during glacial and interglacial periods.

Today the Southern Ocean is characterized by unique environmental features, including a very deep and partly narrow continental shelf, a weakly stratified almost isothermal water column and formation of abyssal waters flowing to other basins. Most of the SO seafloor (27.9 million km^2, Clarke and Johnston 2003) is deep sea (Fig. 2.1). Characteristic for Antarctica is that the shelves are much deeper than the average, 450–500 m, and can exceed 1,000 m in places; the Antarctic shelf comprises 17% of the world's total shelf area. The SO deep-sea faunas are related both to adjacent shelf communities and to those in other deep oceans, and both submergence and emergence processes can occur. Nowadays, natural and anthropogenically driven climate change processes shape the Southern Ocean marine fauna and despite our knowledge of the evolutionary historic developments, it is difficult to anticipate how these processes will drive speciation, extinction or the SO food web in future.

2.2 Programmes and Expeditions to the SO Deep Sea

Since the early exploratory phase of Antarctic research, and in the course of modern programs such as EPOS, EASIZ, ANDEEP, CAML, EBA, LGP (e.g. Dayton 1990; Arntz et al. 1997; Brandt 1991; Arntz and Clarke 2002; De Broyer et al. 2003;

Clarke and Johnston 2003; Clarke et al. 2006; Brandt et al. 2007a, b, c), benthic life of the SO has been a subject of great interest. However, none of the deep-sea biology studies in the Southern Ocean was devoted exclusively to the deep sea. Between 1950 and 1960, during Russian and U.S. expeditions of the RVs *Eltanin*, *Glacier*, Akademik *Kurchatov*, and Akademik *D. Mendeleiev*, deep-sea samples had been taken occasionally. The first deep-sea expedition with HMS *Challenger* sampled the Beagle Channel and several deep-sea stations.

Later, in 1994, the Magellan area was sampled again during the IBMANT programme (Interactions between the Magellan Region and the Antarctic, RV *Polarstern*) (Arntz and Rios 1999). During EPOS (European Polarstern Studies) and EASIZ (Ecology of the Antarctic Sea Ice Zone), some deep-sea data were obtained from the slope but rarely from abyssal sites. However, both programmes have been biologically outstanding and have provided a wealth of benthic data on species of all taxonomic groups and functional guilds of the high Antarctic Weddell Sea and the Antarctic Peninsula (Arntz and Gutt 1997; Arntz and Clarke 2002; Arntz and Brey 2003). Most information on the ecology of benthic deep-sea fauna in the Weddell Sea comes from a few stations sampled during EASIZ II in 1998 (Arntz and Clarke 2002). The ANDEEP (ANtarctic benthic DEEP-sea biodiversity: colonisation history and recent community patterns) expeditions took place in 2002–2005 in order to investigate this little known realm of the SO and provide first insights into the biodiversity and biogeography of the SO deep sea. ANDEEP contributed to the CoML (Census of Marine Life) projects CeDAMar (CEnsus of the Diversity of Abyssal Marine Life) and CAML (Census of the Antarctic Marine Life) and aimed to conduct the first comprehensive survey of megafaunal, macrofaunal and meiofaunal deep-water communities in the Scotia and Weddell Seas.

2.3 Patterns of Biodiversity in the Deep Sea

2.3.1 High Levels of Novel Biodiversity in the Southern Ocean

SO benthic shelf communities display high levels of endemism, gigantism, slow growth, longevity and late maturity. In the deep sea it still remains virtually impossible to obtain animals alive and undamaged from abyssal depths and thus knowledge of the deep-sea faunal composition, particularly in the SO, is still scarce in comparison with that from shelf and upper slope environments and we know almost nothing about the physiology, autecology or life histories of the SO deep-sea organisms. During the ANDEEP project the Scotia and Weddell seas were sampled (Brandt et al. 2004, 2007) and the geological and sedimentological backgrounds of this region were reviewed by Thomson (2004) and Howe et al. (2004). A novel approach of this project was sampling across a broad range of taxonomic groups. This was crucial because large-scale biodiversity and biogeography patterns largely depend on size, biology (feeding mode and reproductive patterns) and mobility of the taxa, their gene flow (compare also Rex et al. 2005), as well as geological history, productivity, predation and the relationship between regional and local species diversity (Witman et al. 2004).

The SO deep-sea communities are unique and highly diverse. The high percentage (often >90%) of new species in most taxa (Brandt et al. 2007a, b, c) and the high degree of endemism of many groups may reflect undersampling of the area. The high benthic diversity can probably be attributed to the constant, cold environment since about the last 34 m.y. The true deep-sea benthic fauna occurred at depths between 1,500 and 2,500 m depending on the taxon (Brandt et al. 2007c). There is a deep-sea affinity with the Antarctic shelf (Lipps and Hickman 1982; Brandt 1991; Brandt et al. 2007b), but some of the species can also be found in adjacent abyssal deep-sea basins (Brandt et al. 2004; Brandt et al. 2007a, b, c). This can be explained by repeated submergence and emergence processes during the evolution of invertebrate taxa and also by the eurybathy of the slope and deep-sea inhabitants (Brey et al. 1996). For example, Foraminifera, revealed close biogeographic links at the species level between deep Weddell Sea localities and those from similar depths in the North Atlantic and Arctic. The interesting biogeography of Arctic and Southern Ocean foraminiferan species might be due to their occurrence in the Weddell Sea, a major source of the world's deep water production, which could enhance the deposition of organic matter to the SO deep-ocean floor (Thomas and Gooday 1996; Brandt et al. 2007b). Isopod and polychaete slope assemblages included species that have invaded from the shelf. In sponges and molluscs the shelf and slope assemblages were more distinct. Abyssal faunas showed stronger links to other oceans, particularly the Atlantic, but mainly within good dispersers such as foraminiferans and polychaetes (Brandt et al. 2007b, c). Poor dispersers, like isopods, ostracods and nematodes are SO species rich and include many "apparently endemic" species. In summary, the ANDEEP data provide a valuable basis for exploring the evolutionary significance of the varied biogeographic patterns observed in this remote environment (Brandt and Hilbig 2004; Brandt et al. 2007b, c; Brandt and Ebbe 2007 and references therein).

The 40 ANDEEP stations based on RV *Polarstern* in the deep Weddell Sea and adjacent areas (748–6,348 m water depth) have revealed high species richness. For instance, 158 live species of Foraminifera were sampled. Within meiofaunal nematodes, 57 species of typical cosmopolitan deep-sea genera were sampled with more than 50% of new species, and more than 100 ostracod species were distinguished, >70% of them new (Brandt et al. 2007b). Macrofaunal isopods were the most diverse benthic invertebrate taxon investigated with 674 species identified and 585 of these species being new to science. Prior to the ANDEEP project 371 isopod species were reported from Antarctica (Brandt et al. 2007c). From 200 polychaete species sampled only 81 were previously known (Schüller and Ebbe 2007). The ANDEEP stations yielded 160 species of shelled gastropods and bivalves; Compared to a total of 279 species known from the shelf (<1,000 m) based on more than a century of Southern Ocean research, numbers of molluscs have also increased remarkably (Brandt et al. 2007b). Megafaunal sponges recognized comprised 76 species, 37 of these were new for the SO and 17 new to science (Brandt et al. 2007b; Janussen and Tendal 2007). Species richness is highest for the peracarid crustaceans, possibly because of the adaptive radiation of several species groups following the extinction of the decapod crustaceans. Some of the other taxa are also divers in other deep-sea areas. On a regional scale, diversity patterns vary strongly between major taxonomic clades (Clarke and Crame 2010).

Bathymetric and biogeographic trends varied among taxa depending on the mobility of the organisms and their reproductive mode. Particularly isopods are distinctive with many species presently unknown outside the SO (−90%). Among the most vagile janiroidean family Munnopsidae > 95% of the ANDEEP species are undescribed. This family comprised 50% of all isopods sampled during ANDEEP at 40 stations. Other important SO isopod families were the Desmosomatidae, the Haploniscidae and the Ischnomesidae. While we know that some species complexes have radiated in the deep SO (e.g. the Haploniscidae), it is unclear whether they have evolved here (and what the drivers of their evolution or potential radiation in the deep sea are) and subsequently spread into other ocean basins. The few known SO deep-sea isopod species show closest biogeographical links to Atlantic faunas. Many ostracod crustaceans are presently unknown outside the SO (Hartmann 1997), e.g. the ostracod family Macrocyprididae was common in the SO material, but usually rare in deep-sea samples of other oceans. The majority (~75%) of mollusc species were also unknown outside the SO, and wide-ranging Atlantic deep-sea species, such as the gastropod *Benthonella tenella,* were not collected (Brandt et al. 2007b). Polychaetes of the families Spionidae, Paraonidae, and Cirratulidae are usually common and species-rich in deep-sea areas of temperate latitudes, but only represented by few species in the SO deep sea. Many nematode species are new in the SO and apparently confined to particular parts of the Weddell Sea (e.g. *Microlaimus*), although some have wider distributions. These biogeographic patterns may be linked to larval ecology. Poor dispersers such as isopods, ostracods and nematodes have a reduced gene flow with restricted species distributions and higher endemism (Brandt et al. 2007b).

Molecular biological studies revealed cryptic species in certain circumantarctic serolid isopods (Held 2003; Raupach et al. 2009) and the bivalve *Lissarca notocardensis* (Linse et al. 2007). On the one hand, the existence of such 'species flocks' in the deep-sea biota may be a more general feature than is currently assumed (e.g. Raupach et al. 2007), on the other, genetic analyses have also revealed the existence of true cosmopolitan species in some planktotrophic taxa (e.g. in polychaetes) and also peracarid brooders (Held pers. communication). The widely distributed SO isopod shelf "species" *Ceratoserolis trilobitoides* (Serolidae), and *Glyptonotus antarcticus* (Valvifera) have been shown to represent complexes of cryptic species (Held 2003). The same proved true for *Betamorpha fusiformis* (Munnopsidae) (Raupach et al. 2007) and *Acanthaspidia drygalskii* (Acanthaspidiidae) (Raupach and Wägele 2006). Other species probably belong to paraphyletic genera such as *Eurycope* (Wilson and Hessler 1987). *Serolis paradoxa* from Patagonia and the Falkland Islands do not show effective gene flow, suggesting that these are also two cryptic species, but *Septemserolis septemcarinata* occurring at different sub-Antarctic islands shows different patterns (Leese et al. 2008; Leese pers. comm.).

Compared to other deep-sea areas, the isopod family Munnopsidae is the most speciose in the SO with 219 species from 3 expeditions and only 40 stations. In the North Pacific or Atlantic deep sea, roughly 50 munnopsid species have been recognized after more than 100 years of deep-sea research (Malyutina and Brandt 2007). It is not known to date which processes drive speciation of this family in the SO deep sea. Stable isotope analyses have revealed that species of Munnopsidae can have completely different diets (Würzberg et al. 2011a, b). Niche partitioning by using different food items could support speciation of the Munnopsidae and might have lead to adaptive radiation of this family; one of their prey items are foraminiferans which are available in high quantity and diversity. Maybe this isopod family radiated in the SO deep sea (due to resource partitioning or dietary specialization) like the Darwin finches on the Galapagos islands?

2.4 Understanding Processes in the Southern Ocean Deep Sea

2.4.1 SYSTCO: From Census to Ecosystem Functioning

While the ANDEEP project has revealed patterns of biodiversity within different faunal groups and documented that these can vary significantly (Brandt et al. 2007), we still know very little about the ecology and the role of deep-sea faunas for trophodynamic coupling and nutrient cycling in oceanic ecosystems. It is still unknown why only few species occur at many stations and with higher abundances, but most species are rare. To fill this knowledge gap, a successor to the ANDEEP project, ANDEEP-SYSTCO (SYSTem COupling) has been started in the Atlantic sector of the Southern Ocean within the framework of the International Polar Year (IPY) with a first expedition ANT XXIV-2 staged from board of RV *Polarstern* between 28.11.2008 and 4.2.2009. This new project seeks to find answers for the questions posed from the biodiversity and biogeography patterns observed during the ANDEEP campaigns. ANDEEP-SYSTCO addresses the processes responsible for the strong differences in biodiversity within and between taxa as well as between areas. SYSTCO aims to investigate the functional biodiversity and the ecology of dominant abyssal species and examine the trophic structure and functioning of abyssal communities of the Atlantic sector of Southern Ocean, focusing on the role and feeding of the abundant key organisms. In the SYSTCO project scientists from a wide variety of disciplines collaborated in the SO in order to try to understand atmospheric-pelagic-benthic coupling processes and gain initial insights into the trophic structure of the SO deep-sea. A second SYSTCO II expedition is planed for 2012 in order to elaborate the first results into a better understanding of the processes and food-web structure.

During the SYSTCO I expedition one station in the polar front at 52°S at about 3,000 m depth was revisited after six weeks and sediment oxygen consumption measurements showed higher values after a phytoplankton bloom. Veith-Köhler et al. (2011) argue that this has been attributed to an enhanced respiratory activity of the living benthic component. The authors assume that low temperatures and ecological strategies are the most important factors for the delayed response of benthic deep-sea copepods. It is worth mentioning that neither meiofauna nor bacteria responded with an increase in individual numbers to the food input from the water column.

Using a Surface and Under Ice Trawl, Flores et al. (2011) showed that in the Austral summer macrozooplankton biomass was dominated by ctenophores in open water and by Antarctic krill under the ice. These authors also emphasize the potential of a number of macro zooplankton and micronecton species to act as energy transmitters between productive sea-ice biota and the pelagic food web.

At Maud Rise (MR), the benthic fauna was investigated with reference to oceanographic features, biogeochemical properties and sediment characteristics, as well as the pelagic, bentho-pelagic and air-breathing fauna. The composition of the deep-sea fauna differed distinctly from surrounding deep-sea basins investigated during previous SO expeditions (ANDEEP) and the overall similarity between MR and adjacent stations was low. The taxon composition was characterised by extremely high abundances of *Vesicomya* spp. (Bivalvia), *Onoba subantarctica wilkesiana* (Gastropoda) and *Thylakogaster* spp. (Isopoda, Haplomunnidae). Members of the bivalve genus *Vesicomya*, characterized by non-reduced guts and no symbiotic chemoautotroph bacteria, have been reported (Krylova and Sahling 2010). Tube dwelling polychaetes occurring at Maud Rise were also not found at the comparison stations. Water-column sampling from the surface to the seafloor, including observations of top predators, indicated the existence of a prospering pelagic food web and local concentrations of top predators and zooplankton were associated with a rich ice-edge blooms located over the northern slope of MR, where the melting of the sea ice might be accelerated by the advection of warm water at intermediate depth. South of Maud Rise, high concentrations of *Euphausia superba* occurred under dense sea ice and attracted *Balaenoptera bonaerensis* and several seabird species. The biological prosperity over MR is likely related to oceanographic as well as sea-ice processes. Downward transport of the organic matter produced in the pelagic realm may be more constant than elsewhere due to low lateral drift over MR.

Investigations of the SO deep-sea food web performed using fatty-acid patterns of peracarid crustaceans revealed that some species feed on a wide variety of different food items including especially foraminiferans (Würzberg et al. 2011a, b), a potential reason why Munnopsidae (Isopoda) are so successful in the SO. Foraminiferans have also been found to play an important role in the diets of tanaidaceans, whereas the amphipods analysed seem to be carnivorous. In general, peracarid crustaceans fall into three dietary groups which are: Mainly phytodetritivore, mainly omnivore (with indications for foraminiferivory), or mainly carnivore. Antarctic demersal fish feed to a high percentage (50–80%) on amphipods, but also other fish, other crustaceans and gastropods (Würzberg et al. 2011a, b). Contrary to peracarid crustaceans, fatty acid analysis of fish indicates that all species except for the Channichthyidae have rather similar diets irrespective of their depth distribution; they mainly feed on benthic amphipods and polychaetes, except for younger (smaller) specimens which seem to feed primarily on zooplankton. Generally, the trophic position estimated based on $\delta^{15}N$ values reflects the assumed feeding habits of the organisms. Narrower trophic ranges compared to Arctic deep-sea systems (Iken et al. 2005) were established. Wide ranges in $\delta^{15}N$ ratios in most benthic taxa of Antarctic food webs indicate feeding across a range of trophic levels and are partly due to a high amount of omnivory (Würzberg et al. 2011a, b).

2.5 Climate Change

Acidification might become one of the largest problems for the long-term stability of the SO ecosystem, in both the pelagic and benthic realms. As a consequence of increasing atmospheric CO_2 most of the SO is expected to become undersaturated by 2,100 in both calcium carbonates, aragonite and calcite, the first being the major component of the skeleton of molluscs and corals, the latter of foraminiferans and coccolithophorids (Orr et al. 2005). The marine realm is at threat in the shallow as well as in deep water because at the sea surface it is directly exposed to increased CO_2, and in deeper water layers saturation principally decreases with increasing pressure. Therefore, the deep sea is undersaturated as is most of the Antarctic shelf. The polar oceans are especially threatened because calcium carbonate saturation is positively correlated with temperature and the uptake of atmospheric CO_2 is above global average. Despite the lack of a general understanding of the ecological consequences for benthic systems, it is generally known that the problems for organisms to build up their skeletons is species-specific. While some suffer, e.g. sea urchins, some even seem to benefit, e.g. tunicates (Dupont and Thorndyke 2009). It has to be considered that such animals are extremely rare in the deep sea. However, on the shelf, echinoderms, hydrocorals and gorgonians will belong to the potentially threatened calcifying organisms. Finally, in the SO deep sea regional abyssal warming and acidification may not be detected for some time due to the buffering effect of the huge volume of water (Kaiser and Barnes 2008; Brandt and Gutt 2011).

Environments have changed continually throughout the Earth's history and the SO marine biota appear to have been remarkably resilient to major, sometimes rapid, temperature and ice changes in the past. For instance, despite the last Ice Age ending only ca. 11,000 years ago, the Antarctic shelf biota has recovered and is exceptionally rich across taxonomic levels (Clarke and Johnston 2003). Nevertheless, recent ecophysiological studies indicate that the strong stenothermy displayed by many Antarctic marine biota makes them vulnerable to ocean warming, as experimental exposure to higher temperatures results in the loss of critical physiological and behavioural functions (Peck et al. 2009). In an ecological context, anthropogenic or natural disturbance comprises temporal and spatial changes in a variety of environmental conditions over different scales, such as for example sea ice, ice scouring, anchor-ice formation, "drop stones", large-scale glacial or pack-ice melt due to temperature increase, sediment instability, CO_2,

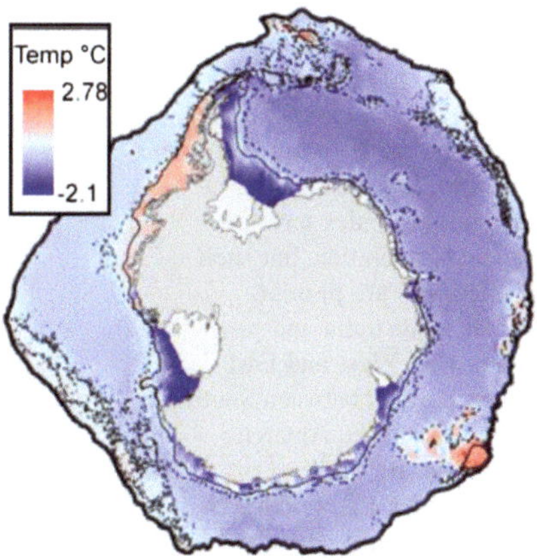

Fig. 2.2 Spatial distribution of bottom (seabed) potential temperatures around Antarctica (after Clarke et al. 2009)

UV-B radiation, and precipitation (Barnes and Conlan 2007). Critical for the SO can be sudden or prolonged temperature changes, and perhaps Milankovitch cyclicity (Clarke and Crame 2010). Amongst the most serious temperature effects of climate change the benthic fauna has to cope with are changes in the extent and quality of the ice sheet which might even lead to new sea-ways. Changes in ice-berg dynamics as well as fresh water flow causing temperature, salinity and stratification changes and near shore sedimentation will influence benthic communities. The potential biological responses are difficult to measure or anticipate because physiological experiments analyzing adaptation and macrophysiological processes are employed at rates of change 10–100,000 times faster than climatically induced oceanic changes (Peck et al. 2009). Changes will first impact molecular levels before determining individual and population fitness and species interactions (e.g. food-web structure) and ultimately influence fundamental ecosystem services such as biogeochemical carbon cycling. Whether the same sensitivity is present in their natural environment, or whether appropriate adaptive responses can occur over decadal to centennial timescales not available to experimenters, are important but as yet unanswered questions.

In the SO, the western Antarctic Peninsula is experiencing one of the fastest rates of regional climate change on Earth, resulting in the collapse of ice shelves, the retreat of glaciers and the exposure of new terrestrial habitat (Barnes and Peck 2008; Meredith and King 2005). Regional atmospheric warming of the Antarctic Peninsula area is linked to oceanographic changes, for example winter sea ice in the Bellingshausen and Amundsen seas has decreased in extent by 10% per decade, and shortened in seasonal duration. Surface waters have warmed by more than 1 K since the 1950s, and even the Circumpolar Deep Water of the Antarctic Circumpolar Current became warmer (Clarke et al. 2007). Clarke et al. (2009) showed a distinct latitudinal gradient in the difference between seabed temperatures on the shelf and in the deep sea. The deep sea is warmer (up to 2 K) at high latitudes and colder (by 2 K) around sub-Antarctic islands (Fig. 2.2). This impact or at least differences of regional climate change will likely have consequences for benthic

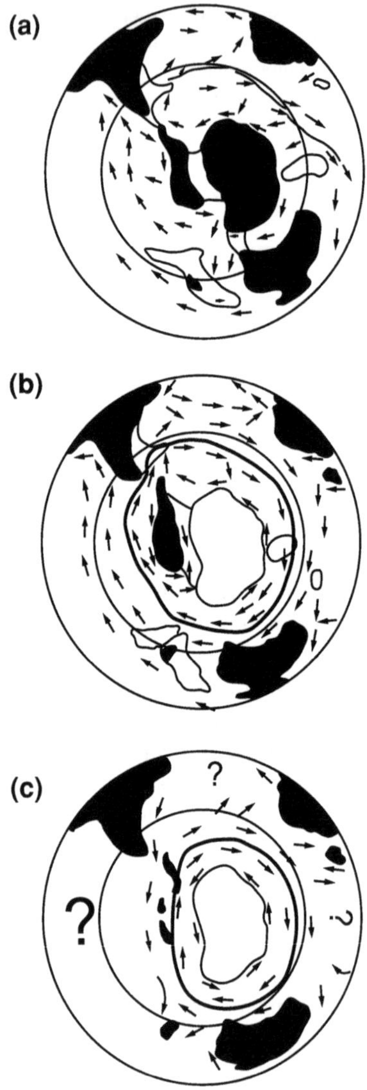

Fig. 2.3 Development of the circum- Antarctic current system; **a**, **b**: redrawn from Crame (1999); **c**: new image of one potential future scenario). **a** Late Eocene, (**a**) plate tectonics have isolated Antarctica, but shelf connections are present between Australia and Antarctica, West and East Antarctica and between South America and the Antarctic Peninsula. **b** Middle Miocene, East Antarctica is isolated and glaciated, a shallow water current system around East Antarctica separates East and West Antarctica. **c** Potential future scenario in 10 million years. Based on (**b**) current measurements (Antarctic Climate Report) and the fact that the West Antarctic Peninsula is rapidly warming while the eastern part of Antarctica is rapidly cooling (compare also geologic scenarios, e.g. Fox 2010; Jamieson et al. 2010) (**c**)

ecology and biogeography of the Antarctic marine biota. However, the complexity of the Southern Ocean food web and species' physiological adaptations as well as interactions make predictions of ecological responses to future changes impossible to date and we can only guess which species will migrate in or out of the SO, which ones will be able adapt to the changing conditions or become extinct.

It is difficult to assess the influence of climate change on deep-sea ecosystems. Shelf communities will be affected in the immediate future and most likely influence deep-sea communities at a much later stage. Smith et al. (2008) have reviewed abyssal food limitation, ecosystem structure and climate change and emphasize the importance of monitoring and modelling efforts. As abyssal food availability will also be driven by climate change, this influence will be especially dramatic in the SO deep sea because of the close coupling of the surface and deep-water layers due to deep-water production especially in the Weddell and Ross Seas (Fig. 2.3).

Based on recent measurements of temperature change one might have to wonder how Antarctica and the current system might develop in the long-term future (e.g. in 10 my). If the warming along the WAP continues and possibly increases and the temperature in the Ross and Weddell seas decrease or stay cold, then the area and islands of the Antarctic Peninsula could be free of sea ice and the Weddell and Ross Sea vanish. In fact, the glacial intensity could resemble that of Miocene times (10–14 my ago; Fox 2010; Jamieson et al. 2010). This in turn would shift and change the Circum Antarctic Current. Consequently, the deep-water production might occur circumantarctically and be less extensive than today and following these changes the world ocean circulation will be governed by different forces than today. If the knowledge of a temperate system during a glacial period could act as a case study for the future of the SO it can be expected that the retreat of the sea-ice in a period of warming and, consequently, a shift in the pelagic community would be mirrored on the sea floor sooner or later, e.g. as recorded in the ostracod and diatom composition (Cronin and Raymo 1997).

A problem with the impact assessment on deep-sea communities is that we do not know the drivers of biodiversity there and how these influence deep-sea assemblages (Kaiser and Barnes 2008).

It is therefore extremely important to study abyssal biodiversity and the key factors generating and maintaining it in order to generate a solid benchmark against which future change can be measured.

Acknowledgments The author is grateful for scientific support to the CoML field projects CAML, CeDAMar and to SCAR for the support of ScarMarBIN. The Alfred-Wegener Institute for Polar- and Marine Sciences is thanked for logistics during the expeditions witha RV *Polarstern*, the German Science Foundation is thanked for financial support of the ANDEEP project (Br 1121/26). Alistair Crame's kind review improved the paper.

References

Arntz WE, Gutt J (eds) (1997) Report of "Polarstern" cruise ANT XIII/3 (EASIZ I) to the eastern Weddell Sea. Ber Polar- u Meeresforsch 249:1–148

Arntz WE, Gutt J, Klages M (1997) Antarctic marine biodiversity: an overview. In: Battaglia B, Valencia J, Walton DWH (eds) Antarctic communities., Species, structure and survival Cambridge University Press, Cambridge, pp 3–14

Arntz WE, Rios C (1999) Magellan-Antarctic: ecosystems that drifted apart. Sci Mar 63(Suppl 1):503

Arntz WE, Clarke A (2002) Ecological studies in the antarctic sea ice zone. Springer, Berlin

Arntz WE, Brey T (2003) The expedition ANTARKTIS XIX/5 (LAMPOS) of RV "Polarstern" in 2002. Ber Polar- u Meeresforsch 462:1–120

Barnes DKA, Peck LS (2008) Vulnerability of Antarctic shelf biodiversity to predicted regional warming. Clim Res 37:149–163

Barnes DKA, Conlan KE (2007) Disturbance, colonization and development of Antarctic benthic communities. In: Rogers A (ed) Antarctic ecology: from genes to ecosystems, vol 362. Royal Society, London, pp 11–38 (Phil Trans Royal Soc B)

Brandt A (1991) Zur Besiedlungsgeschichte des antarktischen Schelfes am Beispiel der Isopoda (Crustacea, Malacostraca). Ber Polarforsch 98:1–240

Brandt A, Brökeland W, Brix S, Malyutina M (2004) Diversity of Antarctic deep-sea Isopoda (Crustacea, Malacostraca)—a comparison with shelf data. Deep-Sea Res II 51(14–16): 1753–1769

Brandt A, Hilbig B (2004) ANDEEP (ANtarctic benthic DEEP-sea biodiversity: colonization history and recent community patterns) - a tribute to Howard L Sanders. Deep-Sea Res 51(14–16):1457–1919

Brandt A, Ebbe B (2007) ANDEEP III Antarctic benthic DEEP-sea biodiversity: colonisation history and recent community patterns. Deep-Sea Res II 54(16–17):1645–1904

Brandt A, Bathmann U, Brix S, Cisewski B, Flores H, Göcke C, Janussen D, Krägefsky S, Kruse S, Leach H, Linse K, Pakhomov E, Peeken I, Riehl T, Sauter E, Sachs O, Schüller M, Schrödl M, Schwabe E, Strass V, van Franeker J, Wilmsen E (2011) Maud rise—a snapshot through the water column. Deep-Sea Res II. doi:10.1016/j.dsr2.2011.01.008

Brandt A, De Broyer C, De Mesel I, Ellingsen KE, Gooday A, Hilbig B, Linse K, Thomson M, Tyler P (2007a) The deep benthos. In: Rogers A (ed) Antarctic ecology: from genes to ecosystems, vol B 362. Royal Society, London, pp 39–66 (Phil Trans Roy Soc)

Brandt A, Gooday AJ, Brix SB, Brökeland W, Cedhagen T, Choudhury M, Cornelius N, Danis B, De Mesel I, Diaz RJ, Gillan DC, Ebbe B, Howe J, Janussen D, Kaiser S, Linse K, Malyutina M, Brandao S, Pawlowski J, Raupach M (2007b) The southern ocean deep sea: first insights into biodiversity and biogeography. Nature 447:307–311

Brandt A, Brökeland W, Choudhury M, Brix S, Kaiser S, Malyutina M (2007c) Deep-sea isopod biodiversity, abundance and endemism in the Atlantic sector of the Southern Ocean—results from the ANDEEP I–III expeditions. Deep-Sea Res II 54:1760–1775

Brandt A, Gutt J (2011) Biodiversity of a unique environment: the southern ocean benthos threat by climate change. In: Zachos F, Habel JC (eds) Biodiversity hotspots. Springer Publishers, Heidelberg. doi: 10.1007/978-3-642-20992-5_25

Brey T, Dahm C, Gorny M, Klages M, Stiller M, Arntz W (1996) Do Antarctic benthic invertebrates show an extended level of eurybathy? Antarctic Sci 8(1):3–6

Brown B, Gaina C, Müller RD (2006) Circum-Antarctic palaeobathymetry: illustrated examples from cenozoic to recent times. Palaeoceanogr Palaeoclimatol Palaeoecol 231:158–168

Clarke A, Johnston NM (2003) Antarctic marine benthic diversity. Oceanogr Mar Biol Ann Rev 41:47–114

Clarke A, Arntz WE, Smith CR (2006) EASIZ: Ecology of the antarctic sea ice zone. Deep-Sea Res II 53:803–1140

Clarke A, Murphy EJ, Meredith MP, King JC, Peck LS, Barnes DKA, Smith RC (2007) Climate change and the marine ecosystem of the western Antarctic Peninsula. Phil Trans Roy Soc London B 362:149–166

Clarke A, Griffiths HJ, Barnes DKA, Meredith MP, Grant SM (2009) Spatial variation in seabed temperatures in the southern ocean: implications for benthic ecology and biogeography. J Geophys Res 114 doi:10.1029/2008JG000886

Clarke A, Crame AJ (2010) Evolutionary dynamics at high latitudes: speciation and extinction in polar marine faunas. Philos Trans R Soc B 365:3655–3666. doi:10.1098/rstb.2010.0270

Crame AJ (1999) An evolutionary perspective on marine faunal connections between southernmost South America and Antarctica. Sci Mar 63:1–14

Cronin TM, Raymo ME (1997) Orbital forcing of deep-sea benthic species diversity. Nature 385:624–627

Dayton PK (1990) Polar benthos. Polar Oceanogr, Part B: Chem, Biol Geol 1:631–683

De Broyer C, Jazdzewski K, Dauby P (2003) Biodiversity in the SO: lessons from Crustacea. In: Huiskes AHL, Gieskes WWC, Rozema J, Schorno RML, van der Vies SM, WJ Wolff (eds) Antarctic biology in a global Context, 201–214

Dupont S, Thorndyke MC (2009) Impact of CO_2-driven ocean acidification on invertebrates early life-history—What we know, what we need to know and what we can do. Biogeosci 6:3109–3131

Flores H, van Franeker J-A, Cisewski B, Leach H, van de Putte AP, Meesters HWG, Bathmann U, Wolff WJ (2011) Macrofauna under sea ice and in the open surface layer of the Lazarev Sea, Southern Ocean. Deep-Sea Res II: 1948–1961

Fox D (2010) Could East Antarctica be headed for big melt? Science 328:1630–1631

Hartmann G (1997) Antarctic and subantarctic podocopa (ostracoda). Theses Zoologicae 26: 1–355

Held C (2003) Molecular evidence for cryptic speciation within the widespread Antarctic crustacean Ceratoserolis trilobitoides (Crustacea, Isopoda). In: Huiskes AHL, Gieskes WWC, Rozema J, Schorno RML, van der Vies SM, Wolff WJ (eds) Antarctic biology in a global context, 135–139

Howe JA, Shimmield TM, Diaz R (2004) Deep-water sedimentary environments of the northwestern Weddell Sea and South Sandwich Islands. Antarctica. Deep-Sea Res II 51(14–16): 1489–1515

Iken K, Bluhm BA, Gradinger R (2005) Food web structure in the high Arctic Canada basin: evidence from δ13C and δ15N analysis. Polar Biol 28:238–249

Janussen D, Tendal OS (2007) Diversity and distribution of Porifera in the bathyal and abyssal Weddell Sea and adjacent areas. Deep-Sea Res II 54(16/17):1864–1875

Jamieson SSR, Sugden DE, Hulton NRJ (2010) The evolution of the subglacial landscape of Antarctica. Earth Planet Sci Lett 239:1–27

Kaiser S, Barnes DKA (2008) Southern Ocean deep-sea responses to climate change. Climate Res 37:165–179

Kaiser S, Barnes DKA, Sands CJ, Brandt A (2009) Biodiversity of the Amundsen Sea (southern ocean): spatial patterns of richness and abundance in shelf isopods. Mar Biodiv 39:27–43

Krylova EM, Sahling H (2010) Vesicomyidae (Bivalvia): current taxonomy and distribution. PLoS ONE5(4). doi:10.1371/journal.pone.0009957

Leese F, Kop A, Wägele J-W, Held C (2008) Cryptic speciation in a benthic isopod from Patagonian and Falkland Island waters and the impact of glaciations on its population structure. Frontiers Zool 5:19. doi:10.1186/1742-9994

Linse K, Cope T, Lörz A-N, Sands C (2007) Some evidence of cryptic speciation in the circum-Antarctic bivalve *Lissarca notorcadensis* (Arcoidea: Philobryidae). Polar Biol 30:1059–1068

Lipps JH, Hickman CS (1982) Origin, age, and evolution of Antarctic and deep-sea faunas. In: Ernst WG, Morin JG (eds) The environment of the deep sea, vol II. Rubey, Prentic Hall, pp 324–254

Malyutina M, Brandt A (2007) Diversity and zoogeography of Antarctic deep-sea Munnopsidae (Crustacea, Isopoda, Asellota). Deep-Sea Res II 54:1790–1805

Meredith MP, King JC (2005) Rapid climate change in the ocean west of the Antarctic Peninsula during the second half of the twentieth century. Geophys Res Lett 32:L19604. doi:10.1029/2005GL024042

Orr JC, Fabry VJ, Aumont O, Bopp L, Doney SC, Feely RA, Gnanadesikan A, Gruber N, Ishida A, Joos F, Key RM, Lindsay K, Maier-Reimer E, Matear R, Monfray P, Mouchet A, Najjar RG, Plattner GK, Rodgers KB, Sabine CL, Sarmiento JL, Schlitzer R, Slater RD, Totterdell IJ, Weirig MF, Yamanaka Y, Yool A (2005) Anthropogenic ocean acidification over the twenty-first century and its impact on calcifying organisms. Nature 437:681–686

Peck LS, Clarke MS, Morley SA, Massey A, Rossetti H (2009) Animal temperature limits and ecological relevance: effects of size, activity and rates of change. Functional Ecol 23:248–256

Poore GCB, Wilson GDF (1993) Marine species richness. Nature 361:597–598

Raupach MJ, Wägele J-W (2006) Distinguishing cryptic species in Antarctic Asellota (Crustacea: Isopoda)—a preliminary study of mitochondrial DNA in *Acanthaspidia drygalskii*. Ant Sci 18(2):191–198

Raupach M, Malyutina M, Brandt A, Wägele JW (2007) Molecular data reveal a highly diverse species flock within the munnopsoid deep-sea isopod *Betamorpha fusiformis* (Barnard 1920) (Crustacea: Isopoda: Asellota) in the SO. Deep-Sea Res II 54(16–17):1820–1831

Raupach MJ, Mayer C, Malyutina M, Wägele J-W (2009) Multiple origins of deep-sea Asellota (Crustacea: Isopoda) from shallow waters revealed by molecular data. Proc R Soc B 276: 799–808. doi:10.1098/rspb.2008.1063

Rex MA, Stuart CT, Hessler RR, Allen JA, Sanders HL, Wilson GDF (1993) Global-scale latitudinal patterns of species diversity in the deep-sea benthos. Nature 365:636–639

Rex MA, Etter RJ, Stuart CT (1997) Large-scale patterns of species diversity in the deep-sea benthos. In: Ormond RFG, Gage JD, Angel MV (eds) Marine biodiversity: patterns and processes. Cambridge University Press, Cambridge, pp 94–122

Rex MA, McClain CR, Johnson NA, Etter RJ, Allen JA, Bouchet P, Warén A (2005) A source-sink hypothesis for abyssal biodiversity. Am Nat 165(2):163–178

Sanders HL (1965) Benthic marine diversity and the stability-time hypothesis. Brookh Symp Biol 22:78–81

Sanders HL, Hessler RR (1969) Ecology of the deep-sea benthos. Science 163:1419–1424

Schüller M, Ebbe B (2007) Global distributional patterns of selected deep-sea polychaeta (Annelida) from the southern ocean. Deep-Sea Res II 54(16–17):1737–1751. doi:10.1016/j.dsr2.2007.07.005

Smith CR, De Leo FC, Bernardino AF, Sweetman AK, Martinez Arbizu P (2008) Abyssal food limitation, ecosystem structure and climate change. Trends Ecol Evol 23(9):518. doi:10.1016/j.tree.2008.05.002

Thomas E, Gooday AJ (1996) Cenozoic deep-sea benthic foraminifera: tracers for changes in oceanic productivity. Geology 24:355–358

Thomson MRA (2004) Geological and palaeoenvironmental history of the Scotia Sea region as a basis for biological interpretation. Deep-Sea Res II 51:1467–1487

Veit-Köhler G, Guilini K, Peeken I, Sachs O, Sauter EJ, Würzberg L (2011) Antarctic deep-sea meiofauna and bacteria react to the deposition of particulate organic matter after a phytoplankton bloom. Deep-Sea Res II 58:1983–1995

Wilson GDF, Hessler RR (1987) Speciation in the deep Sea. Ann Rev Ecol Syst 18:185–207

Witman JD, Etter RJ, Smith F (2004) The relationship between regional and local species diversity in marine benthic communities: a global perspective. Proc Natl Acad Sci USA 101(44):15664–15669

Würzberg L, Peters J, Brandt A (2011a) Fatty acid patterns of Southern Ocean shelf and deep sea peracarid crustaceans and a possible food source, foraminiferans. Deep-Sea Res II 58: 2027–2035

Würzberg L, Peters J, Schüller M, Flores H, Brandt A (2011b) Demersal fishes from the Antarctic shelf and deep sea: a diet study based on fatty acid patterns and gut content analyses. Deep-Sea Res II 58:2036–2042

Zachos JC, Dickens GR, Zeebe RE (2008) An early Cenozoic perspective on greenhouse warming and carbon-cycle dynamics. Nature 451:279–283

Chapter 3
The TUNU-Programme: Euro-Arctic Marine Fishes—Diversity and Adaptation

Jørgen S. Christiansen

3.1 Introduction

A firm focus on the scientific status, the vulnerability and the commercial potential of the Arctic marine fishes is both timely and imperative. Parallel to the ongoing and indisputable retreat of the Arctic summer sea ice (Comiso et al. 2008), human activities increase rapidly into hitherto pristine parts of the Arctic Ocean: petroleum exploitation has begun, commercial fisheries are developing, and shipping routes across the Arctic Ocean are in operation with novel pollutants such as antifouling, ballast water and noise in their wake. Grounds for particular concern are marine bioprospecting, which eagerly extract commercially valuable compounds from otherwise little known Arctic organisms.

The combination of climate and human stressors inevitably changes Arctic marine ecosystems although the magnitude and outcome remains speculative. In the light of ocean warming (Agustí et al. 2010) coming Arctic fisheries will broadly affect two groups of fishes: harvested stocks found north of their traditional distribution areas and non-target fishes native to Arctic waters. It is worrying, therefore, to realise that there is a dearth of real biological data and fundamental knowledge for the latter group of fishes. This is well illustrated by the fact that 97% of the Arctic marine fish species are either data deficient (category DD) or not evaluated (category NE) according to the criteria of the IUCN Red List. The current state of Arctic marine fishes thus raises several key questions ripe for scientific attention.

The ongoing TUNU-Programme at the University of Tromsø was coined 3 October 2002 during the PRE-TUNU-Expedition and became formally endorsed

J. S. Christiansen (✉)
Department of Arctic and Marine Biology, Faculty of Biosciences, Fisheries and Economics, University of Tromsø, NO-9037 Tromsø, Norway
e-mail: jorgen.s.christiansen@uit.no

G. di Prisco and C. Verde (eds.), *Adaptation and Evolution in Marine Environments, Volume 1*, From Pole to Pole, DOI: 10.1007/978-3-642-27352-0_3,

by the International Polar Year (IPY, ID: 318) 25 May 2006. To date about 35 scientists and research students from 10 nations—Denmark, Finland, France, Greenland, Iceland, Italy, Norway (Head), Russia, UK, USA—have been engaged in the research activities.

The research programme addresses the diversity of species, populations and communities of Arctic marine fishes in primarily Northeast Greenland (TUNU-MAFIG, see Sect. 3.3). Genetic and demographic structuring, trophic interactions and physiological adaptations are viewed on a broad evolutionary time scale and in context of novel climate and human stressors. A minor part of the research programme is allocated studies on benthos and plankton communities as well as seal physiology.

The term 'TUNU' is etymologically ambiguous in modern Greenlandic—geographically it refers to East Greenland and, anatomically, to the back or spine. Both interpretations are adopted by the TUNU-Programme—the research activities began in Northeast Greenland with the ambition eventually to grow into a scientific backbone in the study of Euro-Arctic marine fishes at large.

3.2 Arctic Fishes: Past, Present and Prospects

Sea ice is the most conspicuous element of the Arctic Ocean and the TUNU-Programme defines Arctic fishes as those species which are associated with ice-laden seas at any time during their life cycle. Given this restriction, the assumed number of genuine Arctic marine fish species presently known to science may count about 60–70 (Mecklenburg et al. 2011).

3.2.1 Arctic Ichthyology in Retrospect

The roots of Arctic ichthyology can be ascribed to the eminent Danish priest (by trade) and polyhistorian Otto Fabricius (Othonis Fabricii, 1744–1822). Equipped with his volume of *Systema Naturae* (10th edition, Linnaeus 1758), Fabricius initiated the study of the Arctic fauna and he was the first to develop a scientific and annotated account on the Arctic fishes in his treatise *Fauna Groenlandica* (1780, Fig. 3.1).

In this work, Fabricius described no less than 473 invertebrate and vertebrate species of which he categorised 36 as Pisces and 9 cartilaginous fishes (Chondrichthyes) and lumpsuckers (*Cyclopterus*) as Amphibia [sic]. Altogether, *Fauna Groenlandica* contains comprehensive descriptions of 45 fish species. Several of Fabricius' descriptions of the approximately 130 species new to science, however, were published a few years earlier in *Zoologiae Danicae Prodomus* (1776) by his contemporary countryman, the renowned zoologist OF Müller (1730–1784). Fabricius is unquestionably a pioneer of Arctic zoology and many of his scientific contributions are still valid (Kapel 2005). Moreover, Fabricius broke new grounds during his stay in West Greenland (Paamiut district, 1768–1773), since he fully

Fig. 3.1 O Fabricius (1744–1822) and the front page of his *Fauna Groenlandica* (1780)

appreciated the value of and extensively employed traditional ecological knowledge (TEK) in his scientific work (Jensen 1923). Prominent ichthyologists of the nineteenth and the first half of the twentieth century, mainly from Denmark (CF Lütken, 1827–1901, and Ad S Jensen, 1866–1953), Russia and the Soviet Union (LS Berg, 1876–1950, GU Lindberg, 1894–1976, AN Svetovidov, 1903–1985, and AP Andriashev, 1910–2009), invigorated Fabricius' work and developed the classification and zoogeographic knowledge of Arctic fishes even further.

3.2.2 Problems and Prospects for Arctic Ichthyology

Despite the effort of outstanding ichthyologists, knowledge of the Arctic fish fauna remains sparse compared to for example the Antarctic marine fauna which has been extensively studied for decades (Piepenburg 2008). Besides logistic constraints, the most serious hindrance towards a comprehensive understanding of Arctic marine biodiversity lies in the fact that our information, by and large, is qualitative (presence-absence) and based on random and fragmentary snapshots accumulated over periods of time revealing no coherent time series.

The Arctic societies are by far based on living natural resources and the socio-economic progress is inevitably rooted in sound ecosystems. Forecasts of biological responses to climate change require multifaceted actions and empirical gaps can hardly be bridged solely by sophisticated models and unsubstantiated hypotheses. Lack of real data presently represents the most severe shortcoming for a reliable assessment and, ultimately, a sustainable management of the Arctic seas. The scientific rationale behind the TUNU-Programme, therefore, is to advance the biological knowledge of the Euro-Arctic marine fish faunae at large.

3.3 The TUNU-Programme: Structure and Function

3.3.1 The Study Area

A clear-cut definition of the marine Arctic has yet to be developed. One definition simply includes the seas within the Arctic Circle, e.g. AMAP (Arctic Monitoring and Assessment Programme), whereas others are based on major zoogeographic shifts in the fish fauna (Mecklenburg et al. 2011). Geographic boundaries are static and have little ecological meaning. More importantly, they do not account for the ongoing changes in the Arctic sea ice cover and quality, i.e. the decrease in the perennial to annual sea ice ratio (Walsh 2008).

In context of the TUNU-Programme, we define the marine Euro-Arctic as the fjords, shelves and slopes (<1,500 m) associated with the maximum extent of the sea ice cover (usually in March). That embraces areas which are seasonally affected by the *Fast Ice* in the fjords and the drifting *Pack Ice* covering the shelves (Pickard and Emery 1990). This definition introduces a needed dynamic and meaningful ecological dimension to the marine Arctic and inevitably reflects both the temporal variations in sea ice cover and the concomitant responses of the fish faunae.

The geographic area of attention in the TUNU-Programme was initially restricted to the fjords and shelf-slope areas in Northeast Greenland (TUNU-MAFIG). Recently the research area is expanded also to include the marine fish faunae across the entire Euro-Arctic triangle, i.e. from west to east: Northeast Greenland (Arctic water) via Jan Mayen Island (Arctic and Atlantic waters), Svalbard Archipelago (Atlantic and Arctic waters) and, whenever feasible, towards the Franz Joseph Land Archipelago (Arctic water). Those Arctic sites are bordering the Atlantic Arctic Gateway, i.e. the Norwegian Sea and the Barents Sea, and are considered exceptionally appropriate for comparative marine biological studies. They share similar photoperiods (latitudes) but are affected by completely different water masses, thermal conditions and invasion histories (Fig. 3.2).

Larger parts of the Euro-Arctic region are marine protected areas (MPAs) and strict admission permits are required from the respective authorities in Greenland, Norway and Russia. Thus, most studies in the TUNU-Programme have been carried out within the borders of the Northeast Greenland National Park which is the largest in the world (972,000 km^2).

3.3.2 The TUNU Expeditions

The logistic backbone of the TUNU-Programme consists of regular expeditions with the ice strengthened R/V Jan Mayen headed by the University of Tromsø (Table 3.1, Fig. 3.3).

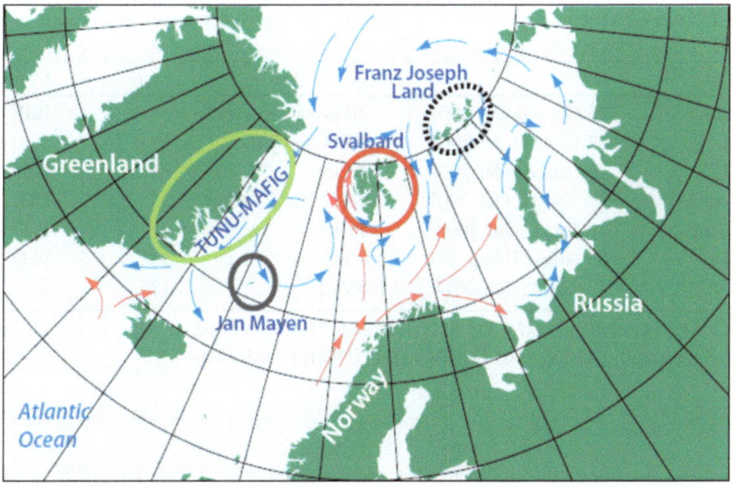

Fig. 3.2 Geographic coverage of the TUNU-Programme. The core areas of investigation are encircled: Northeast Greenland (TUNU-MAFIG, *green*), Jan Mayen Island (*black*) and the Svalbard Archipelago (*red*). The Franz Joseph Land Archipelago is included as a prospect comparative reference site (*dashed*). The arrows indicate the major Arctic (*blue*) and Atlantic (*red*) surface currents

Seven expeditions were conducted successfully in autumns 2002 (PRE-TUNU), 2003 (TUNU-I), 2005 (TUNU-II), 2007 (TUNU-III), summers 2008 (TUNU-Seal) and 2010 (TUNU-IV), and winter 2011 (TUNU-SVAL I). During the TUNU Expeditions, series of biological and hydrographic stations and transects are identified in primarily Northeast Greenland waters (see Sect. 3.3.3). Three expeditions had logistic and scientific collaboration with national research institutions, i.e. Institute of Marine Research and the Dept. of Geology, University of Tromsø. Two PhD courses in Arctic marine biology were carried out in 2005 and 2010 under the auspices of the ARCTOS PhD-School, University of Tromsø, and two PolART performers in visual arts participated in 2008.

3.3.3 Stations and Sampling Procedures

To date, 235 stations *in toto* are covered during the TUNU Expeditions with 170 stations in Northeast Greenland, 30 at Jan Mayen Island, and 35 at Svalbard Archipelago (Table 3.1). The stations embrace habitats from the littoral zone in the innermost parts of the fjords via the shelves to the continental slope (<1,500 m) (Fig. 3.4). Together the stations provide first hand quantitative data and meta-data on inter-annual variations in fish faunae and concurrent hydrographic regimes.

An array of sampling methods is employed during the TUNU Expeditions (Fig. 3.5) although sampling was strongly hampered by exceptionally heavy sea

Table 3.1 Overview of completed TUNU Expeditions and the number of stations examined (N = 235 *in toto*) during the period 2002–2011

Expeditions	Date	Study area	Stations	Comments
PRE-TUNU	25 Sep-18 Oct 2002	NEG and SVAL	38	IMR
TUNU-I	02 Oct-13 Oct 2003	NEG and JMI	50	
TUNU-II	26 Sep-10 Oct 2005	NEG	36	PhD-School
TUNU-III	29 Sep-12 Oct 2007	NEG and SVAL	33	DG
TUNU-Seal	29 Jun-12 Jul 2008	JMI	28	IMR & PA
TUNU-IV	06 Aug-17 Aug 2010	NEG and SVAL	33	PhD-School
TUNU-SVAL I	17 Jan-25 Jan 2011	SVAL	17	

NEG Northeast Greenland, *JMI* Jan Mayen Island, *SVAL* Svalbard Archipelago, *IMR* Institute of Marine Research, *DG* Dept. of Geology, *PA* PolART performers

Fig. 3.3 The *R/V Jan Mayen* in Scoresby Sund Fjord (~70°N), Northeast Greenland, October 2007. On the 17 June 2011, the *R/V Jan Mayen* was renamed *R/V Helmer Hanssen*. Helmer Hanssen (1870, Bjørnskinn-1956, Tromsø) was a polar explorer who participated in three famous expeditions headed by Roald Amundsen: The *Gjøa* expedition through the Northwest passage (1903–1906), the *Fram* expedition to the South pole (1911), and the *Maud* expedition through the Northeast passage (1918–1920)

ice in 2005 and 2007 (Fig. 3.4). Arctic fishes are by far bottom dwelling (Fig. 3.6) and standardized bottom trawls, Campelen Super 1800/96 NOFI, are employed as the main gear for quantitative studies of fish diversity. Whenever required, additional live fishes are sampled at the same sites for experimental studies onboard the R/V Jan Mayen. Fish diversity stations (N = 83) include complementary data on echo-sounding, trawl gear dimensions and speed, and measurements of in situ depth and temperature. Profiles of hydrography and chlorophyll are obtained by CTD-sensors at the same sites as fish sampling (N = 91) and water samples across the entire Northeast Atlantic were obtained in 2005. All fishes are identified to species, counted and weighed. Fresh voucher specimens are labelled, photographed and frozen whole for the TUNU Museum Collection (TMC, see Sect. 3.3.4.4).

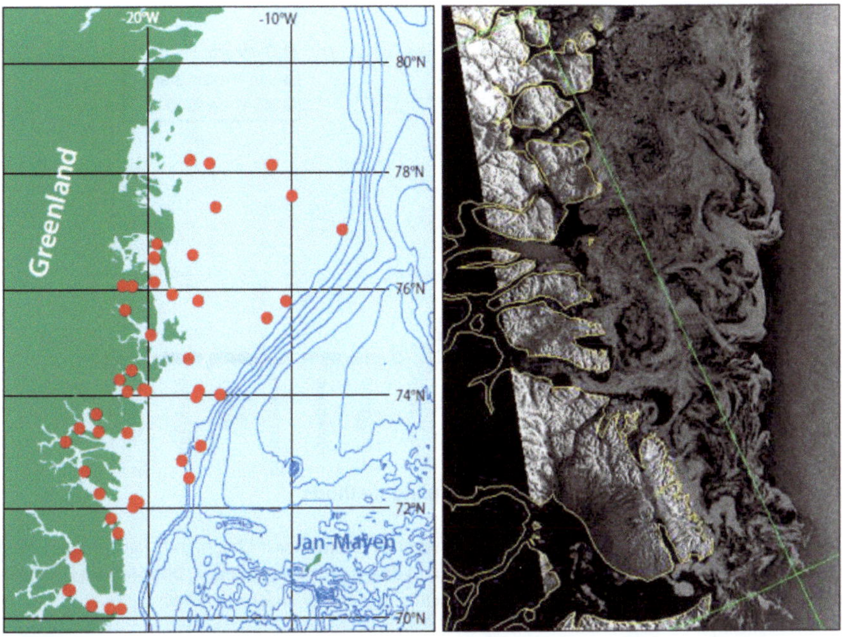

Fig. 3.4 Stations for quantitative studies of fish diversity in Northeast Greenland (N = 56) between latitudes ~70–78°N (2002–2010), cf. Fig. 3.2. The satellite photo shows the pack ice (Greenlandic–'Sikorsuit') between latitudes ~70–74°N, 4 October 2007 (Courtesy: www.seaice.dk). The drifting 'Sikorsuit' in Northeast Greenland is an exceptional marine ecosystem which deserves a research programme of its own

Observations of birds and sea mammals are made regularly during all TUNU Expeditions and they were carried out systematically in 2008 and 2010 (Byrkjedal and Madsen 2008).

3.3.4 Scientific Foci

It is generally assumed that the Arctic marine fish fauna was established relatively recently (2–3 million years ago, mya) compared to its counterpart in the Southern Ocean, Antarctica, which may have evolved under cold environmental conditions for the past 10–17 mya (DeVries and Steffensen 2005; Patarnello et al. 2011). Interestingly, a recent study on ocean floor sediments suggests that the modern circulation in the Arctic Ocean actually dates back about 17 mya and a perennial sea ice cover was formed about 13 mya (Krylov et al. 2008). Given this perspective, Arctic and Antarctic regions display a coeval freeze and the polar faunae should be viewed on the same geological time scale. However, this may not affect our current perception of polar fish evolution in view of the fact that the Arctic and Antarctic regions differ

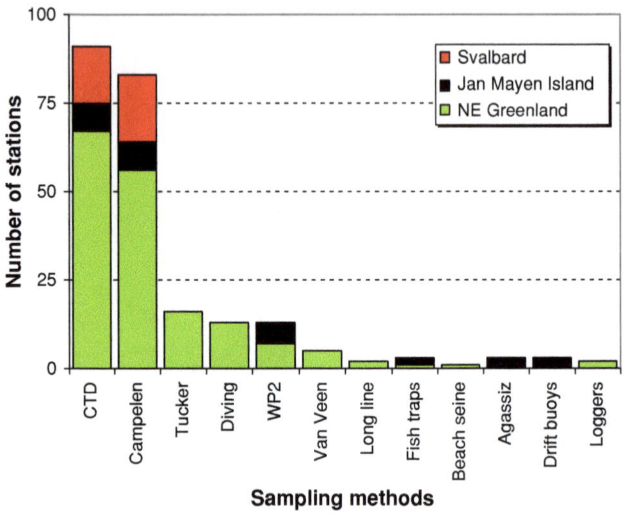

Fig. 3.5 Sampling methods employed at 235 stations in Northeast Greenland, Jan Mayen Island and Svalbard Archipelago, 2002–2011. CTD = hydrography, Campelen = bottom trawl, Tucker and WP2 = zooplankton, Van Veen = sediments, Agassiz = macro-benthos

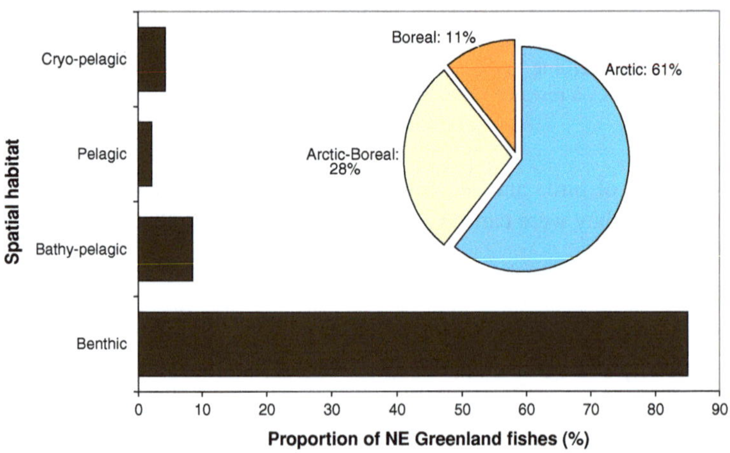

Fig. 3.6 Functional diversity. Spatial habitat and zoogeographic affiliation (inserted pie chart) for the Northeast Greenland fish species (cf. Table 3.2)

significantly in invasion history, biogeographic isolation and environmental stability. Bipolar comparisons, therefore, are particularly valuable to grasp the evolutionary history and prospects of Arctic fishes and several participants in the TUNU-Programme are studying the fish faunae in both polar hemispheres. The TUNU-Programme is broadly organised into four research activities:

Table 3.2 Preliminary checklist of the fish species encountered in Northeast Greenland, 2002–2010. Phylogenetic order and family numbering follows Nelson (2006). The inserted emblem represents the research activities in Northeast Greenland (TUNU-MAFIG)

FISHES OF NORTHEAST GREENLAND 2002-2010	
CLASS ELASMOBRANCHII	
Family 37. SOMNIOSIDAE (sleeper sharks)	
Somniosus microcephalus Bloch-Scheider 1801	
Family 48. RAJIDAE (skates)	
Amblyraja hyperborea Collett 1879	
Amblyraja radiata Donovan 1808	
CLASS ACTINOPTERYGII	
Family 172. OSMERIDAE (smelts)	**Family 327. CYCLOPTERIDAE (lumpfishes or lumpsuckers)**
Mallotus villosus Müller 1776	*Eumicrotremus spinosus* Fabricius 1776
Family 200. MYCTOPHIDAE (lanternfishes)	**Family 328. LIPARIDAE (snailfishes)**
Benthosema glaciale Reinhardt 1837	*Careproctus* idet.
Family 219. PHYCIDAE (phycid hakes)	*Careproctus micropus* Günther 1887
Gaidropsarus argentatus Reinhardt 1837	*Careproctus* see *reinhardti* Krøyer 1862
Family 220. GADIDAE (codfishes)	*Liparis fabricii* Krøyer 1847
Arctogadus glacialis Peters 1872	*Liparis gibbus* Bean 1881
Boreogadus saida Lepechin 1774	*Liparis tunicatus* Reinhardt 1837
Gadus morhua Linnaeus 1758	*Paraliparis bathybius* Collett 1879
Family 291. GASTEROSTEIDAE (sticklebacks)	*Rhodichthys regina* Collett 1879
Gasterosteus aculeatus Linnaeus 1758	**Family 416. ZOARCIDAE (eelpouts)**
Family 304. SCORPAENIDAE (scorpionfishes or rockfishes)	*Gymnelus retrodorsalis* Le Danois 1913
Sebastes mentella Travin 1951	*Lycenchelys kolthoffi* Jensen 1904
Sebastes norvegicus Ascanius 1772	*Lycenchelysmuraena* Collett 1878
Family 320. COTTIDAE (sculpins)	*Lycodes eudipleurostictus* Jensen 1902
Artediellus atlanticus Jordan-Evermann 1898	*Lycodes paamiuti* Møller 2001
Gymnocanthus tricuspis Reinhardt 1831	*Lycodes pallidus* Collett 1879
Icelusbicornis Reinhardt 1840	*Lycodes polaris* Sabine 1824
Myoxocephalus scorpius Linnaeus 1758	*Lycodes reticulatus* Reinhardt 1835
Myoxocephalus quadricornis Linnaeus 1758	*Lycodes rossi* Malmgren 1865
Triglops nybelini Jensen 1944	*Lycodes seminudus* Reinhardt 1837
Triglops pingeli Reinhardt 1831	*Lycodes squamiventer* Jensen 1902
Family 324. AGONIDAE (poachers)	*Lycodonus flagellicauda* Jensen 1902
Leptagonus decagonus Bloch-Scheider 1801	**Family 417. STICHAEIDAE (pricklebacks)**
Family 325. PSYCHROLUTIDAE (fathead sculpins)	*Leptoclinus maculatus* Fries 1838
Cottunculus microps Collett 1875	**Family 493. PLEURONECTIDAE (righteye flounders)**
Cottunculus sadko Esipov 1937	*Hippoglossoides platessoides* Fabricius 1780
Psychrolutes subspinosus Jensen 1902	*Reinhardtius hippoglossoides* Walbaum 1792

3.3.4.1 Taxonomy, Distribution and Diversity

The taxonomy of Arctic marine fishes is both complex and controversial and, in light of the molecular revolution, several genera are ripe for major revisions (Mecklenburg et al. 2011). Barcoding of *mt*DNA and *n*DNA microsatellites have become major tools for identification of fish taxa and populations (Ward et al. 2009) but molecular techniques are no substitute to morphological studies. For example, strong intra-specific phenotypic variations exist (Byrkjedal et al. 2007) and the combination of classic taxonomy (phenotypic plasticity) and molecular studies (the underlying genotype) will provide not only information but also knowledge on the evolution of Arctic marine fishes (Naish and Hard 2008).

The term 'distribution' is ambiguous and we may distinguish between zoogeographic patterns and phylogeographic processes of fish distribution. Whereas zoogeography simply provides the total spatial range for a given taxon (i.e. the pattern), phylogeography reflects the origin, evolutionary history and the putative dispersal routes of taxa through space and time (i.e. the underlying processes). Obviously, the study of phylogeography is much more multifaceted and requires ample information both on the genetic lineages among fish taxa and populations and the temporal dynamics in environmental barriers that give rise to certain zoogeographic patterns (Briggs 1974; Hardy et al. 2011). The study of marine fish distribution within the Euro-Arctic region is of particular actuality due to ocean warming and a concomitant northward shift in the geographic range for a number of boreal species in recent years (Perry et al. 2005; Wienerroither et al. 2011).

Polar seas are usually considered species poor compared to lower latitudes. Recent investigations of marine invertebrates in the Southern Ocean, however, reveal a much higher biodiversity than previously thought (Brandt et al. 2007). The bottom topography in Northeast Greenland is often massively disturbed by moving icebergs and together with the pack ice this poses several logistic problems for sampling (Figs. 3.3 and 3.4). Thus, the numbers of Arctic fish species are either underestimated (i.e. apparent) due to inadequate sampling and/or taxonomic controversies or a biological reality of unknown causes (cf. the idea of "dark diversity" by Pärtel et al. 2011). This clearly calls for further scientific attention.

3.3.4.2 Functional Ecology

Due to environmental constraints such as subzero temperature and seasonal food shortage, Arctic marine fishes are thought to grow and reproduce slowly but even fundamental data on demographic structuring (e.g. Von Bertalanffy Growth Functions), longevity, life history and trophic status are lacking. Trophic interactions inferred from stable isotope profiles and stomach analyses are coupled with studies of pollutants within species (bioaccumulation) and food chains (biomagnification) as well as parasites (Køie et al. 2007). Moreover, most Arctic marine fish species are bottom dwelling and substrate spawners (Fig. 3.6, Christiansen et al. 1998). This would make them particularly vulnerable as unwarranted by-catch and to habitat destruction caused by conventional trawl gears.

3.3.4.3 Thermal Adaptation and Evolution

Polar fishes (Arctic and Antarctic) have evolved an array of exceptional physiological and biochemical adaptations to tackle ice-laden seas (DeVries and Cheng 2005; di Prisco and Verde 2006). This may have resulted in loss of genetic variability at adaptive loci and a limited capacity to overcome novel stressors such as ocean warming (Patarnello et al. 2011). Polar marine fishes are assumed to be extremely temperature sensitive and stenothermal and they display a clear structuring in thermal distribution even within the narrow temperature zone encountered in polar waters. Moreover, physiological key properties in polar fishes may show a sharp curvilinear response by a change from subzero to positive ambient temperatures (Christiansen et al. 1995). For this reason, even a slight rise in sea temperature may have disproportionally large consequences for the overall fitness of polar fishes compared to lower latitude counterparts.

One of the most striking features of polar fishes is their ability to tolerate near-freezing temperatures by means of antifreeze peptides and glycoproteins in tissues and body fluids (DeVries and Cheng 2005). On the other hand, effects of high temperature in polar fishes are much less understood (Hofmann et al. 2000; Bilyk and DeVries 2011) although the Arctic gadoid *Boreogadus saida* may tolerate laboratory temperatures of ~14°C for weeks (Christiansen et al. 1997) and appears to display an intermediate thermal stress response between that of Antarctic and temperate fishes (Whiteley et al. 2006). Studies of whole body metabolism in polar fishes reveal no adaptation to low temperatures (Steffensen 2005), whereas respiratory proteins such as hemoglobins (Hbs) show distinct adaptations to polar environments (Verde et al. 2006, 2012). Fishes devoid of Hbs, however, are yet not known from the Arctic Ocean cf. the Hb-less icefishes Channichthyidae in the Southern Ocean (di Prisco et al. 2002).

3.3.4.4 The TUNU Museum Collection

Natural History collections hold important information to reconstruct long-term data series (Harrison et al. 2011; Lister et al. 2011) and voucher specimens from the TUNU Expeditions are collected systematically for the TUNU Museum Collection (TMC, c/o Senior Curator I Byrkjedal) at Bergen Museum. The TMC contains *pro tem* about 650 voucher specimens from 47 species for further taxonomic studies and corresponding tissues are fixed in ethanol for phylogenetic validation.

3.3.5 Fishes in Northeast Greenland

To date, 3 cartilaginous fish species in 3 families and 44 bony fish species in 14 families are identified from Northeast Greenland during the TUNU Expeditions (Table 3.2). The species encountered are all marine but the diadrome stickleback *Gasterosteus aculeatus*. The diadrome Arctic charr *Salvelinus alpinus* is indeed numerous in the fjords of Northeast Greenland but they return to freshwaters in early summer and were absent at sea at the time of the TUNU Expeditions in August-October.

The Arctic marine fishes are taxonomically complex and further analyses are needed, particularly within the speciose, sensu Hart (2008), families Cottidae, Liparidae, and Zoarcidae.

The spatial habitat and zoogeographic affiliation of the Northeast Greenland fishes are shown in Fig. 3.6. More than 85% of the species display a benthic life style and only two are considered cryo-pelagic, sensu Andriashev (1970), as they utilize the sea ice as feeding grounds, refuge and spawning substrate (see Sect. 3.3.6).

More than 60% of the species are considered genuinely Arctic and spawn mainly at subzero temperatures. On the other hand, 11% are boreal and spawn solely at positive ambient temperatures (Mecklenburg et al. 2011). In the light of ocean warming, the number and proportion of pelagic and boreal species (Fig. 3.6) is expected to increase in the Euro-Arctic region (Wienerroither et al. 2011).

3.3.6 Arctic Codfishes and How (Not) to Name Them

Two cryo-pelagic gadoids (family Gadidae) are endemic to the Arctic Ocean with each genus being represented by only a single species: *Arctogadus glacialis* and *Boreogadus saida* (Table 3.2, Fig. 3.7). The *B. saida* is beyond doubt an Arctic key-stone species both in terms of abundance and ecological significance and being the northernmost occurring fish species it is widespread throughout the Arctic Seas (Christiansen and Fevolden 2000; Christiansen et al. 2010 and references therein). The *A. glacialis*, on the other hand, is much less abundant and it is primarily associated with the Arctic fjords and shelves (Aschan et al. 2009). The two species are taxonomically distinguished by simple phenotypic features for adults (Fig. 3.7) and genetic markers for the larval stages (Madsen et al. 2009).

The scientific nomenclature is unsurpassed to vernacular names in scientific communication. There are several unfortunate examples from the scientific literature where the identification of these Arctic codfishes is mixed up due to lack of consistency in the use of vernacular names. Hence, *B. saida* is known both as 'Polar cod' and 'Arctic cod' and vice versa for *A. glacialis*. To add further confusion, the migrating population of Atlantic cod *Gadus morhua* (Norwegian–*skrei*) is also known as 'Northeast Arctic cod' by ICES (International Council of Exploration of the Seas) and it has been mistaken for *B. saida*.

Fig. 3.7 Endemic gadoids of the Arctic Ocean—the cryo-pelagic 'Polar cod' and 'Ice cod'

The scientific name is conclusive and should always follow the vernacular name for these species. The vernacular names 'Ice cod' for *Arctogadus glacialis* and 'Polar cod' for *Boreogadus saida* are employed by the TUNU-Programme.

3.4 Dissemination and Communication

The TUNU-Programme has communicated results in various ways both to the general public and the scientific community. Thus far, studies are mainly published within the fields of thermal adaptation and genetics. Studies combining quantitative fish diversity and the physical environment are presently being undertaken to examine inter-annual variations and putative trends since 2002. The scientific communication comprises: Technical Expedition Reports (TERs), peer-reviewed scientific journals, popular science, MSc and PhD theses, internal workshops and international conferences. In addition, international courses for PhD-students and young scientists are given during selected expeditions. Live Arctic fishes are collected regularly for display at the public aquarium (POLARIA) in Tromsø and an art exhibition was presented by PolART at the Tromsø Art Society in 2009. A web-site with general information, dissemination and access to meta-data from the TUNU Expeditions is planned operative from spring 2012.

The TUNU-Programme is affiliated with CAFF (Conservation of Arctic Flora and Fauna), the IUCN Redlist (International Union for Conservation of Nature), the EU COST-Action: Conservation Physiology of Marine Fishes, EBA (Evolution and Biodiversity in Antarctica), and the ARCTOS network at the University of Tromsø.

3.5 Prospects for the TUNU-Programme

Basic transects for long-term monitoring and studies on ocean climate and the marine fish faunae are now established in Northeast Greenland and key-stations are adopted by CBMP (Circumpolar Biodiversity Monitoring Program) commissioned by CAFF. More investigations are needed in the littoral zone and in deep waters (>1,500 m) and an array of gears such as beach seines, traps and long lines should be employed to a larger extend to cover also difficult and little accessible habitats. Studies on the fish fauna during the Arctic winter are practically absent and coming TUNU Expeditions should emphasise this issue to discriminate between seasonal and resident fish species (cf. TUNU-SVAL I, Table 3.1).

The TUNU-Programme was initially confined to Northeast Greenland (TUNU-MAFIG: **MA**rine **FI**shes of NE **G**reenland—Diversity and Adaptation). With the expanded **TUNU**-Programme: **E**uro-**A**rctic **M**arine **Fish**es—Diversity and Adaptation (acronym: TEAM-Fish), we aim to continue our work also in the remaining parts of the Euro-Arctic region.

Acknowledgements I sincerely thank the members of the TUNU-Programme, the crew onboard the R/V Jan Mayen, and the Government of Greenland. I further thank G. di Prisco and C. Verde for the opportunity to write this summary, F. Strand for preparation of maps, M. Jobling for useful references, and S. Adams at Blå Rock for suitable working conditions.

References

Agustí S, Sejr MK, Duarte CM (2010) Impacts of climate warming on polar marine and freshwater ecosystems. Polar Biol 33:1595–1598

Andriashev AP (1970) Cryopelagic fishes of the Arctic and Antarctic and their significance in polar ecosystems. In: Holdgate MW (ed) Antarctic ecology. Academic, New York, pp 297–304

Aschan M, Karamushko OV, Byrkjedal I, Wienerroither R, Borkin I, Christiansen JS (2009) Records of the gadoid fish *Arctogadus glacialis* (Peters, 1814) in the European Arctic. Polar Biol 32:963–970

Bilyk KT, DeVries AL (2011) Heat tolerance and its plasticity in Antarctic fishes. Comp Biochem Physiol A 158:382–390

Briggs JC (1974) Operation of zoogeographic barriers. Syst Zool 23:248–256

Brandt A, Gooday AJ, Brandão SN et al (2007) First insights into the biodiversity and biogeography of the Southern Ocean deep sea. Nature 447:307–311

Byrkjedal I, Madsen J (2008) Autumn bird observations in the Northeast Greenland sea ice. Dansk Orn Foren Tidsskr 102:325–330

Byrkjedal I, Rees DJ, Willassen E (2007) Lumping lumpsuckers: molecular and morphological insights into taxonomic status of *Eumicrotremus spinosus* (Fabricius 1776) and *Eumicrotremus eggvinii* Koefoed, 1956 (teleostei: cyclopteridae). J Fish Biol 71:111–131

Christiansen JS, Fevolden S-E (2000) The polar cod of Porsangerfjorden, Norway; revisited. Sarsia 85:189–193

Christiansen JS, Chernitsky AG, Karamushko OV (1995) An Arctic teleost with a noticeably high body fluid osmolality—a note on the navaga, *Eleginus navaga* (Pallas 1811), from the white sea. Polar Biol 15:303–306

Christiansen JS, Karamushko LI, Nahrgang J (2010) Sub-lethal levels of waterborne petroleum may depress routine metabolism in polar cod *Boreogadus saida* (Lepechin, 1774). Polar Biol 33:1049–1055

Christiansen JS, Schurmann H, Karamushko LI (1997) Thermal behaviour of polar fish: a brief survey and suggestions for research. Cybium 21:353–362

Christiansen JS, Fevolden S-E, Karamushko OV, Karamushko LI (1998) Maternal output in polar fish reproduction. In: di Prisco G, Pisano E, Clarke A (eds) Fishes of Antarctica, a biological overview. Springer, New York, pp 41–52

Comiso JC, Parkinson CL, Gersten R, Stock L (2008) Accelerated decline in the Arctic sea ice cover. Geophys Res Lett. doi:10.1029/2007GL031972

DeVries AL, Cheng C-HC (2005) Antifreeze proteins and organismal freezing avoidance in polar fishes. In: Farrell AP, Steffensen JF (eds) The physiology of polar fishes. Academic, New York, pp 155–201

DeVries AL, Steffensen JF (2005) The Arctic and Antarctic polar marine environments. In: Farrell AP, Steffensen JF (eds) The physiology of polar fishes. Academic, New York, pp 1–24

di Prisco G, Verde C (2006) Predicting the impacts of climate change on the evolutionary adaptations of polar fish. Rev Environ Sci Biotech 5:309–321

di Prisco G, Cocca E, Parker SK, Detrich HW III (2002) Tracking the evolutionary loss of hemoglobin expression by the white-blooded Antarctic icefishes. Gene 295:185–191

Fabricius O (1780) Faunae groenlandica. http://www.archive.org/details/ faunagroenlandi 00fabrgoog

Hardy SM, Carr CM, Hardman M, Steinke D, Corstorphine E, Mah C (2011) Biodiversity and phylogeography of Arctic marine fauna: insights from molecular tools. Mar Biodiv 41:195–210

Harrison IJ, Chakrabarty P, Freyhof J, Craig JF (2011) Correct nomenclature and recommendations for preserving and cataloguing voucher material and genetic sequences. J Fish Biol 78:1283–1290

Hart MW (2008) Speciose versus species-rich. TREE 23:660–661

Hofmann GE, Buckley BA, Airaksinen S, Keen JE, Somero GN (2000) Heat-shock protein expression is absent in the Antarctic fish *Trematomus bernacchii* (family nototheniidae). J Exp Biol 203:2331–2339

Jensen AdS (1923) Naturforskeren Otto Fabricius. Meddelelser om Grønland 42:331–399 (in Danish)

Kapel FO (2005) Otto Fabricius and the seals of Greenland. Meddelelser om Grønland, Biosci 55:1–150

Køie M, Steffensen JF, Møller PR, Christiansen JS (2007) Parasite fauna of *Arctogadus glacialis* (Peters) (Gadidae) from western and eastern Greenland. Polar Biol 31:1017–1021

Krylov AA, Andreeva IA, Vogt C et al (2008) A shift in heavy and clay mineral provenance indicates a middle Miocene onset of a perennial sea ice cover in the Arctic Ocean. Paleoceanography 23:PA1S06. doi:10.1029/2007/PA001497

Lister AM, Climate change research group (2011) Natural history collections as sources of long-term datasets. TREE 26:153–154

Madsen ML, Fevolden S-E, Christiansen JS (2009) A simple molecular approach to distinguish between two Arctic gadoid fishes *Arctogadus glacialis* (Peters, 1874) and *Boreogadus saida* (Lepechin, 1774). Polar Biol 32:937–939

Mecklenburg CW, Møller PR, Steinke D (2011) Biodiversity of arctic marine fishes: taxonomy and zoogeography. Mar Biodiv 41:109–140

Naish KA, Hard JJ (2008) Bridging the gap between the genotype and the phenotype: linking genetic variation, selection and adaptation in fishes. Fish Fisheries 9:396–422

Nelson JS (2006) Fishes of the world, 4th edn. Wiley, New York

Patarnello T, Verde C, di Prisco G, Bargelloni L, Zane L (2011) How will fish that evolved at constant sub-zero temperatures cope with global warming? Notothenioids as a case study. Bioessays 33:260–268

Pärtel M, Szava-Kovats R, Zobel M (2011) Dark diversity: shedding light on absent species. TREE 26:124–128

Perry AL, Low PJ, Ellis JR, Reynolds JD (2005) Climate change and distribution shifts in marine fishes. Science 308:1912–1915

Pickard GL, Emery WJ (1990) Descriptive physical oceanography an introduction, 5th edn. Butterworth-Heinemann, Oxford

Piepenburg D (2008) As time goes by: polar biology over the years 1982–2008. Polar Biol 32:3–7

Steffensen JF (2005) Respiratory systems and metabolic rates. In: Farrell AP, Steffensen JF (eds) The physiology of polar fishes. Academic, New York, pp 203–238

Verde C, Lecointre G, di Prisco G (2006) The phylogeny of polar fishes and the structure, function and molecular evolution of haemoglobin. Polar Biol 30:523–539

Verde C, Giordano D, Russo R, di Prisco G (2012) The adaptive evolution of polar fishes. Lessons from the function of hemoproteins. In: di Prisco G, Verde C (eds) Adaptation and evolution in marine environments—The impacts of global change on biodiversity, vol 1. Series "From Pole to Pole". Springer, Berlin, pp 197–213

Walsh JE (2008) Climate of the Arctic marine environment. Ecol Appl 18:S3–S22

Ward RD, Hanner R, Hebert PDN (2009) The campaign to DNA barcode all fishes, FISH-BOL. J Fish Biol 74:329–356

Whiteley NM, Christiansen JS, Egginton S (2006) Polar cod, *Boreogadus saida* (Gadidae), show an intermediate stress response between Antarctic and temperate fishes. Comp Biochem Physiol 145:493–501

Wienerroither R, Nedreaas KH, Uiblein F, Christiansen JS, Byrkjedal I, Karamushko O (2011) The marine fishes of Jan Mayen Island, NE Atlantic—past and present. Mar Biodiv 41:395–411 doi: 10.1007/s12526-010-0055-y

Chapter 4
Sea-Ice Interactions with Polar Fish: Focus on the Antarctic Silverfish Life History

Marino Vacchi, Philippe Koubbi, Laura Ghigliotti and Eva Pisano

4.1 Introduction

The sea ice, extending seasonally over a vast area of the circumpolar Oceans, is a component of the polar cryosphere, playing major roles in earth global systems. The extent and dynamics of the sea-ice cover influence the ocean circulation patterns (Brandon et al. 2010), the sea surface/atmosphere exchanges and the level of reflected radiation (Perovich et al. 2007).

At seasonal scale, it is a dynamic biogeochemical system (Thomas et al. 2010) that deeply influences marine ecosystems (Massom and Stammerjohn 2010); at geological scale, changes in sea ice cover influence rates of climate change (Keeling and Stephen 2001) with impact on marine ecosystems (e.g. Fraser et al. 2009).

From initial formation, the sea ice is shaped by the interaction of physical, chemical and biological processes. The large-scale cycles of sea-ice formation and melting in turn influence the oceanography, govern global climate (Schmitz et al. 2003) and act on the life cycles of marine plants and animals, from micro-organisms

M. Vacchi (✉)
ISPRA (Istituto Superiore per la Protezione e la Ricerca Ambientale),
c/o MNA, University of Genoa, I-16132 Genoa, Italy
e-mail: m.vacchi@unige.it

P. Koubbi
Laboratoire d'Océanographie de Villefranche, Université Paris 06,
UMR 7093, 06230 Villefranche-sur-mer, France
e-mail: koubbi@obs-vlfr.fr

L. Ghigliotti · E. Pisano
DipTeRis, University of Genoa, I-16132 Genoa, Italy
e-mail: laura.ghigliotti@unige.it

E. Pisano
e-mail: pisano@unige.it

G. di Prisco and C. Verde (eds.), *Adaptation and Evolution in Marine Environments, Volume 1*, From Pole to Pole, DOI: 10.1007/978-3-642-27352-0_4,

to vertebrates. Rich and diverse biological communities (ice biota) live in association with sea ice in both polar seas (reviews in Horner 1985; Horner et al. 1992; Legendre et al. 1992; Garrison 1991; Knox 1994; Thomas and Dieckmann 2010). In addition to hundreds of viruses, bacteria, fungi, algae, and a wide range of animals living inside, on top of, or attached to the bottom of the sea ice, several organisms live in close relation, including crustaceans and fish of primary importance as biomass producers and key species in the polar ecosystems. The rich and diverse ice-related biological assemblages in polar seas, indicates that numerous taxa that met the ice over evolutionary time scale have been able to cope with it, through metabolic, biochemical and ecological adaptations.

Due to its complexity, the sea-ice science spans from physical science, e.g. geophysics, oceanography, glaciology, geology and chemistry, to life science, e.g. ecology and biology. However, despite the long history of research on sea ice, only from the early 1990s the wide community of scientists involved in sea-ice research started to interact, thus making significant advances towards a more comprehensive understanding of the sea-ice system. An important step in this interdisciplinary effort was made during the International Symposium of the International Glaciology Society, held in Fairbanks (Alaska) in June 2000: "Sea Ice and Its Interactions with the Ocean, Atmosphere and Biosphere" (Jeffries and Eicken 2000). Sea-ice research, as a collaborative enterprise among scientists with different expertise, including both Arctic and Antarctic, had been a major feature and merit of scientific and outreach programs during the Fourth International Polar Year (IPY 2007–2009). Under the IPY umbrella, specialists from various fields had the opportunity to integrate their expertise in a joint effort for a better understanding of the sea ice as a fundamental component of the Earth system. Such interaction among assorted players was also solicited by the large public as well as by decisions makers, requiring sound advice and science-based support to the discussions on environmental changes and predictions of future climate conditions.

The sea ice, as part of the global climate system, and the implications of changes in the sea ice regime, were dominant themes during IPY, either as objectives of scientific projects or as focus of multidisciplinary meetings and polar schools. Some IPY international schools provided platforms for interdisciplinary field activities in the polar areas, mostly performed in the Arctic (simply for logistic reasons), and involving researchers, teachers and students. As an example concerning polar fish, which is the theme of this report, the TUNU-MAFIG Project (Marine Fishes of North East Greenland—Diversity and Adaptation, IPY activity #318) involved scientists with various expertise, and students, in surveys on fish biodiversity along the coasts of Greenland, a key site to study effects of climate change on marine biota (see Chap. 3, Christiansen 2012). The field IPY course held in Barrow (Alaska, May 2008) led to the publication of the book "Field techniques for sea ice research" (Eicken et al. 2009). The volume has become a major reference for polar field activities because it provides detailed information on field sampling techniques and approaches the integration of measurements, to address questions that go beyond disciplinary boundaries. A significant multidisciplinary approach to various issues on sea-ice science is provided in the second edition of the volume "Sea Ice" (Thomas and

Dieckmann 2010). Beside the information on sea-ice structure, growth and dynamics, five of the fifteen chapters update the information on diversity and complexity of the biological communities associated to sea ice (from bacteria and viruses, to algae and heterotrophic protists and metazoans), the role of the biological communities in structuring the sea-ice system itself, and their function as mediators between sea ice and pelagic/benthic habitats.

The present chapter will focus on the relation between polar sea ice and fishes. Such a wide and complex topic is still far from being exhausted, mainly due to logistic constraints. In particular, the possibility to sample those species living close to land-based sea ice is mostly seasonal, due to the limits in field activities all year round, even from winter-over open scientific stations. The fragmented sampling makes it even more difficult to draw the life cycle of a species, as well as to understand the nature of the its relationship with sea ice during the seasons and at different ontogenetic stages. The possibility to sample in diverse areas and in diverse periods of the year is a great advantage for research on polar fishes. For example, the opportunity to work in the field earlier than usual (in Spring and Summer, instead of Summer only) close to the Italian Mario Zucchelli Station (Ross Sea) allowed the first discovery of a nursery area of the Antarctic silverfish *Pleuragramma antarcticum*. Millions of eggs and developing embryos were found encased within, and floating among platelet ice dislodged under the fast ice in a site named "Silverfish Bay" (see below). This finding greatly contributed to the ecological knowledge of silverfish, and stimulated international and multi-disciplinary investigations to study the environmental features of the nursery area, to understand the role of the sea ice in the life cycle, as well as to approach basic aspects related to the physiology of embryos and larvae surviving in a such a harsh icy environment.

In this review, we first provide background information on the features of polar sea-ice formation and basic aspects concerning sea ice and polar fish. In the second part, we focus on *P. antarcticum* by integrating biological and ecological information from recent advancements, mostly acquired during IPY. By updating information on the Antarctic silverfish from various Antarctic regions, it appears how geographical, oceanographic, and environmental conditions are all important features in the life history of this key polar fish, with coastal sea ice playing major roles for the early life phases and reproduction. The very new hypothesis of homing as spawning behaviour for the silverfish stresses the importance of long term monitoring to make sound predictions on the possible effects of environmental changes on silverfish life cycle and related impacts at ecosystem level.

4.2 Which Types of Sea Ice Do Fish Meet in Polar Waters?

The geographic characteristics and distinct glacial histories of the Arctic and Antarctic lead to fundamental differences between the sea ice of the two regions. Thickness, amplitude of the annual seasonal extent, texture characteristics, salt regime, influence of terrestrial input, and composition of associated communities are important features (reviewed in DeVries and Steffenson 2005; Dieckmann and Hellmer 2010). At large scale, some differences are critical for cryosphere-system functioning and must be taken into account in evaluating the consequences of current changes in the sea-ice mass regimes at the two poles (e.g. Notz 2009).

The features of sea ice in the Arctic and Antarctic were recently updated and compared (Dieckmann and Hellmer 2010). Therefore in the following sections we provide an overview of the distinct types of sea ice encountered by polar fishes: sea-ice cover (fast ice and pack ice), platelet ice, anchor ice, and ice crystals.

4.2.1 Sea-Ice Cover

The ice in shallow areas adjacent to land is called fast ice (e.g. Garrison 1991; Knox 1994; Thomas and Dieckmann 2010), also with reference to official nomenclature (Gutt 2001 and references therein). The pack ice is the part of ice cover extending from fast ice seaward, that will break into free-floating pack ice late in the sea-ice cycle (e.g. Knox 1994).

In Antarctica the extent of land-fast ice has been estimated as 5% of the total cover (0.8×10^6 km^2) (Fedotov et al. 1998). No estimate is reported for the Arctic. Despite a structural continuum between fast and pack ice, the differences in physical regimes of coastal and oceanic areas covered by ice result in large scale heterogeneity of ice floes as habitats (Garrison 1991).

At specific level, only pelagic fish can theoretically have direct relationships with ice cover all around the life cycle. In contrast, benthic species can directly relate to the ice cover during their pelagic stage as larvae or juveniles. At biogeographic level, the latitudinal extent of the ice cover strongly influences the composition and diversity of the fish community living in Polar Regions. In Antarctica, the ice cover has been used, together with frontal zones and distribution of phytoplankton and zooplankton, as a parameter to separate, from north to south, the Ice-Free Zoogeographic Zone, the Seasonal Pack-Ice Zone and the Permanent Ice or High-Antarctic Zone (Kock 1992; Hureau 1994). In the High Antarctic Zone, where waters are ice-covered for most of the year, the fish fauna includes ca. 80 species and endemic notothenioids are dominant. In the Seasonal Pack-ice Zoogeographic Zone, extending between the maximum limit of pack ice in winter/spring and the minimum in autumn/summer, the biomass is higher but species diversity is lower, with 65 fish species (Eastman and McCune 2000).

4.2.2 Platelet Ice and the Under-Ice Water Layer

Although there is evidence of the occurrence of platelet ice in the Arctic sea (Jeffries et al. 1995), platelet ice is considered a typical Antarctic feature (Dieckmann and Hellmer 2010). It consists of various-sized flat plate-like crystals up to above 10 cm in diameter, randomly oriented, mostly occurring under the coastal cover (Gow et al. 1998; Leonard et al. 2006). Platelet ice can be incorporated into fast ice and it is found as a component of fast-ice cores, at depths larger than 1 m (Smith et al. 2001). More often, it loosely aggregates under fast or pack ice, making a semi-solid layer from a few cm to several meters in thickness. Due to its prominent occurrence under the fast ice, platelet ice is considered a coastline high-latitude feature, although its real extent is still unknown. The mechanisms of growth of platelet ice have implications for the dynamics of the communities associated with it (Dieckmann et al. 1986; Bluhm et al. 2010), but still remains a puzzle (McGuinness and Langhorne 2006; Crook 2010).

Among the number of hypotheses proposed to explain platelet-ice formation, the presence of adjacent ice shelves seems to be important (Holland and Jenkins 1999). In fact, platelet ice may be the result of frazil ice growing in supercooled plumes forming near the ice shelf; here frazil ice collides, provides secondary nucleation, and rises out of suspension due to buoyancy. Alternatively, the platelet ice may grow entirely at the fast-ice interface by turbulent heat transfer to supercooled water flowing by; combination of these two mechanisms may also occur. The brine rejection from growing ice cover can also play a role in the dynamics of platelet ice by affecting the rate and timing of its incorporation in the bottom levels of fast ice (McGuinness and Langhorne 2006). The biological role of platelet ice is widely recognized. As intermediate between the relatively impermeable congelation ice above and the water column below, it is characterized by rapid nutrient exchanges (Arrigo and Thomas 2004; Thomas et al. 2010). The platelet ice offers a stable surface for algal attachment and growth and is a site of high algal production and accumulation (Mangoni et al. 2009 and references therein). The standing stock of algae and bacteria provides habitat for pelagic zooplankton species such as a copepods and amphipods, associated in large amount with the platelets (Bluhm et al. 2010). Some sea-ice inhabitant are preferentially found in the platelet ice in some stages of their cycle life, such as the Antarctic calanoid copepod *Stephos longipes* that possibly uses ice platelets as substrate for its eggs (Kurbjeweit et al. 1993; Bluhm et al. 2010). The platelet-ice environment conditions seem to be favourable for nauplii hatch, by providing food and protection from predation (Schnack-Schiel et al. 1998; Hubold and Ekau 1990; Hammer and Hammer 2000). Although the extent of platelet-ice formation is unknown, platelet ice is expected to occur around the Antarctic continent wherever there are floating ice shelves or glacier tongues, and represents an important source of primary and secondary production available also for the benthos on the coastal shelf. In areas where there is no platelet-ice layer, the under-ice water layer (the boundary layer extending few meters from the bottom of the ice cover) plays an equally important ecological role (e.g. Gradinger 2001; Werner and Martinez Arbizu 1999).

Such an under-ice water layer is a special habitat, strongly influenced by the ice freezing–melting cycle (different in the Arctic and Antarctic) as well as by water current direction and speed. Regional or local changes in thickness and morphology of the ice cover also add variability to the biological properties of this boundary layer. Organisms are either sympagic (originating from the ice) or pelagic, and they use this habitat as shelter and feeding and nursing grounds. The dominant under-ice macrofauna living close to sea ice, at least for some time,

Fig. 4.1 Anchor ice mat at 20 m depth, in the proximity of Mario Zucchelli Station, Terra Nova Bay, November 2010. (Photograph by R. Palozzi)

includes euphausiids in Antarctica (e.g. Nicol 2006) and gammarid amphipods in the Arctic (e.g. Macnaughton al. 2007; Arndt and Swadling 2006). Within euphausiids, *Euphausia superba* is one of the most important members of the Antarctic sub-ice community, spending part of its life cycle at the ice–water interface, feeding on ice-algal biomass (Nicol 2006). An Arctic/Antarctic comparative picture of the macrofauna living in such an habitat is provided by Bluhm et al. (2010). Fish species of the cryopelagic community (see below) directly interact with the underside of the ice cover and platelet ice, either as adults and juveniles.

4.2.3 Anchor Ice

The anchor ice is submerged ice attached to the bottom. Although there is still some uncertainty concerning the physical conditions under which this type of ice forms, it is commonly described as the result of growing masses of platelet ice attached to the bottom in shallow waters (Dayton et al. 1969).

In the Ross Sea the anchor ice forms in abundance to depths of ca. 30 m and prevents setting of sessile organisms in many areas (DeVries and Steffenson 2005 and references therein). In reviewing the direct impact on marine benthic communities in Antarctica, Gutt (2001) stressed the effects of anchor ice as the main factor determining the upper limit of a generally rich coastal benthic assemblage.

Near the Italian MZ station (Fig. 4.1), the occurrence of anchor ice is a frequent feature from few to 30 m until spring (Vacchi, personal observation).

Over evolutionary time, several benthic fish living in the coastal shallow waters have learned to cope with exposure to this ice mat. For instance, benthic notothenioids that are likely to contact anchor ice, such as the trematomids *Trematomus bernacchii* and *T. hansoni*, in addition to antifreeze molecules (see below), seem to have specific resistance to ice propagation across their surface tissues (DeVries and Cheng 2005). Moreover, Cziko et al. (2006) suggested high levels of protection for the eggs laid in bottom nests by the bathydraconid *Gymnodraco acuticeps* which have to face anchor ice growth on the shallow benthos of McMurdo Sound during the 10 month development.

Based on its origin from platelet ice, which has been observed at 250 m depth (Dieckmann et al. 1986), anchor ice may also form in deeper waters under ice shelves. However, present information on the extent of anchor ice in the Antarctic is scattered and limited, most of the records being based on diving surveys. In the Arctic, anchor ice seems to be common (Gutt 2001).

4.2.4 Free Sea-Ice Crystals

In addition to ice cover, platelet layer and anchor ice, the nucleation of ice crystals in polar waters can theoretically occur also in the water column, wherever the local equilibrium among salinity, pressure and temperature keeps the sea water at its freezing point. For instance, when the advection of super-cold water from underside of ice shelves upwards reduces the hydrostatic pressure, a rising in its freezing point leads to formation of many small ice crystals in the upper pelagic environment (DeVries and Steffenson 2005). These minute ice crystals are all potential ice-nucleating agents for fish that make contact with them during swimming. The sound evidence that nucleation of undercooled water and free growth of ice crystals can occur at depth of 225 m (Dieckmann et al. 1986) stresses the possibility of contact of fish with ice in a wide vertical and spatial range.

In the Arctic, the lack of substantial ice shelves results in a lower amount of available frazil ice, thus in reduced formation of platelet ice. This is an additional difference between the two polar environments that has led to apparently less severe freezing condition for Arctic species with respect to Antarctica. As biological response, although the two faunas have distinct evolutionary histories, Arctic fish are expected to have a generally lower resistance to freezing if compared to the Antarctic counterparts, and they actually do (DeVries and Cheng 2005).

4.2.5 Sea Ice Crystals Inside the Polar Fish Body

The most striking evidence of the exposure of polar fish to ice crystals in the environment over evolutionary time is the fact that antifreeze proteins (APs) are the most important biological mechanism allowing fish not to freeze. In fact APs exert their action by inhibiting growth of ice crystals to which they bind, therefore implying the presence of ice crystals in the body fluids to be functionally and physiologically pertinent (DeVries 1971, 1986). Antarctic fish exposed to ice crystals all year, such as cryopelagic *Pagothenia borchgrevinki*, contain ice in their body, entered via contact with surface tissues or ingestion through diet and seawater drinking (exogenous ice); they can also store ice in the spleen (endogenous ice).

How would fish remove endogenous ice in order to avoid physiological problems is still not understood. However the mechanism is possibly different in the Arctic and Antarctic. In the Arctic, the favourable seasonal temperature regime, should assure melting of endogenous ice possibly acquired by fish in the colder months. In the Antarctic, low water temperature in the coastal environment over the seasonal cycle would not assure melting, and a biological disposal mechanism involving macrophages has been suggested: ice that entered the cells via endocytose would melt in endosomes where ion transport raises the melting point. For more details, see DeVries and Cheng (2005).

4.3 The Cryopelagic Community

Our knowledge of composition and ecological role of the polar sea-ice biota has greatly increased in recent years. We now know better how rich and diverse group of autotrophs, bacteria and protozoans are harboured by sea ice and how several metazoans also use sea ice as habitat, feeding ground, refuge, breeding and/or nursery ground (Deming 2010; Arrigo et al. 2010; Caron and Gast 2010; Bluhm et al. 2010; Tynan et al. 2010).

The term cryopelagic, originally assigned to epipelagic fish found in association with the ice cover (Andriashev 1968; Parin 1968) is more generally used to describe the community of metazoans found at the ice-water interface. Such a community includes taxonomically diverse species of invertebrates and fish, either adults or larvae and juveniles, trophically related to ice-associated biological assemblages, as primary or secondary consumers (Knox 1994).

The most common cryopelagic species, in the Arctic and Antarctic, are listed as under-ice and sub-ice species in recent reviews (see Bluhm et al. 2010).

Despite their similar ecological role, these sea-associated communities are obviously different in species composition and diversity at the two poles. As a consequence of the different features under the sea cover, the cryopelagic communities are mostly found in platelet-rich environments in the Antarctic, whereas the Arctic ones have been described in platelet-free environments. The cycles of temporal melting and formation (annual for the majority of ice cover in Antarctica and multiannual in the Arctic) is reflected in the complex life cycles of the ice-associates organisms that have to cope with different temporal changes in both availability of substrate and trophic conditions (Bluhm et al. 2010). For most species, being cryopelagic does not imply to spend the entire life cycle near the ice.

Among invertebrates, the dominant metazoans algal grazers in the cryopelagic community are euphausiids in Antarctica and gammarid amphipods in the Arctic. Within euphausiids, *E. superba*, the key species in the Southern Ocean depends on sea–ice food at some stage of its life cycle; its important interactions with the sea cover were extensively studied in the last years (Brierley et al. 2002). Copepods can be highly concentrated under the ice both in the Arctic and Antarctica showing complex and various life cycles (e.g. Kurbjeweit et al. 1993).

In Antarctica, fish species living in the High-Antarctic Zone (Kock 1992; Hureau 1994) are essentially benthic notothenioids, with only very few pelagic or benthopelagic species (Eastman and McCune 2000); therefore, a minority of fish are found in cryopelagic communities as adults; in contrast, pelagic, benthopelagic and benthic species can be components of the cryopelagic community as larvae and juveniles. The Antarctic silverfish, the only notothenioid in which all development stages are pelagic, enters the cryopelagic community at various phases of its life history (see below).

The nototheniids *P. borchgrevinki* and *P. brachysoma* were first recognized as cryopelagic by Andriashev (1970, 1987). *P. borchgrevinki* is possibly the most studied species with respect to its specializations for life in the cryopelagic environment during adult and juvenile stages (Eastman and DeVries 1985); in addition to trophic relationships (Knox 1994; La Mesa et al. 2004), *P. borchgrevinki* uses sea ice as refuge against predation (Brierley 2002), and so does the most abundant cryopelagic fish in the Arctic, *Boreogadus saida* (Gradinger and Bluhm 2004). As a significant adaptive response to exposure to ice crystals, *P. borchgrevinki* has more antifreeze than benthic species confined to ice-free deep water (Eastman and DeVries 1985), and it can manage the occurrence of both exogenous and endogenous ice inside its body tissues (DeVries and Cheng 2005).

The notothenioid *Trematomus newnesi* is also considered a component of the cryopelagic community at adult and juvenile stages (Williams 1988); its food spectrum changes according to the ice-cover (La Mesa et al. 2000). Although not studied in detail, larvae and juveniles of other trematomid benthic or benthopelagic species are common in cryopelagic communities (Knox 1994).

A direct relationship with ice cover was observed for the larvae of icefish such as *Chionodraco rastrospinosus*, that uses the ice canopy as hunting ground for ice-associated furcilia and fish larvae (Moline at al. 2008 and references therein).

Among Arctic fish, the gadids *B. saida* and *Arctogadus glacialis* live in close association with the ice underside (Carey 1985; Melnikov 1997). At young age, the polar cod *B. saida* utilizes cavities and narrow wedges of sea water along the edges of melting ice floes (Carey 1985; Gradinger and Bluhm 2004). As adult, the polar cod forms aggregations to spawn under the ice during the winter months (Bouchard and Fortier 2011, Geoffroy et al. 2011).

4.4 Ice As Evolutionary Driver

Despite the very low water temperatures, sometimes reaching values well below the equilibrium freezing point (f.p.) of the body fluids, modern polar fishes (both Arctic and Antarctic) live and thrive in frigid sea waters and do not freeze. In most polar fishes the equilibrium f.p. is significantly depressed by higher blood salt content. However this change in blood osmolarity, by itself, would not be sufficient to avoid freezing in waters where very low temperature is often accompanied by sea ice crystals that can act as nucleating agents, providing a template for water molecules to convert from liquid to solid phase. Environmental ice is usually separated from body fluids as long as the fish is protected by physical barriers, such as the skin, but as soon as a single crystal comes into contact with the undercooled body fluids of a fish, and it can happen due to damages in the physical barriers or through ingestion, the organism will freeze and die (DeVries and Cheng 2005).

In an environment characterized by the combination of frigid temperature and ice-crystal richness, one of the main challenge for polar fishes is, and has historically been, freezing avoidance. One could argue that the presence of ice crystals, in such a context, could have driven evolution of key biological features in polar fishes, one of the most important being AP genes and function.

Since the finding of the first fish AP in Antarctic notothenioids (DeVries and Wohlschlag 1969; DeVries 1970, 1971) and identification of the first antifreeze peptide in the winter flounder (Duman and DeVries 1974), dating back to the early 1970s, much effort has been devoted to study these molecules, leading to the discovery of a multiplicity of APs, including antifreeze glycoproteins (AFGPs) and four types of antifreeze peptides (AFPs) in many polar fishes (Cheng 1998a). Despite structural diversity, the role and mechanism of action of all APs seem to be the same: all recognize and bind the same substrate, ice crystals. Structural, evolutionary and molecular biologists have been intrigued for decades by APs.

AFGPs, made of Ala/Pro-Ala-Thr repeats (Shier et al. 1972) ranging from 4 to 56 (Cheng et al. 1996), occur in both Antarctic notothenioids and Arctic gadids. Detailed analyses of AFGP genes from these taxonomically distinct groups revealed that genes were evolved independently from different genomic origins: a rare case of convergent evolution (Cheng 1998a).

AFPs have been found in some Arctic fishes and in two non-notothenioid Antarctic species. These molecules differ in their primary and higher order structures, and are conventionally classified in four types (Fletcher et al. 2001). In the case of AFPs, the structural heterogeneity of proteins having same ice-binding and growth-inhibition properties epitomize functional convergence (Cheng 1998b).

The mechanisms of genesis of the novel genes coding for APs from functionally unrelated ancestor genes have been long investigated in both Arctic and Antarctic fishes. The AFGP gene in the Antarctic notothenioid *Dissostichus mawsoni* is suggested to have evolved from de novo amplification of a 9-nt Thr-Ala-Ala coding region in a duplicated trypsinogen-like protease (TLP) gene, leading first to a chimeric AFGP/TLP intermediate and then to the novel AFGP gene. The recent description of the genomic organization of the AFGP locus in *D. mawsoni* (Nicodemus-Johnson et al. 2011) has provided new data that question the role of the chimeric AFGP/TLP gene as evolutionary intermediate. At present, the escape from adaptive conflict (EAC) model of evolution seems to be the most likely mechanism that underlies evolution of APs. According to this model, the ancestor gene, driven by adaptive constraints (i.e. freezing habitat), would have acquired an emergent function additional to and in conflict with its primary function. Gene duplication could have resolved the conflict, allowing gene duplicates to improve one of the other functions separately. In an Antarctic zoarcid, *Lycodichthys dearborni*, EAC-driven evolution of type III AFP (AFPIII) coding gene from an old sialic acid synthase (SAS) gene has been recently proven through molecular and functional evidences (Deng et al. 2010).

4.5 Focus on the Antarctic Silverfish

The notothenioid *P. antarcticum* (Antarctic silverfish) is the most abundant fish in coastal regions of Antarctica. In all areas sampled to date it has been found in waters between the shelf break and the continental margin (DeWitt et al. 1990; Miller 1993; Trunov 2001). This fish is the dominant species in permanent pack-ice zones (Hureau 1994), but it has also been found in the zone of seasonal pack ice and in the shelf waters of the West Antarctic Peninsula (Donnelly and Torres 2008). Among notothenioids, *P. antarcticum* is the only known holopelagic fish, living all stages of development throughout the water column, at depths from 0 to 900 m (Gerasimchuk 1986; DeWitt et al. 1990; Fuiman et al. 2002; Knox 1994). Ontogenetic shifts in vertical/spatial distribution characterize the silverfish life cycle (Hubold and Ekau 1987; Granata et al. 2009; La Mesa et al. 2010; O'Driscoll et al. 2011; Koubbi et al. 2011). This fish reaches sexual maturity after 6–7 years, at a size larger than ca. 130 mm SL (Faleyeva and Gerasimchuk 1990), and lays pelagic eggs (Vacchi et al. 2004) with absolute fecundity values ranging between 4,315 and 17,774 eggs/female (Kock and Kellermann 1991).

The capability to pelagic life has occurred in evolutionary time as a result of adaptive radiation within notothenioids, possibly driven by trophic competition among demersal species and by the largely unexploited food sources available in the pelagic realm (Eastman 1993; Klingenberg and Ekau 1996). Specialization to pelagic life has been accomplished through adaptations that include reduction of skeletal ossification and lipid storage (Eastman 1997; Wöhrmann et al. 1997; Near et al. 2009; Albertson et al. 2010; La Mesa and Eastman 2011).

As the prevalent plankton-feeder of the intermediate trophic level, this small fish plays a pivotal role in the trophic structure of the High-Antarctic coastal system and its patterns of energy flow. Therefore *P. antarcticum* is considered a keystone species, much like *E. superba* (Antarctic krill) is for waters beyond the continental shelf (Guglielmo et al. 1998; Fuiman et al. 2002) and *E. crystallorophias* (ice krill) is for the neritic zone (Vallet et al. 2011).

The ecological position of the silverfish in the Antarctic coastal system, together with its adaptive features, justifies the interest on this species by a wide community of researchers and the number of studies carried out in the last years, before, during and after the 4th International Polar Year. In particular, during the Census of Antarctic Marine Life (EBA), and under the umbrella of CCAMLR (Convention on the Conservation of Antarctic Marine Living Resources), some country members developed large-scale ecological and acoustic surveys on pelagic fish and plankton, in which the silverfish was one of the main target. Azzali et al. (2010), performed *ex situ* experiments of acoustic target strength (TS) of the silverfish to make a model to estimate the abundance of *P. antarcticum* based on previous acoustic surveys in the Ross Sea. The New Zealand survey onboard R/V Tangaroa, in February–March 2008, focused on distribution, abundance and acoustic properties of silverfish in the Ross Sea (O'Driscoll et al. 2011). A substantial bulk of data on biology, ecology, physiology and structure of populations of *P. antarcticum* were collected during the Collaborative East Antarctic Marine Census (CEAMARC) project of CAML (Giraldo et al. 2011; Koubbi et al. 2011; Moteki et al. 2011; Mayzaud et al. 2011; Vallet et al. 2011). In addition to publications, several recent CCAMLR documents and reports update the data on ecology, population structure and life cycle of the silverfish (www.ccamlr.org).

4.6 The Sea Ice and the Antarctic Silverfish

4.6.1 Sea-Ice Cover as a Key Environmental Feature for the Silverfish Life Cycle

Due to its distribution in mid-level waters, at the adult stage the silverfish unlikely faces contact with the ice cover. However, the capability to produce an array of AFGPs (Wöhrmann et al. 1997) even if in small amount (Jin and DeVries 2006) suggests that during adult life this fish could be exposed to ice crystals in the water column. By contrast, larval stages of *P. antarcticum* cope with direct exposition to sea ice cover, as they are well known component of the cryopelagic community (Knox 1994 and references herein).

Recent evidence of direct interaction between the silverfish and the ice cover dates back to spring 2002, when the first spawning ground of this species was localized in the Ross Sea. Silverfish eggs were found floating in huge amounts among the platelet ice under the fast-ice in Terra Nova Bay (TNB) from late October to early December (Vacchi et al. 2004). The monitoring of TNB, by drilling 225 sea-ice stations, during spring surveys in the subsequent years confirmed that embryonated silverfish eggs (at the last developmental stage), mixed with just hatched larvae, were distributed in an area of almost 270 km^2 (Vacchi et al. 2012). The higher amounts of eggs were found in a roughly triangular body of water northward the Campbell Ice Tongue, after named "Silverfish Bay" (SCAR Gazetteer Ref. No 18082, http://data.aad.gov.au/aadc/gaz/scar/) where up 1,328 fish eggs/larvae per litre of water/platelet ice were counted. The egg abundance appeared correlated with a sea-ice thickness (in prevalence over 2.4 m), and with the occurrence of platelet ice underneath.

The features of such nursery area led to significant progress in the knowledge of the early life cycle, and provided answers to some important questions. For instance, which features make TNB a suitable spawning site for the silverfish? Are the geographical or environmental features making an area suitable for spawning? These issues are crucial to a) find other possible reproduction sites along the Antarctic coast; b) understand the role of sea ice, as environmental feature, for reproduction and nursery of the silverfish; c) evaluate the influence of sea ice dynamics on the silverfish spawning and life cycle.

All the above points are relevant to our understanding and predictions about possible impacts of environmental changes on the coastal Antarctic ecosystem where the silverfish plays a central role (e.g. La Mesa et al. 2004; Smith et al. 2007; Moline et al. 2008).

Geographically, the spawning area of silverfish is located at TNB, a well defined coastal region of Victoria Land (Western Ross Sea), limited north by Cape Washington and south by the Drygalski Ice Tongue. The TNB coastline is indented with numerous embayments including Gerlache Inlet and Silverfish Bay in its northern part. The Drygalski Ice Tongue, the Nansen Ice Sheet and the Campbell Glacier Tongue flow down from the continent into the TNB. The bottom topography is rather irregular and it is characterized by steep seabeds and by the Drygalki depression, a deep pit, elongated along shore, reaching more than 1,100 m (the greatest depth of the Ross Sea). Summer circulation in TNB shows a prevailing northward direction in the upper layer along the coast with a clockwise rotation with depth. The coastal area is characterized by warmer and more saline water, while the lowest temperature values are found in the central area of TNB, probably due both to local eddies and to upwelling processes determined by katabatic winds (Budillon and Spezie 2000; Buffoni et al. 2002). A large polynya (maximum extension 80 km) is maintained during winter by combination of persistent katabatic winds, coupled with the barrier effect of the Drygalski Ice Tongue on the pack-ice advection coming from south and south-west (Kurtz and Bromwich 1983, 1985).

The TNB polynya acts as an ice factory during winter and has a primary function in the sea-ice local dynamics (Van Woerst 1999). As a result, a major environmental feature of TNB is seasonal sea ice, bordering the coastal areas for almost 9 months per year. High amounts of platelet ice is present under the annual sea ice, possible resulting from frazil ice growing in the supercooled plumes forming near the ice shelves and glacier tongues.

Data from the Weddell Sea (Hubold 1984; Kellermann 1986), the Ross Sea (Maes et al. 2006; Granata et al. 2009; La Mesa et al. 2010), and the Dumont D'Urville Sea region (Koubbi et al. 2011 and references therein) indicated that geographic and oceanographic features (e.g. close ice shelf or glacier tongues, canyons, water masses stratification, polynyas and katabatic winds) are important for the early life history of the silverfish: combinations of the above features contribute to retain larvae in a favourable environment in water close to shore, and to gradually carry them towards the inner-shelf depressions and banks as they increase in size. The fact that this unique spawning site for the silverfish is covered by ice stresses the role that ice, as an environmental feature, can also play.

The coastal location of a spawning area, together with the consistent pattern of ontogenetic segregation among larvae, juveniles and adults, confirms predictions about migrations to coastal areas to spawn (Kellermann 1986). On the other hand, the observations on hatching within the platelet ice contrast with earlier speculations relating to *P. antarcticum* in the Weddell Sea, in which it was suggested that the embryonated eggs drift over the shelf slope, and hatch at depths below 500 m giving rise to larvae that then swim to the surface (Wöhrmann et al. 1997).

How do silverfish adults choose a suitable environment to spawn? A possible answer has been suggested by Koubbi et al. (2011), indicating "homing" as a possible strategy. According to homing mechanisms, spawning areas are geographically determined when fish return to spawn at the same place where they were born, or spawning areas are environmentally determined when fish return to spawn in environmental conditions they experienced at the larval stage (Koubbi et al. 2011). Does a combination of geographic and environmental homing behaviour help adult silverfish to recognize their spawning sites during their winter migration to coastal areas? Due to environmental inter-annual variations, homing strategy could not lead to larvae being released in optimal areas each year, but it might ensure a good larval survival rate over the long term (Petitgas et al. 2006). On the other hand, occurrence of such a homing mechanism may create a time lag in the detectable impact of long-term environmental change on spawning distribution (Corten 2002), making it necessary to perform long-term monitoring surveys in order to predict the possible effects of environmental changes.

4.6.2 The Sea Ice Cover as a Challenging Feeding Ground and Refuge

Apart from the hypothetical role of sea ice for homing, it is important to address whether the sea cover is mostly an environmental constraint to cope with, waiting for the onset of production cycle following ice retreat, or whether it provides favourable conditions and contributes to success of the early life history.

Several records in the 1980s in the Ross and Weddell Seas in waters adjacent to the continental ice shelves indicate that *P. antarcticum* is associated with sea ice early in life history (Kellermann 1986; Knox 1994). As seen in previous sections, the underside of sea ice and the platelet-ice layer are sites of enhanced primary production, and provide favourable conditions for hatching and protection for several grazers whose life cycle is coupled with sea-ice formation and melting (Arrigo et al. 2010; Caron and Gast 2010; Bluhm et al. 2010). In addition, Granata et al. (2009) have found that some ice-associated grazers are preferentially eaten by silverfish post-larvae in the platelet ice in TNB in spring, thus indicating a direct trophic role of sea ice in the early life history of the silverfish.

The occurrence of huge amounts of developing eggs within the platelet ice suggested a possible function of platelet ice as a refuge, reducing egg predation by other fish, commonly reported in benthic feeders in the Ross Sea (La Mesa et al. 2010). However, such kind of brooding opens several questions dealing with the biological and physiological needs that allows survival in this challenging environment. For instance, the early developmental stages lack significant antifreeze capacity, however underdeveloped gill structures certainly minimize exposition to ice (Cziko et al. 2006; Bottaro et al. 2009). Early larvae of *P. antarcticum* seem to possess responsiveness toward increase of prooxidant conditions naturally occurring in the platelet ice in early spring (Regoli et al. 2005), nevertheless, soon after

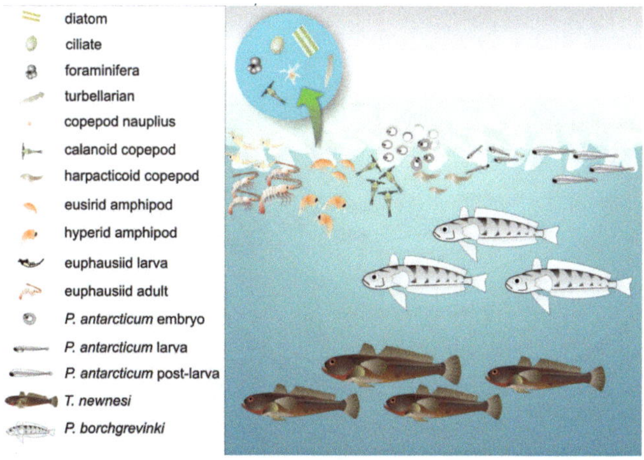

Fig. 4.2 The Antarctic cryopelagic community in Northern Terra Nova Bay in spring

hatching, they are negatively photo tactic in their swimming behaviour and sink in the water column (Evans et al. 2012; Vacchi et al. 2012).

By considering the data on the nursery ground in the Ross Sea, and the former information on spatial and vertical distribution of larval phases, juveniles and adults, we can draw an updated picture of the relationship between the sea ice and this key Antarctic fish. Several points still remain uncertain and new data need to be acquired by long-term monitoring surveys, possibly including winter periods. However, by combining evidence from available information and some hypotheses, the interactions between this fish and sea ice during early life and reproduction emerge as major features in its life history, as shown in Figs. 4.2 and 4.3.

As shown in Fig. 4.2, when hatching occurs, in spring, eggs and newly hatched larvae of silverfish become locally important component of the cryopelagic community at TNB. The eggs are of small size (2.2–2.5 mm in diameter) and float in the platelet ice under the sea cover (Vacchi et al. 2004). Newly hatched larvae, although appearing finely tuned to local conditions, do not seems to feed in situ, and soon sink in the water column (Vacchi et al. 2012). Post-larvae are also present the cryopelagic community in various Antarctic regions (Hubold 1985; Knox 1994; Granata et al. 2009).

In winter/spring juveniles and adults have not been reported in proximity of the sea ice cover. Juveniles are largely pelagic and progressively occupy a wide area of the continental shelf, often in close proximity to the slopes surrounding the banks and near the shelf break (Hubold 1984, 1985; Kellermann 1986; White and Piatkowski 1993; La Mesa et al. 2010). Adults are widely distributed both pelagically and demersally in the shelf and slope areas around the continent (DeWitt et al. 1990; Miller 1993; Trunov 2001; Donnelly et al. 2004; O'Driscoll 2011). Nevertheless, some adult *Pleuragramma* have been recorded shoaling and feeding beneath the ice along the Antarctic Peninsula (Daniels 1982; Daniels and Lipps 1982). In the pelagic zone of the ice-covered McMurdo Sound *Pleuragramma* is

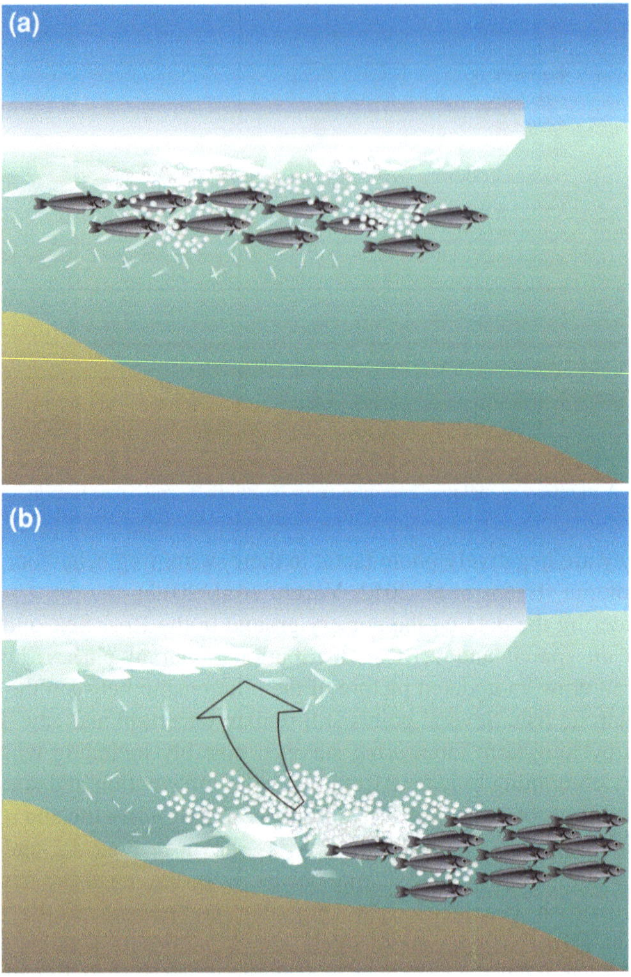

Fig. 4.3 Two alternative hypotheses of spawning mode for *P. antarcticum*. Eggs could be released close to the sea ice cover and soon embedded by the growing platelet ice (**a**), or they could be released in deeper waters and lifted up, along with ice crystals and platelets, by local hydrodynamic flows (**b**)

very abundant and serves as major food item for a wide variety of organisms (Eastman 1985). Moreover, some dead adult specimens were collected close to the ice cover near in the spawning area of TNB (Vacchi et al. 2012).

Although the relationship between the silverfish and the ice cover in winter largely remains to be investigated, a direct trophic interaction for early phases, and a possible interaction for reproduction can be suggested. The hypothesis of trophic relationship during early phases is supported by the evidences that larvae do not accumulate large energy reserves for the winter season (Koubbi et al. 2007), and that the foraging activity of young *Pleuragramma* does not change substantially during the year (Hubold and Hagen 1997). Moreover, post-larvae actively feeding on sea-ice associated copepods have been observed in TNB, thus suggesting that similar trophic interaction can also occur during winter at the ice–water interface (Granata et al. 2009).

Direct interaction of the adult silverfish with the ice can occur for spawning. Evidence supports spawning events occurring in winter. Adults migrate to cold in-shore areas and possibly recognize their spawning sites by geographical and/or environmental characteristics, according to the homing behaviour. However, spawning events have not yet been directly observed in the field, and where the eggs are released remains to be elucidated. As a first possibility (Fig. 4.3a) the eggs could be released and fertilized close to the sea ice cover, remaining in a challenging environment for all the long time of embryonic development (3–4 months). Alternatively (Fig. 4.3b), the eggs could be laid in deeper, ice–free seawater, and lifted up under to the sea ice by local hydrodynamic flows. In this second case, eggs reach the nursery environment within the platelet ice after fertilization.

At that stage, the micropyle, potentially large enough to allow entry of ice, has been occluded and an intact chorion can protect developing embryos from freezing (Cziko et al. 2006). Given the prediction of spawning in July and August (Vacchi et al. 2012), when the platelet ice forms in the water column in the Ross Sea, we cannot exclude that ice platelets could act as vectors for lifting eggs from a deeper deposition site to overlying waters covered by sea ice, according to mechanism already hypothesized for explaining colonization of sea ice cover by invertebrates (Dieckmannn et al. 1986; Bluhm et al. 2010).

References

Albertson C, Yi-Lin Yan YL, Titus TA, Pisano E, Vacchi M, Yelick PC, Detrich HW, Postlethwait JH (2010) Molecular pedomorphism underlies craniofacial skeletal evolution in Antarctic notothenioid fishes. BMC Evol Biol 10:4

Andriashev AP (1968) The problem of life community associated with Antarctic fast ice. In: Currle RI (ed) Symp Antarctic oceanography. Scott Polar Res Inst, Cambridge, pp 147–155

Andriashev AP (1970) Cryopelagic fishes of the Arctic and Antarctic and their significance in polar ecosystems. In: Holdgate MW (ed) Antarctic ecology 1. Academic Press, London, pp 297–304

Andriashev AP (1987) A general review of the antarctic bottom fish fauna. In: Kullander SO, Fernholm B (eds) Proceedings of 5th Congress European Ichthyol. Swedish Mus Nat Hist, Stockholm, pp 357–372

Arndt CE, Swadling KM (2006) Crustacea in Arctic and Antarctic sea ice: distribution, diet and life history strategies. Adv Mar Biol 51:197–315

Arrigo KR, Thomas DN (2004) Large scale importance of sea ice biology in the southern ocean. Antarctic Sci 16:471–486

Arrigo KR, Mock T, Lizotte MP (2010) Primary producers and sea ice. In: Thomas DN, Dieckmann GS (eds) Sea ice, 2nd edn. Blackwell, Oxford, pp 283–325

Azzali M, Leonori I, Biagiotti I, De Felice A, Angiolillo M, Bottaro M, Vacchi M (2010) Target strength studies on Antarctic silverfish (*Pleuragramma antarcticum*) in the Ross Sea. CCAMLR Science 17:75–104

Bluhm BA, Gradinger RR, Schnack-Schiel SB (2010) Sea ice meio- and macrofauna. In: Thomas DN, Dieckmann GS (eds) Sea ice, 2nd edn. Blackwell, Oxford, pp 357–393

Bottaro M, Oliveri D, Ghigliotti L, Pisano E, Ferrando S, Vacchi M (2009) Born among the ice: first morphological observations on two developmental stages of the Antarctic silverfish *Pleuragramma antarcticum*, a key species of the southern ocean. Rev Fish Biol Fish 19: 249–259

Bouchard C, Fortier L (2011) Circum-arctic comparison of the hatching season of polar cod *Boreogadus saida*: a test of the freshwater winter refuge hypothesis. Prog Oceanogr 90: 105–116

Brandon MA, Cottier FR, Nilsen F (2010) Sea ice and oceanography. In: Thomas DN, Dieckman GS (eds) Sea Ice. Wiley-Blackwell, Oxford, pp 79–111

Brierley AS, Fernandes PG, Brandon MA, Armstrong F, Millard NW, McPhail SD, Stevenson P, Pebody M, Perrett J, Squires M, Bone DG, Griffiths G (2002) Antarctic krill under sea ice: elevated abundance in a narrow band just south of ice edge. Science 295:890–1892

Budillon G, Spezie G (2000) Thermohaline structure and variability in Terra Nova Bay polynya, Ross Sea. Antarct Sci 12:493–508

Buffoni G, Cappelletti A, Picco P (2002) An investigation of thermohaline circulation in Terra Nova Bay polynya. Antarct Sci 14:83–92

Carey AG (1985) Marine ice fauna: Arctic. In: Horner R (ed) Sea ice biota. CRC Press, Boca Raton, Florida, pp 173–190

Caron DA, Gast RJ (2010) Heterotrophic protists associated with sea ice. In: Thomas DN, Dieckmann GS (eds) Sea ice, 2nd edn. Blackwell, Oxford, pp 327–356

Cheng C-HC (1996) Genomic basis for antifreeze glycopeptide heterogeneity and abundance in Antarctic notothenioid fishes. In: Ennion SJ, Goldspink G (eds) Gene expression and manipulation in aquatic organisms, vol 58. Cambridge University Press, Cambridge, pp 1–20

Cheng C-HC (1998a) Origin and mechanism of evolution of antifreeze glycoproteins in polar fishes. In: di Prisco G, Pisano E, Clarke A (eds) Fishes of Antarctica. A biological overview. Springer, Milano, pp 311–328

Cheng C-HC (1998b) Evolution of the diverse antifreeze proteins. Curr Opin Genet Dev 8:715–720

Corten A (2002) The role of “conservatism” in herring migrations. Rev Fish Biol Fish 11:339–361

Crook J (2010) Ice growth and platelet crystals in Antarctica. PhD Thesis, Victoria University of Wellington, New Zealand

Christiansen JS (2012) The TUNU-programme: Euro-Arctic marine fishes—diversity and adaptation. In: di Prisco G, Verde C (eds) Adaptation and evolution in marine environments, vol 1. Series “From Pole to Pole”. Springer-Verlag, Berlin Heidelberg, pp 35–50

Cziko PA, Evans CW, Cheng C-HC, DeVries AL (2006) Freezing resistance of antifreeze-deficient larval Antarctic fish. J Exp Biol 209:407–420

Daniels RA (1982) Feeding ecology of some fishes of the Antarctic Peninsula. Fish Bull 80:575–588

Daniels RA, Lipps JH (1982) Distribution and ecology of fishes of the Antarctic Peninsula. J Biogeogr 9:1–9

Dayton PK, Robilliard GA, Devries AL (1969) Anchor ice formation in McMurdo sound, Antarctica, and its biological effects. Science 163:273–274

Deming JW (2010) Sea ice bacteria and viruses. In: Thomas DN, Dieckmann GS (eds) Sea ice. Wiley-Blackwell, Oxford, pp 247–282

Deng C, Cheng C-HC, Ye H, He X, Chen L (2010) Evolution of an antifreeze protein by neofunctionalization under escape from adaptive conflict. Proc Nat Acad Sci U S A 107:21593–21598

DeVries AL (1970) Freezing resistance in Antarctic fishes. In: Holdgate MW (ed) Antarctic ecology. Academic Press, New York

DeVries AL (1971) Glycoproteins as biological antifreeze agents in Antarctic fishes. Science 172:1152–1155

DeVries AL (1986) Glycopeptide and peptide antifreeze-interaction with ice. In: Colowick SP, Kaplan NO (eds) Methods in enzymology, vol 127. Academic Press, New York, pp 293–303

DeVries AL, Cheng C-HC (2005) Antifreeze proteins and organismal freezing avoidance in polar fish. In: Farrell AP, Steffensen JF (eds) The physiology of polar fishes, vol 22. Elsevier Inc, San Diego, pp 155–201

DeVries AL, Steffenson JF (2005) The Arctic and Antarctic polar marine environments. In: Farrell AP, Steffensen JF (eds) The physiology of polar fishes, vol 22. Elsevier Inc, San Diego, pp 1–24

DeVries AL, Wohlschlag DE (1969) Freezing resistance in some Antarctic fishes. Science 163: 1073–1075

DeWitt HH, Heemstra PC, Gon O (1990) Nototheniidae. In: Gon O, Heemstra PC (eds) Fishes of the southern ocean. JLB Smith Institute of Ichthyology, Grahamstown, pp 279–331

Duman JG, DeVries AL (1974) Freezing resistance in winter flounder, *Pseudopleuronectes americanus*. Nature 247:237–238

Dieckmann GS, Hellmer HH (2010) The importance of sea ice: an overview. In: Thomas DN, Dieckmann GS (eds) Sea ice. Wiley-Blackwell, New York, pp 1–22

Dieckmann GS, Rohardt G, Hellmer H, Kipfstuhl J (1986) The occurrence of ice platelets at 250 m near the Filchner Ice Shelf and its significance to sea ice biology. Deep Sea Res 33:141–148

Donnelly J, Torres JJ (2008) Pelagic fishes in the Marguerite Bay region of the West Antarctic Peninsula shelf. Deep Sea Res II 55:523–539

Donnelly J, Torres JJ, Sutton TT, Simoniello C (2004) Fishes of the eastern Ross Sea, Antarctica. Polar Biol 27:637–650

Eastman JT (1985) *Pleuragramma antarcticum* (Pisces, Nototheniidae) as food for other fishes in McMurdo sound, Antarctica. Polar Biol 4:155–160

Eastman JT (1993) Antarctic fish biology: evolution in a unique environment. Academic Press, San Diego

Eastman JT (1997) Phyletic divergence and specialization for pelagic life in the Antarctic notothenioid fish *Pleuragramma antarcticum*. Comp Biochem Physiol A 118:1095–1101

Eastman JT, DeVries AL (1985) Adaptations for cryopelagic life in the Antarctic notothenioid fish *Pagothenia borchgrevinki*. Polar Biol 4:45–52

Eastman JT, McCune AR (2000) Fishes on the Antarctic continental shelf: evolution of a marine species flock? J Fish Biol 57:84–102

Eicken H, Gradinger R, Salganek M, Shirasawa K, Perovich D, Leppäranta M (2009) Field techniques for sea ice research. University of Alaska Press, Alaska, p 566

Evans CW, Williams DE, Vacchi M, Brimble MA, DeVries AL (2011) Metabolic and behavioural adaptations during early development of the Antarctic silverfish, *Pleuragramma antarcticum*. Polar Biol. doi: 10.1007/s00300-011-1134-7

Faleyeva TI, Gerasimchuk VV (1990) Features of reproduction in the Antarctic sidestripe, *Pleuragramma antarcticum* (Nototheniidae). J Ichthyol 30:67–79

Fedotov VI, Cherepanov NV, Tyshko KP (1998) Some features of the growth, structure and metamorphism of East Antarctic landfast ice. In: Jeffries MO (ed) Antarctic sea ice: physical processes, interactions and variability. antarctic research series, vol 74. American Geophysical Union, Washington DC, pp 343–354

Fletcher GL, Hew CH, Davies PL (2001) Antifreeze proteins of teleost fishes. Ann Rev Physiol 63:359–390

Fraser CI, Nikula R, Spencer HG, Waters JM (2009) Kelp genes reveal effects of subantarctic sea ice during the last glacial maximum. Proc Nat Acad Sci U S A 106:3249–3253

Fuiman LA, Davis RW, Williams TM (2002) Behavior of midwater fishes under Antarctic ice: observations by a predator. Mar Biol 140:815–822

Garrison D (1991) Antarctic sea ice biota 1. Am Zool 31:17–33

Geoffroy M, Robert D, Darnis G, Fortier L (2011) The aggregation of polar cod (*Boreogadus saida*) in the deep Atlantic layer of ice-covered Amundsen Gulf (Beaufort Sea) in winter. Polar Biol 27:595–603

Gerasimchuk VV (1986) Characteristics of Antarctic silverfish, *Pleuragramma antarcticum* (Nototheniidae), from Olaf-Pruds Bay (Commonwealth Sea, Eastern Antarctica) with notes on the identification of the species. J Ichthyol 26:10–17

Giraldo C, Cherel Y, Vallet C, Mayzaud P, Tavernier E, Moteki M, Hosie G, Koubbi P (2011) Ontogenic changes in the feeding ecology of the early life stages of the Antarctic silverfish (*Pleuragramma antarcticum*) documented by stable isotopes and diet analysis in the Dumont d'Urville Sea (East Antarctica). Polar Sci 5:252–263

Gow AJ, Ackley SF, Govoni JW (1998) Physical and structural properties of land-fast ice in McMurdo Sound, Antarctica. In: Jeffries MO (ed) Antarctic sea ice: physical processes, interactions and variability, Antarctic research series, vol 74. American Geophysical Union, Washington DC, pp 355–374

Gradinger RR (2001) Adaptation of Arctic and Antarctic ice metazoa to their habitat. Zoology 104:339–345

Gradinger RR, Bluhm BA (2004) In situ observations on the distribution and behaviour of amphipods and Arctic cod (*Boreogadus saida*) under the sea ice of the high Arctic Canada basin. Polar Biol 27:595–603

Granata A, Zagami G, Vacchi M, Guglielmo L (2009) Summer and spring trophic niche of larval and juvenile *Pleuragramma antarcticum* in the western Ross Sea, Antarctica. Polar Biol 32: 369–382

Guglielmo L, Granata A, Greco S (1998) Distribution and abundance of postlarval and juvenile *Pleuragramma antarcticum* (Pisces, Nototheniidae) off Terra Nova Bay (Ross Sea, Antarctica). Polar Biol 19:37–51

Gutt S (2001) On the direct impact of ice on benthic communities, a review. Polar Biol 24: 553–564

Holland DM, Jenkins JA (1999) Modelling thermodynamic ice-ocean interactions at the base of an ice shelf. J Phys Oceanogr 29:1787–1800

Hammer WM, Hammer PP (2000) Behavior of Antarctic krill (*Euphausia superba*): schooling, foraging, and antipredatory behavior. Can J Fish Aquat Sci 57(Suppl 3):192–202

Horner RA (1985) Sea ice biota. CRC Press, Boca Raton

Horner RA, Ackley SF, Dieckmann GS, Gulliksen B, Hoshiai T, Legrendre L, Melnikov IA, Reeburgh WS, Spindler M, Sullivan CW (1992) Ecology of sea ice biota. 1 habitat, terminology, and methodology. Polar Biol 12:417–427

Hubold G (1984) Spatial distribution of *Pleuragramma antarcticum* (Pisces: Nototheniidae) near the Filchner and Larsen ice shelves (Weddell Sea, Antarctica). Polar Biol 3:231–236

Hubold G (1985) The early life history of the high Antarctic silverfish, *Pleuragramma antarcticum*. In: Siegfried WR, Condy PR, Laws RM (eds) Antarctic nutrient cycles and food webs. Springer-Verlag, Heidelberg, Berlin, pp 445–451

Hubold G, Ekau W (1987) Midwater fish fauna of the Weddell Sea, Antarctica. In: Kullander SO, Fernholm B (eds) Proceedings of 5th congress European ichthyol Society. Swedish Mus Nat Hist, Stockholm, pp 391–396

Hubold G, Ekau W (1990) Feeding patterns of post-larval and juvenile notothenioids in the southern Weddell Sea (Antarctica). Polar Biol 10:255–260

Hubold G, Hagen W (1997) Seasonality of feeding and lipid content in juvenile *Pleuragramma antarcticum* (Pisces: Nototheniidae) from the southern Weddell Sea. In: Battaglia B, Valencia J, Walton WH (eds) Antarctic communities. Species, structure and survival. Cambridge University Press, New York, pp 277–283

Hureau JC (1994) The significance of fish in the marine Antarctic ecosystems. Polar Biol 14:307–313

Jeffries MO, Eicken H (2000) Internatl symp on Sea Ice and its Interactions with the Ocean, Atmosphere and Biosphere. Internatl Glaciol Soc, Ann Glaciol 33:591

Jeffries MO, Schwartz K, Morris K, Veazey AD, Krouse HR, Gushing S (1995) Evidence for platelet ice accretion in Arctic sea ice development. J Geophys Res 100:10905–10914

Jin Y, DeVries A (2006) Antifreeze glycoprotein levels in Antarctic notothenioid fishes inhabiting different thermal environments and the effect of warm acclimation. Comp Biochem Physiol B 144:290–300

Keeling RF, Stephens BB (2001) Antarctic sea ice and the control of pleistocene climate instability. Paleoceanogr 16:112–131

Kellermann A (1986) On the biology of early life stages of notothenioid fishes (Pisces) off the Antarctic Peninsula. Berichte zur Polarforschung 31:1–155

Klingenberg CP, Ekau W (1996) A combined morphometric and phylogenetic analysis of an ecomorphological trend: pelagization in Antarctic fishes (Perciformes: Nototheniidae). Biol J Linn Soc 59:143–177

Knox GA (1994) The biology of the Southern Ocean. Cambridge University Press, Cambridge

Kock KH (1992) Antarctic fish and fisheries. Cambridge University Press, Cambridge

Kock KH, Kellermann A (1991) Reproduction in Antarctic notothenioid fish. Antarctic Sci 3:125–150

Koubbi P, O'Brien C, Loots C, Giraldo C, Smith M, Tavernier E, Vacchi M, Vallet C, Chevallier J, Moteki M (2011) Spatial distribution and inter-annual variations in the size frequency distribution and abundances of *Pleuragramma antarcticum* larvae in the Dumont d'Urville Sea from 2004 to 2010. Polar Sci 5(2):225–238

Koubbi P, Vallet C, Razouls S, Grioche A, Hilde D, Courcot L, Janquin MA, Vacchi M, Hureau JC (2007) Condition and diet of larval *Pleuragramma antarcticum* from Terre Adélie (Antarctica) during summer. Cybium 31:67–76

Kurbjeweit F, Gradinger R, Weissenberger J (1993) The life cycle of *Stephos longipes*—an example for cryopelagic coupling in the Weddell Sea (Antarctica). Mar Ecol Prog Ser 98: 255–262

Kurtz DD, Bromwich DH (1983) Satellite observed behaviour of the Terra Nova Bay polynya. J Geophys Res 88:9717–9722

Kurtz DD, Bromwich DH (1985) A recurring, atmospherically forced polynya in Terra Nova Bay. In: Jacobs SS (ed) Oceanology of the Antarctic continental shelf. American Geophysical Union, Washington, pp 177–201

La Mesa M, Eastman JT (2011) Antarctic silverfish: life strategies of a key species in the high-Antarctic ecosystem. Fish Fish. doi:10.1111/j.1467-2979.2011.00427.x

La Mesa M, Catalano B, Russo A, Greco S, Vacchi M, Azzali M (2010) Influence of environmental conditions on spatial distribution and abundance of early life stages of Antarctic silverfish, *Pleuragramma antarcticum* (Nototheniidae), in the Ross Sea. Antarct Sci 22:243–254

La Mesa M, Eastman JT, Vacchi M (2004) The role of notothenioid fish in the food web of the Ross Sea shelf waters: a review. Polar Biol 27:321–338

La Mesa M, Vacchi M, Zunini Sertorio T (2000) Feeding plasticity of *Trematomus newnesi* (Pisces, Nototheniidae) in Terra Nova Bay, Ross Sea, in relation to environmental conditions. Polar Biol 23:38–45

Legendre L, Ackley SF, Dieckmann GS, Gulliksen B, Horner R, Hoshiai T, Melnikov IA, Reeburgh WS, Splinder M, Sullivan CW (1992) Ecology of sea ice biota: 2 global significance. Polar Biol 12:429–444

Leonard GH, Purdie CR, Langhorne PJ, Haskell TG, Williams MJM, Frew RD (2006) Observations of platelet ice growth and oceanographic conditions during the winter of 2003 in McMurdo sound, Antarctica. J Geophys Res 111 C04012 doi:10.1029/2005JC002952

Macnaughton MO, Thormar J, Berge J (2007) Sympagic amphipods in the Arctic pack ice: redescriptions of *Eusirus holmii* Hansen, 1887 and *Pleusymtes karstensi* (Barnard 1959). Polar Biol 30:1013–1025

Maes J, Van de Putte A, Hecq J-H, Volckaert FAM (2006) State dependent energy allocation in the pelagic Antarctic silverfish *Pleuragramma antarcticum*: trade-off between winter reserves and buoyancy. Mar Ecol Prog Ser 326:269–282

Mangoni O, Saggiomo M, Modigh M, Catalano G, Zingone A, Saggiomo V (2009) The role of platelet ice microalgae in seeding phytoplankton blooms in Terra Nova Bay (Ross Sea, Antarctica): a mesocosm experiment. Polar Biol 32:311–323

Massom RA, Stammerjohn SE (2010) Antarctic Sea ice change and variability–physical and ecological implications. Polar Sci 4:149–186

Mayzaud P, Chevalier J, Tavernier E, Moteki M, Koubbi P (2011) Lipid composition of the high Antarctic fish *Pleuragramma antarcticum*, influence of age class. Polar Sci 5:104–117

McGuinness MJ, Langhorne P (2006) A platelet puzzle in Antarctica. In: Proceedings of the KSIAM 2006 annual meeting, Konkuk University, Seoul, Korea, 24–25 Nov

Melnikov IA (1997) The Arctic sea ice ecosystem. Gordon Breach, Amsterdam

Miller RG (1993) A history and Atlas of the fishes of the Antarctic ocean. Foresta Institute for Ocean and Mountain Studies, Carson City, Nevada

Moline MA, Karnovsky NJ, Brown Z, Divoky GJ, Frazer TK, Jacoby CA, Torres JJ, Fraser WR (2008) High latitude changes in ice dynamics and their impact on polar marine ecosystems. Ann N Y Acad Sci 1134:267–319

Moteki M, Koubbi P, Pruvost P, Tavernier E, Hulley PA (2011) Spatial distribution of pelagic fish off Adélie and George V Land, East Antarctica in the austral summer 2008. Polar Sci 5(2):211–224

Near TJ, Jones CJ, Eastman JT (2009) Geographic intraspecific variation in buoyancy within Antarctic notothenioid fishes. Antarctic Sci 21:123–129

Nicodemus-Johnson J, Silic S, Silic L, Pisano E, Cheng C-HC (2011) Assembly of the antifreeze glycoprotein/trypsinogen-like protease genomic locus in the Antarctic fish *Dissostichus mawsoni* (Norman). Genomics 98:194–201

Nicol S (2006) Krill, currents, and sea ice: *Euphausia superba* and its changing environment. Bioscience 56:111–120

Notz D (2009) The future of ice sheets and sea ice: between reversible retreat and unstoppable loss. Proc Nat Acad Sci U S A 106:20590–20595

O'Driscoll RL, Macaulay GJ, Gauthier S, Pinkerton M, Hanchet S (2011) Distribution, abundance and acoustic properties of Antarctic silverfish (*Pleuragramma antarcticum*) in the Ross Sea. Deep-Sea Res II 58:181–195

Parin NV (1968) Oceanic ichthyologeography: an attempt to review the distribution and origin of pelagic and bottom fishes outside continental shelves and neritic zones. Arch Fisch Wis 1:5–41

Perovich DK, Light B, Eicken H, Jones KF, Runciman K, Nghiem SV (2007) Increasing solar heating of the Arctic ocean and adjacent seas, 1979–2005: attribution and role in the ice-albedo feedback. Geophys Res Lett 34:L19505. doi:10.1029/2007GL031480

Petitgas P, Reid D, Planque B, Nogueira E, O'Hea B, Cotano U (2006) The entrainment hypothesis: an explanation for the persistence and innovation in spawning migration and life cycle patterns. In: ICES CM documents 2006/B:07, p 9

Regoli F, Nigro M, Benedetti M, Fattorini D, Gorbi F (2005) Antioxidant efficiency in early life stages of the Antarctic silverfish, *Pleuragramma antarcticum*: responsiveness to pro-oxidant conditions of platelet ice and chemical exposure. Aquat Toxicol 75:43–52

Schmitz OJ, Post E, Burns CE, Johnston KM (2003) Ecosystem responses to global climate change: moving beyond color mapping. Bioscience 53:1199–1205

Schnack-Schiel SB, Thomas DN, Dahms HU, Haas C, Mizdalski E (1998) Copepods in Antarctic sea ice. In: Lizotte MP, Arrigo K (eds) Antarctic sea ice biological processes, interactions and variability, Antarctic research series, vol 73. American Geophysical Union, Washington, pp 173–182

Shier WT, Lin Y, DeVries AL (1972) Structure and mode of action of glycoproteins from Antarctic fishes. Biochim Biophys Acta 263:406–413

Smith WO, Ainley DG, Cattaneo-Vietti R (2007) Marine ecosystems: the Ross Sea. Antarctic ecology: from genes to ecosystems. Phil Trans Royal Soc B 362:95–111

Smith IJ, Langhorne PJ, Haskell, Trodahl HJ, Frew R, Venner R (2001) Platelet ice and the land-fast sea ice of McMurdo sound, Antarctica. Ann Glaciol 33:21–27

Thomas DN, Dieckmann GS (2010) Sea Ice 2. Wiley-Blackwell, Oxford

Thomas DN, Papadimitriou S, Michel C (2010) Biogeochemistry of sea ice. In: Thomas DN, Dieckmann GS (eds) Sea ice. Wiley-Blackwell, West Sussex, pp 425–467

Trunov IA (2001) Occurrence of *Pleuragramma antarcticum* (Nototheniidae) off South Georgia Island and the South Sandwich Islands (Antarctica). J Ichthyol 41:549–550

Tynan CT, Ainley DG, Stirling I (2010) Sea ice: a critical habitat for polar marine mammals and birds. In: Thomas DN, Dieckmann GS (eds) Sea ice. Wiley- Blackwell, West Sussex, pp 395–423

Vacchi M, La Mesa M, Dalù M, Macdonald J (2004) Early life stages in the life cycle of Antarctic silverfish, *Pleuragramma antarcticum* in Terra Nova Bay, Ross Sea. Antarct Sci 16: 299–305

Vacchi M, DeVries AL, Evans CW, Bottaro M, Ghigliotti L, Pisano E (2012) The nursery area of the Antarctic silverfish *Pleuragramma antarcticum* at Terra Nova Bay (Ross Sea): first estimate of spatial distribution and abundance of eggs and larvae under the seasonal fastice.

Vallet C, Labat JP, Smith M, Koubbi P (2011) Interannual variations in euphausiid life stage distribution in the dumont d'Urville sea from 2004 to 2008. Polar Sci 5:166–178

Van Woert ML (1999) Wintertime dynamics of the Terra Nova Bay polynya. J Geophys Res 104:7753–7769

Werner I, Martinez Arbizu P (1999) The sub-ice fauna of the Laptev Sea and the adjacent Arctic Ocean in summer 1995. Polar Biol 21:71–79

White MG, Piatkowski U (1993) Abundance, horizontal and vertical distribution of fish in eastern Weddell Sea micronekton. Polar Biol 13:41–53

Williams R (1988) The inshore marine fishes of the Vestfold Hills region, Antarctica. Hydrobiol 165:161–167

Wöhrmann APA, Hagen W, Kunzmann A (1997) Adaptations of the Antarctic silverfish *Pleuragramma antarcticum* (Pisces: Nototheniidae) to pelagic life in high Antarctic waters. Mar Ecol Prog Ser 151:205–218

Chapter 5
Connectivity and Molecular Ecology of Antarctic Fishes

Filip A. M. Volckaert, Jennifer Rock and Anton P. Van de Putte

5.1 Introduction

The international program on Evolution and Biodiversity in the Antarctic (Anonymous 2005) focused on the influence of evolution and diversity of life on the properties and dynamics of the Southern Ocean (SO) biome. It also wanted to predict how communities and organisms respond to environmental change. A component of the program aimed at understanding micro-evolutionary processes and dynamics during the Pleistocene and Holocene. The past three million years have shaped the "shallow" evolution of genes, organisms and ecosystems through major climate changes and short period earth periodicities. Fish, a major source of ecosystem goods, play a key role in the ecosystem. However, it is only since relatively recently that the fish communities of the SO started to reveal their characteristics. Before summarizing the current understanding of their connectivity and molecular ecology, we introduce the reader to those aspects that have affected their recent evolution so much.

F. A. M. Volckaert (✉) · A. P. Van de Putte
Laboratory of Biodiversity and Evolutionary Genomics, Katholieke Universiteit Leuven, Charles Deberiotstraat 32, 3000 Leuven, Belgium
e-mail: filip.volckaert@bio.kuleuven.be

J. Rock
Department of Zoology, University of Otago, Dunedin 9054, New Zealand
e-mail: jennifer.rock@otago.ac.nz

A. P. Van de Putte
Belgian Biodiversity Platform, Royal Belgian Institute for Natural Sciences, Vautierstraat 27, 1000 Brussels, Belgium
e-mail: anton.vandeputte@bio.kuleuven.be

G. di Prisco and C. Verde (eds.), *Adaptation and Evolution in Marine Environments, Volume 1*, From Pole to Pole, DOI: 10.1007/978-3-642-27352-0_5,

A most outstanding observation is that environmental conditions at high latitudes are extreme. The water temperature hovers amazingly stable close to the freezing point and has high oxygen saturation values (see Chap. 8 Marshall 2012). Local fish species have adapted to these conditions through amongst others slow metabolism, restricted thermal tolerance (Pörtner 2006), long generation times, above average values of blood osmolarity through lacing with anti-freeze proteins (see Chap. 8 Marshall 2012) and the loss of gene functions, e.g. heat shock and hemoglobin (di Prisco et al. 2002; see Chap. 11 Verde et al. 2012). The strong seasonal photoperiod leads to time-constrained primary production with the microbial loop playing an important role throughout the year (Thomson et al. 2010).

Also the conditions of the SO proper are peculiar. The upper water column has been isolated from the world oceans for some 31 million years through the presence of a stable frontal system, the Antarctic Polar Front (APF) (Lawver and Gahagan 2003). South of the APF flows the Antarctic Circumpolar Current (ACC) clockwise around the Antarctic continent up to 2,000 m depth. It might favour large-scale dispersal of Antarctic organisms and lead to homogenised communities. The physical and hydrodynamic continuity between ocean surface and bottom suggests continuity between the faunas of the surface waters and adjacent deep-sea (Gutt et al. 2010). On a regional scale the nearshore East Wind Drift and clockwise gyres (such as the Weddell and Ross Sea) may have lead to evolutionary isolation. The average depths of the continental shelves are larger than most shelves (500 m) and widths are rather narrow except for the large embayments in the Ross and Weddell Seas. Sea ice covers 60% of the continental shelf in winter and 20% in summer. In the past the continental ice mass has shaped the profile of the continental shelf, especially through the recurrence of grounded ice during Glacial Periods (Thatje et al. 2005). Benthic biodiversity is locally very high, but unlike first thought not uniformly high (Gutt et al. 2010).

Global change has been dramatic in the SO with the loss of many animal taxa since the Cenozoic (Rogers 2007). The disappearance of many fish taxa created new opportunities (niches) for the surviving taxa. Notothenioidea (129 species) and Liparidae expanded and diversified into novel habitats. Today the SO harbours a remarkably high (88%) level of endemism (Eastman 2005; Chap. 6 Lecointre 2012). Ancestral notothenioids had a demersal life-style and gradually colonized the water column. The absence of a swim bladder lead to alternative paths to neutral buoyancy, for example through lipid deposition and reduced skeletal mineralisation (Eastman and Barrera-Oro 2010; Patarnello et al. 2011). Pleistocene glaciations have led to the extrusion of continental shelf fauna during cold epochs and recolonisation from refugia during warm epochs (Thatje et al. 2005). As the shelves were narrow, taxa were forced either to occupy greater depths or to adapt a pelagic life-style. SO fish has stenothermal traits (Patarnello et al. 2011), which contrasts with the broader phenotypic plasticity of smaller and more active invertebrates (Peck et al. 2009). Although fish may acclimate up to 4°C (Bilyk and DeVries 2010), overall the current scale of change outpaces their evolutionary potential. Their typically long generation times do not allow fast adaptation.

Considerable conceptual and technical progress have stimulated research on the ecosystems and biota of the SO. It includes the launching of ice-strengthened research vessels and satellites equipped with platforms for remote sensing, tools operational under extreme conditions, hydrodynamic models (Webb and de Cuevas 2007) and novel molecular approaches (see below). They all promoted our understanding of the recent evolution of the SO.

5.2 Patterns of Genetic Diversity in Past and Present

Genetic diversity integrates as well historical events (the accumulation of mutations through expanding populations combined with systematic loss due to extinction of populations) as contemporary events of population migration and fitness (the potential to adapt). On average, marine populations have high genetic diversities because of high effective population sizes (and hence reduced chances for genetic drift and inbreeding).

5.2.1 Accessing Genetic Diversity

Remarkable technological progress originating from the drive to supply affordable single genomes (1,000 US$ genome) has created new opportunities to study biological evolution (Clark et al. 2011). Low-density genotyping is increasingly complemented by high density and high-throughput genotyping and profiling. There are now suitable tools available for the correct estimation of genetic origin at the between and within species level. At the population level large numbers of individuals can be typed, regardless of size. The transition has lead to remarkable and unique outcomes. Here we list some of the realized and potential applications of genomic tools to screen populations of Antarctic fishes.

Metagenomics, the analysis of the genomes of organisms recovered from the environment, started on picoplankton communities of the open ocean (Rusch et al. 2007). It is now generally implemented and used for screening prokaryotes and eukaryotes of complete biological communities (microbiome). Applications in the SO are in progress. An equally massive but more targeted approach is metagenetics, the *en masse* characterization of operational taxonomic units (OTU) at a single locus, for example the nuclear small subunit 18S (nSSU). Logical targets are soft bottom communities (Creer et al. 2010), but also plankton and fish qualifies. Ward et al. (2009) analyzed bacterial 16S rRNA gene sequences from the intestinal tract of *Notothenia coriiceps* and *Chaenocephalus aceratus*. Simple Sequence Repeats (SSRs) or DNA microsatellites, which are tandem repeats from one to five base pairs, have proven their usefulness in detecting contemporary patterns. Progress from the first microsatellites of notothenioids isolated with some difficulty from genomic DNA libraries (Reilly and Ward 1999; Susana et al. 2007;

Van Houdt et al. 2006; Van de Putte et al. 2009) to the current large scale mining of EST-SSRs (albeit not yet in fishes of the SO, but see Vogiatzi et al. 2011) has lead to more balanced genome-wide sampling and hence characterization of populations. However, SSRs suffer from poor standardization and transportability between platforms. Point mutations in the genome, known as Single Nucleotide Polymorphisms (SNPs), are the most common type of genetic marker (Brumfield et al. 2003). They have the advantage of being platform independent and hence standardized. Their genome wide occurrence has lead to quick acceptance (Kuhn and Gaffney 2008). Large-scale applications on fishes of the SO are under development. Outlier SNPs seem perfectly suitable to detect weak population structure, which is a most typical feature of marine populations (Nielsen et al. 2009). Also Copy Number Variation (CNV), although occurring at a lower frequency throughout the genome, harbours a high information content.

Much is expected from the genotyping of complete genomes (Hohenlohe et al. 2010); it represents a major change in approach, as the full blueprint can be mined for information. Also, gene transcripts contain lots of information to understand genome evolution, population patterns and molecular physiology. Excellent high resolution and high throughput tools (from microarrays to next generation sequencing) are operational (Peck et al. 2005; Clark et al. 2008, see Chap. 9 Peck and Clark 2012).

From this survey it is clear that genomic tools are well suited to analyse the biodiversity of the SO. It is expected that sequencing access and hence the means to scrutinise genomes at the population level will continue to improve and that the statistical tools (e.g. coalescence analysis) will continue to expand. It should make it possible to monitor in real time and in situ genetic patterns.

5.2.2 Barcoding Genetic Diversity in the Southern Ocean

Through a number of large-scale projects and the use of standardised methods DNA barcoding has promoted routine molecular identification. DNA barcoding is a molecular technique that uses a standardized region (e.g. cytochrome oxidase 1 gene) as biomarker for species identification (Hebert et al. 2003). One of the foremost projects is the Barcode of Life project (www.barcodeoflife.org). It relies on the sequencing of standardized regions (e.g. cytochrome oxidase 1, cytochrome b (cyt *b*) and rhodopsin) of voucher specimens. Identification of undetermined samples can be obtained by comparing one or more of these regions with a publically available reference library (Barcode of Life Data System–BOLD). Like any other method barcoding has both advantages and disadvantages (Grant and Linse 2009). Nevertheless barcoding allows for identifying a large number of unknown taxa, which would be an immense undertaking solely by morphological methods. It is especially useful in cases of cryptic speciation, missing morphological information (e.g. degraded species or fragments, stomach contents and larval stages), population differentiation, speciation and phylogeography. Barcoding is a valuable tool for assessing known and unknown biodiversity (Blaxter 2003).

Over the last decade the Census of Antarctic Marine Life (www.caml.aq), the Antarctic component of Census of Marine Life (www.coml.org), focused on determining the distribution of Antarctica's vast biodiversity in the SO and around the sub-Antarctic Islands. Within the framework of the International Polar Year CAML has taken the initiative to contribute to the Barcode of Life project and to barcode as many marine species as possible. It resulted in a strong increase in the number of publicly available sequences and species. The output of this initiative is impressive. No studies on Antarctic life were published before 2007. Rock et al. (2008) used just 124 sequences from 34 putative fish species and Dettaï et al. (2011) included already 538 sequences from 68 putative fish species.

Several authors (Rock et al. 2008; Lautrédou et al. 2010; Dettaï et al. 2011; Allcock et al. 2011) testify that COI barcoding appears to be a good choice for the identification of Antarctic fish. However, a number of issues remain. Identification using BOLD relies on the availability of a complete database. Species for which no voucher specimens are available will not be indentified using the system. Also, some groups would benefit from additional markers. For instance two species pairs of trematomids *Trematomus loennbergii–T. lepidorinus* and *T. bernacchii–T. vicarius* could not differentiated using COI markers. Ultravariable SSR markers were suggested for full species delineation (Dettaï et al. 2011; Van de Putte et al. 2009). It is expected that as the database grows, its applicability in combination with other databases (geographical coverage with SCAR-MarBin/AntaBIF and ecohydrodynamic models) will increase considerably.

5.2.3 Patterns of Genetic Diversity in the Past

Over the last 800 kYA the Antarctic has experienced a cyclic climatic pattern with a period of about 100 kYA resulting in eight glacial cycles causing dramatic climatic changes (Petit et al. 1999). Models suggest that during the last interglacial (ca 120 kYA) the Antarctic ice sheet was somewhat smaller than today. During the subsequent cooling phase the ice sheet first attained an intermediate coverage. The timing of the Last Glacial Maximum (LGM) with the maximum extent of the continental ice sheet is situated around 15 kYA. This Holocene retreat took about 6–8 kYA to reach the present extent; the ice sheet will reach a minimum coverage, similar to that of the last interglacial, in the "near" future (Huybrechts 1993).

The glacial cycles had a different impact on benthic and pelagic species in the Antarctic compared to the Arctic. In the northern hemisphere a number of population expansions of pelagic species were associated with the Pleistocene ice ages. The presence of land masses in combination with glacial expansion resulted in the isolation of stenothermal populations in refuges across the Atlantic Ocean and the Mediterranean Sea; the formation of allopatric lineages lead to complex haplotype networks (Zane et al. 2006). In the SO there are no such land barriers; the APF and Antarctic planktonic ecosystem shifted northwards during periods of glaciation and southward during interglacials (Flores and Sierro 2007). Such northward shifts would allow for population expansions; the lack of land barriers would not lead to allopatric differentiation as observed in the northern hemisphere and result in a relatively simple demographic history (Zane et al. 2006).

During warm interglacial periods (such as the current epoch) Antarctic ice sheets retreated towards the coast and sea ice only covered the Antarctic continental margin seasonally. Occasionally icebergs, breaking off from the ice shelf, ran aground on the Antarctic shelf. Iceberg scouring locally eradicated the benthic community. During cold glacial periods, such as the LGM or the late Eocene–early Oligocene, ice sheet build-up caused grounded ice masses to advance across the shelf (Anderson et al. 2002). In order to escape extinction organisms either had to occupy the deeper slope areas or survived in refuges. Submergence might account for the present depth distribution of notothenioid fishes (Andriashev 1987) and the eurybathy of many invertebrate taxa (Brey et al. 1996). During the LGM the shelf was not completely covered with grounded ice; there were refuges on the shelf (Thatje et al. 2005). Benthic and demersal organisms with a pelagic larval phase, such as observed in most of the notothenioids, would be able to disperse to these refuges (Thatje et al. 2005). Such a mechanism would promote stronger geographic partitioning and also and more recent population expansions.

The most common genetic marker allowing for comparison of historical genetic diversity between fish taxa is mitochondrial DNA. In the SO cyt *b* has been used most commonly. Diversity values are strongly influenced by the marker of choice. For example, average mtDNA diversity was found for *Champsocephalus gunnari* (nucleotide sequence diversity 0.19–0.22, Williams et al. 1994), but lower levels were indicated by allozyme studies (Duhamel et al. 1995). MtDNA diversity for *Dissostichus mawsoni* was found to be very low, and some thirty times less than that observed for the conspecific *D. eleginoides* (e.g. nucleotide diversity of 0.0002 versus 0.0052; Smith and Gaffney 2005). Low genetic variation was also suggested from nuclear RAPD analysis, indicative of a historical bottleneck or physiologically low rates of microevolutionary change (Parker et al. 2002).

5.2.4 Contemporary Patterns of Genetic Diversity

Patterns of genetic diversity during the Holocene, up to today, are lodged in genome loci with a higher mutation rate than the allozymes and mitochondrial cyt *b* locus. SNPs are the most commonly used markers for this purpose and provide information on diversity, population mixing and population structure. Microsatellite diversity measured as observed heterozygosity ($H_o = 0.79$) is considerably higher than anadromous ($H_o = 0.68$) and freshwater populations ($H_o = 0.46$) of fish (DeWoody and Avise 2000). The values observed in the SO seem comparable.

Microsatellite analysis for *Chionodraco rastrospinosus* indicated relatively high levels of genetic diversity across all sampling locations (0.70–0.79) (Papetti et al. 2012). Genetic diversity for *C. aceratus* from the same region was comparable (0.69–0.72; Papetti et al. 2009), and significantly higher in allelic richness (16.43–18.39) than in *C. rastrospinosus* (14.52–15.77).

5.3 Historical Patterns: Vicariance or Dispersal?

Mitochondrial DNA haplotypes are a prime source of information on the spatial and temporal dynamics of Pleistocene populations (Avise 2000). Although the current distribution does not directly allow to identify past ranges of taxa, comparisons between haplotype networks allow to estimate the location of refuges and the nature of the population dynamics. The interpretation of these patterns is complicated by the limitations of markers and the difficulty to calibrate the molecular clock. A means to understand the historic population dynamics is the comparison of pairwise mutations of mtDNA haplotypes (so called mismatch distribution analysis). In case of a sudden population expansion a vast number of mutations occur in a short time and hence haplotypes resemble each other (Avise 2000).

C. gunnari showed the strongest support for expansion in populations from South Georgia, Shag Rocks and South Shetland (Kuhn and Gaffney 2006; Rock, pers. comm.), whereas no evidence of expansion was found for populations such as Dallman Bay and Kerguelen (Rock, pers. comm.). Timing of the population expansion/selective sweep at Shag Rocks is suggested at 71 kYA, with the South Shetland demographic event more recent, and the ancient divergence of both populations suggested at around 120 kYA. In the Indian Ocean sector, a lack of detectable substructure, attributed to lowered sea levels as recently as 18 kYA, is thought to have created a homogeneous population (Williams et al. 1994).

Within the trematomid group demersal and pelagic species reacted differently to the glacial cycles. For the demersal species, there was a recent population expansion that was attributed to the last glacial cycle. The expansion of *Trematomus pennellli* was dated to 25–30 kYA ago coinciding with the onset of the last deglaciation. The population expansion of *T. bernacchii* started a little earlier during the last glacial period, indicating that these demersal species had suffered a population bottleneck during the last glacial maximum (Janko et al. 2007). In contrast the expansions of the pelagic species *T. newnesi*, *Pagothenia borchgrevinki* and *Pleuragramma antarcticum* predated the last interglacial (Janko et al. 2007; Zane et al. 2006). *Pleuragramma antarcticum* shows a population expansion between 111 and 126 mYA and a remarkable weak population structure.

The congeneric demersal icefishes *Chionodraco hamatus*, *C. myersi* and *C. rastrospinosus* showed the onset of population expansion at 90, 47 and 25 kYA respectively (Patarnello et al. 2003). *C. hamatus* showed a complex network of expansions and heterogenous distribution (Patarnello et al. 2003) and *Gobionotothen gibberifrons* a sudden expansion some 24–32 kYA ago (Matschiner et al. 2009).

From these cases, a close association between environment and population dynamics appears. Under influence of periodic fluctuations during the Quaternary fishes of the SO expanded or reduced in population sizes and natural range. During cold glacial periods they occupied a vastly expanded range while during relaxed interglacial periods they retreated to refuges. Narrow continental shelves led to limited opportunities for refuges. Taxa largely reacted in line with their life-history traits. Therefore the spatial patterns of demersal species are generally fragmented while pelagic species have a broad historical range. Similar to other taxa worldwide, vicariance and dispersal have shaped the historical patterns, albeit with lower constraints by the continental landmasses.

5.4 Connectivity and Gene Flow

Population connectivity relates to the extent to which marine populations in different parts of the range are connected by the exchange of individuals (larvae, juveniles or adults) (Palumbi 2003). It is regulated by the environment (distance, retention, recruitment/transport and mortality) while biological traits (behaviour and life history) modify the connectivity patterns. The multi-generational outcome of connectivity is quantified through gene flow, a genetic measure of connectivity. In the following paragraphs the impact of several of these factors on the fishes of the SO have been summarized, beginning with the effect of distance on spatial structure across large, intermediate and small scales.

5.4.1 Distance and Genetic Structure

Geographic distance plays a significant role in population differentiation at both circumpolar and regional levels for many species. Indeed, the general picture of connectivity for most species is restricted gene flow. There is, however, much variation in the effect of distance. Shelf areas, the habitat of most species, are far and few between and bathymetry may often preclude migration between proximate island populations. For example, short stretches of deep water may form powerful barriers to gene flow by adult dispersal for many demersal notothenioids. Directional ocean flows associated with the ACC place further constraints on the possible transport routes between isolated populations in the SO. Local variability in current directionality and flow rate, combine with biological factors including post larval development (PLD) and species- or region-specific life history and behaviour characteristics, to determine transport across spatial scales.

5.4.1.1 Spatial Structure Across Large and Intermediate Distances

For a few notothenioids, a lack of genetic divergence has been observed across enormous spatial scales, with panmictic assemblages on even a circumpolar scale (i.e. *C. hamatus* (Patarnello et al. 2003), *P. antarcticum* (Zane et al. 2006), *T. nicolai* (Kuhn et al. 2009), *N. rossii* (Rock, pers. comm.) and *E. antarctica* (Van de Putte et al. 2012)). For most, however, clear genetic differentiation is found at larger scales of spatial separation. To date, these have most often been assessed by comparisons between the Scotia Sea and Indian Ocean sector of the SO.

For example, isozyme analysis of *Lepidonotothen squamifrons* populations indicated divergence between these regions (Kerguelen versus South Georgia, South Orkney and Elephant Islands; Schneppenheim et al. 1994). Although Duhamel et al. (1995) detected no isozyme polymorphism between populations of *C. gunnari* in the Kerguelen versus South Orkney Islands, recent genetic comparisons of Scotia Sea populations with those in the Indian Ocean sector (Kerguelen and/or Heard Island) were highly divergent (Kuhn and Gaffney 2006; Rock, pers. comm.). This is supported by morphometric and meristic measurements, parasite infestation patterns and reproductive characteristics (Duhamel et al. 1995). Similarly, significant differentiation was found between three oceanic sectors for populations of *D. mawsoni* (Kuhn and Gaffney 2008; Parker et al. 2002), and for *D. eleginoides* (Smith and McVeagh 2000). For the latter species, Scotia Sea populations were found to be genetically distinct from those in the Indian Ocean sector (for microsatellite though not mtDNA markers; Rogers et al. 2006), as well as from Macquarie and Macdonald Islands (Appleyard et al. 2002; microsatellite markers for the same sites curiously did not reveal significant heterogeneity).

Seamounts and islands providing suitable shelf habitat for demersal fishes are regionally clustered in discrete ridges or archipelagos. Geographic distance between populations is often at the extremes of large or small, with little representation in between. However, one such intermediary location occurs between the Scotia Sea and Indian Ocean sector at Bouvet Island. Situated some 1,800 km east of the Scotia Arc and with no intervening seamounts breaking water generally over 3,000 m deep, it is separate from the Scotia Sea, but notably downstream in the west–east flow of the ACC. In a preliminary analysis (of small sample sizes), three of four notothenioid species sampled at Bouvet showed genetic homogeneity with the Scotia Sea (Jones et al. 2008).

5.4.1.2 Spatial Structure Across Small Distances

At smaller distances, there is great variation in the extent of population structuring. Overall, evidence of gene flow suggests that across time scales producing evolutionary patterns, populations upstream in the ACC represent important source populations for those downstream (e.g. within the Scotia Sea from the Antarctic Peninsula (AP) region to South Georgia), with intermediary populations sometimes acting as stepping stones. However, it is also clear that many populations rely on local recruitment. Connectivity within two regions has been more fully assessed, including between islands on the Kerguelen Plateau and surroundings in the Indian Ocean sector, and between islands within the Scotia Sea and AP region.

Peculiar is the evidence from mitochondrial and nuclear markers for heterogeneity in species currently targeted by fisheries, *Dissostichus* spp. and *C. gunnari*, indicating distinct stocks within regions. For the Patagonian toothfish, *D. eleginoides*, microsatellite markers indicated population structure between the Ross Sea, Macquarie, Heard, and Prince Edward Island (in contrast to the homogeneity revealed by allozymes; Smith and McVeagh 2000), although between Marion,

Crozet and Kerguelen islands no consistent differentiation was found. Significant homogeneity amongst seamount populations connected by oceanic ridge systems within the Bouvet region was also suggested by both mtDNA and microsatellite markers (Rogers et al. 2006). At smaller spatial scales, however, no mtDNA genetic differentiation was found within regions for Heard and Macdonald Islands, or for South Georgia and Shag Rocks (Appleyard et al. 2002, Appleyard and Williams Appleyard et al. 2004). Microsatellite analyses also resolved no significant differentiation between South Georgia, Shag Rocks and the northern North Scotia Ridge (Shaw et al. 2004), although significant partitioning between South Georgia/Shag Rocks and populations to the north west was suggested by mtDNA (including the North Scotia Ridge and Falkland Islands; Shaw et al. 2004, Rogers et al. 2006) and attributed to their separation by both the Antarctic Polar Front (APF) and deep water of ~2,000 m. However short stretches of deep water (tens of kilometers) appear not to form a complete barrier to gene flow in *D. eleginoides* (Shaw et al. 2004). Overall, structuring appears to be a case of regional differentiation with local patchiness in gene flow that is not necessarily correlated with distance (Smith and McVeagh 2000). For the Antarctic toothfish, *D. mawsoni*, mtDNA and nuclear (SNP) markers have, to date, distinguished differentiation between populations across large distances (e.g. South Shetlands, Ross Sea and Australian Antarctic Territory; Kuhn and Gaffney 2008) and nuclear (RAPD) markers revealed divergence between the Ross and Bellingshausen Sea (Parker et al. 2002).

For populations of mackerel icefish *C. gunnari* within the Indian Ocean, no significant differentiation has been found (Williams et al. 1994, Rock, pers. comm.). However in the Scotia Sea both mtDNA and microsatellite data have revealed significant genetic structure at small spatial scales (Kuhn and Gaffney 2006, Rock, pers. comm.). Hierarchical genetic analyses revealed significant separation between northern (South Georgia and Shag Rock) and southern groups (South Shetlands, Kuhn and Gaffney 2006; or Dallman Bay, Elephant Island and South Orkney). Microsatellite hierarchical analysis also revealed distinct groupings in this region separating South Georgia and Shag Rocks from Dallman Bay and Elephant Island, with South Orkney exhibiting additional differentiation from the other groups. Genetic differentiation was particularly strong between the southernmost populations vs. northern populations (i.e. Dallman Bay vs. South Georgia), although it was also observed between southwestern populations (including Elephant Island) and South Orkney, as well as between the latter population and northern populations (Kuhn and Gaffney 2006, Rock, pers. comm.). In contrast, connectivity was found between Dallman Bay and Elephant Island. Differentiation between South Georgia and Shag Rocks was also non-significant for mtDNA markers (Kuhn and Gaffney 2006; Rock, pers. comm.) although weakly significant for nuclear (SNP and microsatellite) markers (Kuhn and Gaffney 2006; Shaw et al. 2004, Rock, pers. comm.). The correlation between mtDNA genetic and geographic distance included aberrations; for example, a higher connectivity was found between South Orkney with Dallman Bay than with Elephant Island. In the Scotia Sea generally, a stepping-stone isolation-by-distance pattern was observed, dominated by uni-directional transport.

Genetic substructure within the Scotia Sea was also found for *C. aceratus*, with significant differentiation between South Orkney and both Elephant and South Shetland Islands (Papetti et al. 2009). In accord with previous studies of parasite infestation patterns, microsatellite data showed that South Shetland and Elephant Islands were, however, a single panmictic population, although significant genetic differentiation was detected between individual year–classes of fish (Papetti et al. 2007). Temporal differentiation was also found between South Shetland and Elephant Island between sampling years (Papetti et al. 2009). Mantel tests resolved an isolation by distance pattern and gene flow barrier analysis indicated a large genetic break between the AP populations and South Orkney (a distance of about 1,500 km). Whilst dominant gene flow was found to follow the ACC, counter-current migration was suggested between South Orkney and Elephant Island. The amount of gene flow was not sufficient to have a homogenising effect. Although Jones et al. (2008) suggested no significant divergence in mtDNA between the AP region and the northern Scotia Sea, this analysis was limited by small sample sizes. Data from otolith chemistry for *C. aceratus* has confirmed fine-scale heterogeneity between certain sites in the AP region (King George and Elephant island), as well as strong divergence between this region and South Georgia (Ashford et al. 2010).

For several species, the divergence of populations despite close proximity suggests a role of complex oceanography, in particular at the Weddell-Scotia Sea confluence. While significant allozyme divergence was reported for channichthyids *Neopagetopsis ionah* and *Chionodraco myersi* populations at large spatial scales (Prydz Bay *vs* Weddell Sea), differentiation was also found across spatial scales of less than 100 km (Clement et al. 1998). Even pelagic *P. antarcticum* has mtDNA divergence between the Weddell Sea and the Western AP (Zane et al. 2006).

In contrast to general isolation-by-distance patterns described above, several notothenioids are notable for exhibiting genetic homogeneity across similar spatial scales. In *N. rossii*, parasite infestation (Zdzitowiecki and White 1992), egg morphology (White et al. 1996) and other morphological and meristic measurements (Gon and Heemstra 1990) indicated regional differentiation and population structuring. However, genetic structure has not been found at either large or small scales. Isozyme analysis revealed no significant differences between Kerguelen Island and Skif Bank populations (Duhamel et al. 1995). Analysis of mtDNA and microsatellite data revealed no significant differentiation within or between sites in the Scotia Sea and/or Indian Ocean sector, with no evidence for correlation between genetic and geographic distance (Rock, pers. comm.). Similarly, for *C. rastrospinosus* microsatellite data revealed no detectible divergence on geographic or temporal scales in the southern Scotia Sea, and hierarchical AMOVAs indicated no substructuring by geographic local or year (i.e. no significant differentiation between cohorts in 1996, 2002 and 2006; Papetti et al. 2012). Mantel tests also indicated no significant correlation between genetic and either geographic or temporal distance and Isolation-by-migration models revealed dominant westward migration events, suggesting a single panmictic population in the South

Scotia Sea region maintained by gene flow from west to east with the ACC. Similar to the results of coalescent model predictions for *N. rossii*, these analyses suggest that few migration events/year (e.g. ~5 migrations between the AP and South Orkney Islands for *C. rastrospinosus*) appear to suffice to maintain genetic homogeneity between populations. For *Gobionotothen gibberifrons* a similar pattern of homogeneity in the Scotia Sea is evident. Neither microsatellite nor mtDNA data revealed significant pairwise differentiation, and although mtDNA separated AP populations from South Georgia and the South Sandwich Islands, microsatellites did not (Matschiner et al. 2009). There was unidirectional connectivity across the Scotia Sea from AP to South Sandwich Islands, most likely via South Orkney, with an average of 10 migration events per year. This points to a panmictic population spanning as much as 1,900 km, comprising a large part of the species' distribution range. Preliminary mtDNA analyses have also revealed no significant divergence between populations of *N. coriiceps,* and *L. larseni* in the NE Scotia Sea versus those along the AP (Jones et al. 2008).

5.4.2 *Effect of Life History on Connectivity*

Many notothenioids have an extended larval duration (1–2 years), although for most developmental stages dispersal away from the shelf habitat equates to loss in the open ocean. Consequently it has been considered a 'paradox' that notothenioids should possess an extended planktonic larval phase, as such dispersive propagules might be expected to have significant loss of early life stages (White 1998). However, while the predictions of oceanographic models have affirmed that species with extended pelagic dispersive phase (*N. rossii*) exhibit relatively low levels of retention, they have also indicated scenarios of long distance successful transport due to high dispersal rates with the ACC. Pervasive (if low-level) long-distance connectivity has been predicted by biophysical models and confirmed by genetic analysis for both *C. rastrospinosus* and *N. rossii* indicating that a lack of population substructure may be due to mobility of at least one life stage within the ACC. As adults are thought to be restricted to shelf waters and incapable of long migrations, particularly across deep ocean waters (Eastman 1993), larval dispersal is expected to be the key to connectivity, with low level exchange sufficient to maintain mitochondrial homogeneity. A pattern of high gene flow was also observed for *G. gibberifrons* in the Scotia Sea, in accord with predictions of extensive larval dispersal derived from drifter tracks (Matschiner et al. 2009).

Neither PLD nor other basic features like being a demersal dweller can singularly predict the level of population connectivity of a notothenioid (Appleyard et al. 2002; Kellerman 1990; Everson et al. 2001; North 2001; North and White 1987). Indeed the effect of life history on the timing and position of larvae entering the water column is a critical factor. Differences in gene flow will result from variation in spawning site, proximity to oceanic retention features and fine differences in the timing of larval phases and behaviour. Nesting icefishes that delay the pelagic phase until late winter may facilitate connectivity to the south and east, whereas species that release pelagic propagules in summer are likely to reach regions further north including South Georgia (Ashford et al. 2010).

Cross-shelf migration is believed to be critical for downstream recruitment (White et al. 1982). Ontogenic timing of such larval behaviour as vertical diurnal migration (known for multiple notothenioid species including *G. gibberifrons* and *C. gunnari*; North and Murray 1992) coupled with oceanographic variability in the velocity and directionality of shelf break front and transitional zones (Meredith et al. 2005) will have profound effects on effective larval dispersal. Behavioural mechanisms including natal fidelity have been suggested as the cause of population subdivision for *D. mawsoni* (Parker et al. 2002), although juveniles remain pelagic with the potential to be highly dispersive (Smith and Gaffney 2005). Other life history characteristics will also interact to affect connectivity. For instance, higher fecundity may allow partial compensation for low successful larval dispersal or retention. *N. rossii*, with its low retention but high dispersal, produces roughly five to ten times as many eggs as *C. gunnari*.

5.5 Seascape Genetics, Exploitation of Living Resources and Marine Protected Areas

Understanding the evolutionary dynamics of the ocean needs accounting for the impact of physical features on the population genetic features. Although marine barriers to dispersal are often weak, they do influence the connectivity triangle of fishes between spawning, nursery and feeding grounds (Selkoe et al. 2010; Galindo et al. 2010). In addition, long generation times of SO fishes have made populations respond quickly to fishing pressure. As collapsed stocks are slow to recover, marine protected areas (MPAs) are a strong management option.

5.5.1 Corroboration of Connectivity Patterns with Oceanographic Models

Dispersal being waterborne makes that hydrodynamics play a crucial role in connectivity patterns. One approach to understand its dynamics is the simulation of transport for passive particles, released at depths, localities and times of year relevant to the life cycle. For example larvae of *C. aceratus* are likely to be transported further east along the Scotia Ridge and fail to reach South Georgia (Ashford et al. 2010). It was corroborated by evidence for significant population differentiation at South Georgia. In another case, drifter data was used as a proxy for passive larval dispersal and showed that passive particles cross the Scotia Sea between the AP region and South Georgia within the larval duration of

G. gibberifrons of four months. It confirmed unidirectional connectivity from west to east following the ACC and resolved a mechanism for the population homogeneity (Matschiner et al. 2009). Particle transport models for *C. gunnari* and *N. rossii* also pointed to the potential for trans-Scotia Sea recruitment, although the magnitude differed between species, source locations and notably, across years encapsulating significant ENSO variation (Rock, pers. comm.). While exceptionally high connectivity values were indicated by genetic data for *N. rossii* (Rock, pers. comm.), oceanographic models also generally predicted transport of a greater number of propagules than for *C. gunnari*. It suggests that the low level but geographically broad connectivity predicted for *N. rossii* is key to its marked gene flow (vs. the more sporadic transport of high numbers characteristic of *C. gunnari*, at least in some years). It is expected that the integration of ecological data (e.g. prey and predator field) in the physical model, will provide a closer understanding of the mechanisms affecting dispersal.

5.5.2 Fisheries Dynamics and Genetics

Although fish resources of the SO may not be diverse, several stocks have been exploited for seafood and fish meal. Following ocean wide exploratory surveys, fishing started in the 70s (Kock et al. 2007). The exploitation of low Antarctic fishes focused on top predators such as mackerel icefish *C. gunnari* during a period of some 15 years (Kock et al. 2007), Patagonian toothfish *D. eleginoides* (Parker et al. 2002; Kock et al. 2007) and various icefishes (Kock 2005). Several stocks were quickly depleted and some collapsed. For example, *C. gunnari* has not recovered yet in the vicinity of the AP (Kock et al. 2005). High Antarctic fishes were much less exploited because of harsh environmental conditions (Antarctic toothfish *D. mawsoni*—Kock et al. 2007). Since the late eighties the pelagic species *D. mawsoni*, *D. eleginoides*, *C. gunnari* and some demersal fishes are managed in a multi-species and ecosystem context by CCAMLR (www.ccamlr.org/fisheries), which has been lauded for its management (Cullis-Suzuki and Pauly 2010). The SO has not escaped overexploitation; particularly illegal, unregulated and unreported (IUU) fishing is of great concern in the Indian Ocean.

As heavy fishing has clear ecological consequences, long lasting ecological regime shifts may be expected, such as reported for example in the Northwestern Atlantic Ocean (Scheffer et al. 2001). Particularly krill fishing has had dramatic impacts, largely through bottom up effects. It is a key prey species for many fishes (Kock et al. 2007) and vertebrates (Trivelpiece et al. 2011); depressed stocks immediately affect recruitment and survival at higher trophic levels of the Antarctic food web. Fishing has measurable evolutionary impacts at the population level. Genetic structure may be modified through extirpated populations, effective population size may diminish because of collapsed populations and adaptive shifts are anticipated when fishing pressure is exerted over many generations (Hauser and Carvalho 2008).

5.5.3 Marine Protected Areas

As effort based management (e.g. quota) measures are complex to enforce and have largely failed worldwide (Worm et al. 2009; FAO 2010), consensus grows to integrate more spatial and temporal management measures. An important tool to enforce spatial measures are Marine Protected Areas (MPAs), an umbrella term referring to a wide range of options for the spatial management to human interventions in the marine environment. They may serve two mutually non-exclusive purposes: (1) Conservation goals (genetic, species and ecosystem biodiversity) and (2) Fisheries (sustainable exploitation) goals. In the latter case, MPAs represent areas designated to enhance conservation of marine resources (Lubchenko et al. 2003) and include no-take fisheries zone, restricted access zones, and zones assigned for specific or targeted fishing practices. Important characteristics are their spillover (larvae and adults: Williams et al. 2008) and seeding capacities (larvae and juveniles: Christie et al. 2010). The overall effectiveness of MPAs from a connectivity and evolution perspective depends on the architecture, namely size, size structure, spacing and coverage (Gaines et al. 2010).

Size is an important feature to allow for minimum viable population sizes. Below a threshold, populations may be too small and inbreeding may erode genetic diversity (Frankham et al. 2002), which may lead to extinction in the long term. The range in size structures reflects the diversity of species, life-histories and reproduction strategies; one size won't fit for all taxa and ecosystems. Most organisms have very specific requirements throughout their life-cycle. For example many fish spawn at determinate hydrodynamical zones in the ocean, named larval retention zones (Patagonian toothfish: Rogers et al. 2006). Size and size structure are only effective if the spacing between the habitats is guaranteed for the taxa inhabiting them. An important ecological determinant to delineate MPAs is the dispersal of larval and adult organisms in time and space (connectivity between populations). Of special interest is that the low metabolic rate of Antarctic fishes allows for larger spacing between MPAs to accommodate for larger dispersal ranges (Laurel and Bradbury 2006). The size of naturally functioning ecosystems and hence the area accessible to breed, grow and mature is determined by coverage, the cumulative requirements of individual taxa and the communities they harbour. They have to buffer the current requirements of maintaining viable populations but also the long term requirements of opportunities to evolve.

Criteria to select areas for conservation in the absence of a global perspective on marine biodiversity are of great relevance (Harris et al. 2007; Koubbi et al. 2011). With the initially small knowledge portfolio of the SO, sea birds have been proposed as a surrogate measure because of their relative ease to monitor (Harris et al. 2007). Also molluscs delineate three biogeographical areas on the shelf (Clarke and Johnston 2003; Linse et al. 2006) and qualify for assigning areas of particular interest. Most significantly, the SCAR-MarBin/AntaBIF data base has reached a high content of information such that mapping and prioritization exercises become increasingly feasible (De Broyer et al. 2011; Griffiths et al. 2011).

The CCAMLR bioregionalisation approach (Grant et al. 2006) relies on the physical characterization of the SO (depth, temperature, salinity, sea ice and nutrients) and uses chlorophyll *a* as a surrogate for biotic parameters. A more promising tool to develop conservation strategies is ecoregionalisation, where not only abiotic factors are used to design MPAs but above all biotic and ecological factors. At a mesoscale level, Koubbi et al. (2011) showed for species assemblages of myctophids, typical inhabitants of the oceanic zone, that regional breaks in distribution patterns correspond with latitudinal gradients. They confirm previous expert judgement. So far the only designated MPA is the South Orkney Island group (UK) covering 94,000 km^2. No fishing activities and no discharge or refuse disposal from fishing vessels are allowed in the area. In addition, CCAMLR is discussing the implementation of a mosaic of MPAs in the SO.

5.6 Perspectives

Ancestral fish diversity of the SO has experienced dramatic bottlenecks and adaptations of body characteristics. Species reductions are obvious from the limited number of taxa at the family level. Austral fishes have high endemism and hence are unique and different from conspecifics elsewhere. The modifications are reflected in the physiology and niche preferences, linked to discrete modifications of the genome (Patarnello et al. 2011; Chap. 9 Peck and Clark 2012). The geographical patterns of the few demersal taxa were shaped by periodic demographic contractions to refuges (glacial periods) and expansions across habitats (interglacials). Orbital forcing and subdecadal variability will continue to impact life in the oceans. However, the pace of current ecological and evolutionary change is likely to be higher than ever because of worldwide anthropogenic impact. Five key threats are remodelling appreciably the ecosystems, and with them the fish communities, of the SO. They include overfishing, global change, alien species, habitat loss and pollution.

Projections of fish biomass show dwindling trends across the SO (Kock et al. 2007; Tittensor et al. 2010). Regional stocks do not support high fishing pressure because of K-selected life histories, hence the great need for adaptive management. Chance for extinctions of local populations without even being noticed are considerable (see in the North Atlantic Ocean; Hutchinson et al. 2003). Of great concern is the evidence that heavily fished stocks show an evolutionary ('Darwinian') debt of delayed maturation and reduced growth rate (Jørgensen et al. 2007).

The current rates of global warming, which are particularly felt in the seas bordering the Antarctic Peninsula (Clarke et al. 2007), are of major concern. The changing abiotic conditions may lead to strong biotic responses. For example, the adaptability of fish to temperature and acidity in view of global change is limited (Patarnello et al. 2011). There is no other option than to translocate to suitable habitats. Coupled physical—biological/genetic models may allow to estimate the impact on connectivity and population structure (Galindo et al. 2006). At the ecosystem level, such changes may lead to regime shifts (Scheffer et al. 2005).

Globalization facilitates intercontinental translocations; terrestrial and freshwater evidence is overwhelming, especially on some subarctic islands and the AP (Chown et al. 2008). Evidence at sea remains limited. For example northern species such as chinook salmon have become established in the coastal zone of the sub-Antarctic (Becker et al. 2007); the North Atlantic spider crab *Hyas araneus* is one of the first benthic species reported in the SO (Tavares and De Melo 2004).

The Antarctic coastal zone is continuously reshaped by glacial scouring and freeze/thaw cycles. Unlike mid- and low latitude zones, direct habitat loss through coastal engineering and dredging of mineral aggregates is limited. The threat of hydrocarbon extraction and deep-sea mining, especially of polymetallic nodules, remains (Pettis and Forest 1979). Trawling for demersal fish on the delicate benthic communities of sponges, corals and echinoderms is of concern.

Although local sources of pollution are limited (Tin et al. 2009), most pollution is water- and airborne. The Great Ocean Conveyor Belt carries contaminated water masses world-wide while atmospheric fractionation is an important redistributor of organic substances. As fish occupy mid- to high ecosystem levels, they are effective concentrators of organic and metallic pollutants. High tissue loads may lead to genotype-dependent survival and hence may affect overall populations fitness and ecosystem resilience. In addition, pollution and above average UV radiation make that mutational load and skewed selection pressures increasingly threaten the genetic integrity of the local fauna and flora (Walther et al. 2002).

Impact abatement of these threats necessitates the further development of a conservation policy for the SO, which should pay attention to genetic diversity, connectivity and evolutionary processes. Science has been developing tools to provide information on conservation (Frankham 2005; Allendorf et al. 2010) and management strategies. The integration of knowledge in marine governance requires international vision and commitment.

Acknowledgments Funding is acknowledged from the Belgian Federal Agency for Science Policy (BELSPO; programs PELAGANT, PADI and ANTABIF), the Research Foundation Flanders (Belgian network for DNA barcoding; W0.009.11N) and UK NERC AFI 6/16 grant. We thank G. di Prisco for excellent guidance in the preparation of the chapter.

References

Allcock AL, Barratt I, Eléaume M et al (2011) Cryptic speciation and the circumpolarity debate: a case study on endemic Southern Ocean octopuses using the COI barcode of life. Deep-Sea Res II 58:242–249

Allendorf FW, Hohenlohe PA, Luikart G (2010) Genomics and the future of conservation genetics. Nat Rev Genet 11:697–709

Anderson RF, Chase Z, Fleisher MQ, Sachs J (2002) The Southern Ocean's biological pump during the last glacial maximum. Deep-Sea Res II 49:1909–1938

Andriashev AP (1987) A general review of the Antarctic bottom fish fauna. In: Kullander SO, Fernholm BO (eds) Proceedings 5th Congress European Ichthyol 1985, pp 357–372. Swedish Museum of Natural History, Stockholm

Anonymous (2005) Evolution and biodiversity in the Antarctic, Science Plan SCAR 2004. SCAR report 24:37–47
Appleyard S, Ward R, Williams R (2002) Population structure of the Patagonian toothfish around heard, McDonald and Macquarie Islands. Antarctic Sci 14:364–373
Appleyard SA, Williams R, Ward RD (2004) Population genetic structure of Patagonian toothfish in the west Indian sector of the Southern Ocean. CCAMLR Sci 11:12–32
Ashford J, La Mesa M, Fach B, Jones C, Everson I (2010) Testing early life connectivity using otolith chemistry and particle-tracking simulations. Can J Fish Sci 67:1303–1315
Avise JC (2000) Phylogeography. Harvard University Press, Cambridge
Becker LA, Pascual MA, Basso NG (2007) Colonization of the Southern Patagonia ocean by exotic chinook salmon. Cons Biol 21:1347–1352
Blaxter ML (2003) Molecular systematics: counting angels with DNA. Nature 421:122–124
Brey T, Dahm C, Gorny M, Klages M, Stiller M, Arntz WE (1996) Do Antarctic benthic invertebrates show an extended level of eurybathy? Antarctic Sci 8:3–6
Brumfield RT, Beerli P, Nickerson DA, Edwards SV (2003) The utility of single nucleotide polymorphisms in inferences of population history. Trends Ecol Evol 18:249–256
Chown SL, Sinclair BJ, Jansen van Vuuren B (2008) DNA barcoding and the documentation of alien species establishment on sub-Antarctic Marion Island. Polar Biol 31:651–655
Christie MR, Tissot BN, Albins MA et al (2010) Larval connectivity in an effective network of marine protected areas. PLoS One 5:e15715
Clark MS, Fraser KPP, Burns G, Peck LS (2008) The HSP70 heat shock response in the Antarctic fish *Harpagifer antarcticus*. Polar Biol 31:171–180
Clark MS, Thorne MAS, Toullec J-Y et al (2011) Antarctic krill 454 pyrosequencing reveals chaperone and stress transciptome. PLoS One 6:e5919
Clarke A, Johnston NM (2003) Antarctic marine benthic diversity. Oceanogr Mar Biol Annu Rev 41:47–114
Clarke A, Murphy EJ, Meredith MP et al (2007) Climate change and the marine ecosystem of the western Antarctic Peninsula. Phil Trans R Soc B 362:149–166
Clement O, Ozouf-Costaz C, Lecointre G, Berrebi P (1998) Allozyme polymorphism and phylogeny of the family Channichthyidae. In: di Prisco G, Pisano E, Clarke A (eds) Fishes of Antarctica. Springer, Milan, pp 299–309
Creer S, Fonseca VG, Porazinska DL et al (2010) Ultrasequencing of the meiofaunal biosphere: practice, pitfalls and promises. Mol Ecol 19(Suppl 1):4–20
Cullis-Suzuki S, Pauly D (2010) Failing the high seas: a global evaluation of regional fisheries management organizations. Mar Pol 5:1036–1042
De Broyer C, Danis B et al (2011) How many species in the Southern Ocean? Towards a dynamic inventory of the Antarctic marine species. Deep-Sea Res II 58:5–17
Dettaï A, Adamowizc SJ, Allcock L, et al. (2011) DNA barcoding and molecular systematics of the benthic and demersal organisms of the CEAMARC survey. Polar Sci 38:298–312
DeWoody JA, Avise JC (2000) Microsatellite variation in marine, freshwater and anadromous fishes compared with other animals. J Fish Biol 56:461–473
di Prisco G, Cocca E, Parker SK, Detrich HW III (2002) Tracking the evolutionary loss of hemoglobin expression by the white-blooded Antarctic icefishes. Gene 295:185–191
Duhamel G, Ozouf-Costaz C, Cattaneo-Berrebi G, Berrebi P (1995) Interpopulation relationships in two species of Antarctic fish *Notothenia rossii* and *Champsocephalus gunnari* from the Kerguelen Islands: an allozyme study. Antarctic Sci 7:351–356
Eastman JT (2005) The nature of the diversity of Antarctic fishes. Polar Biol 28:93–107
Eastman JT, Barrera-Oro E (2010) Buoyancy studies of three morphs of the Antarctic fish *Trematomus newnesi* (Nototheniidae) from the South Shetland Islands. Polar Biol 33:823–831
Everson I, North A, Paul A, Cooper R, McWilliam N (2001) Spawning locations of mackrel icefish at South Georgia. CCAMLR Science 8:1–12
FAO (2010) The state of world fisheries and aquaculture. FAO Fisheries and Aquaculture Department. Food and Agriculture Organization of the united Nations, Rome

Flores JA, Sierro FJ (2007) Pronounced mid-Pleistocene southward shift of the Poar Front in the Atlantic sector of the Southern Ocean. Deep Sea Res II 54:2432–2442
Frankham R (2005) Genetics and extinction. Cons. Biol 126:131–140
Frankham R, Ballou JD, Briscoe DA (2002) Introduction to conservation genetics. Cambridge University Press, Cambridge
Gaines SD, White C, Carr MH, Palumbi SR (2010) Designing marine reserve networks for both conservation and fisheries management. Proc Natl Acad Sci USA 107:18286–18293
Galindo H, Olson D, Palumbi S (2006) Seascape genetics: a coupled oceanographic-genetic model predicts population structure of Caribbean corals. Curr Biol 16:1622–1626
Galindo HM, Pfeiffer-Herbert AS, McManus MA et al (2010) Seascape genetics along a steep cline: using genetic patterns to test predictions of marine larval dispersal. Mol Ecol 19: 3692–3707
Gon O, Heemstra P (1990) Fishes of the Southern Ocean. JLB Smith Institute of Ichthyology, Grahamstown, p 462
Grant S, Constable A, Raymond B et al (2006) Bioregionalisation of the Southern Ocean: report of the experts workshop. WWF Australia and ACE CRC, Hobart
Grant RA, Linse K (2009) Barcoding Antarctic biodiversity: current status and the CAML initiative, a case study of marine invertebrates. Polar Biol 32:1629–1637
Griffiths HJ, Danis B, Clarke A (2011) Quantifying Antarctic marine biodiversity: The SCAR-MarBin data portal. Deep-Sea Res II 58:18–29
Gutt J, Hosie G, Stoddart M (2010) Marine life in the Antarctic. In: McIntyre AD (ed) Life in the World's Oceans. Diversity, distribution and abundance. Chapter 11:203–220
Harris J, Haward M, Jabour J, Woehler EJ (2007) A new approach to selecting Marine Protected Areas (MPAs) in the Southern Ocean. Antarctic Sci 19:189–194
Hauser L, Carvalho GR (2008) Paradigm shifts in marine fisheries genetics: ugly hypotheses slain by beautiful facts. Fish Fisheries 9:333–362
Hebert PDN, Cywinska A, Ball SL, deWaard JR (2003) Biological identifications through DNA barcodes. Proc Royal Soc London B 270:313–321
Hohenlohe PA, Bassham S, Etter PD et al (2010) Population genomics of parallel adaptation in threespine stickleback using sequenced RAD tags. PLoS Genet 6:e1000862
Hutchinson WF, van Oosterhout C, Rogers SI, Carvalho GR (2003) Temporal analysis of archived samples indicates marked genetic changes in declining North Sea cod (*Gadus morhua*). Proc Royal Soc London B 270:2125–2132
Huybrechts P (1993) Glaciological modeling of the late Cenozoic East Antarctic ice sheet: stability or dynamism? Geogr Ann 75A:221–238
Janko K, Lecointre G, DeVries AL et al (2007) Did glacial advances during the Pleistocene influence differently the demographic histories of benthic and pelagic Antarctic shelf fishes? Inferences from intraspecific mitochondrial and nuclear DNA sequence diversity. BMC Evol Biol 7:220
Jones C, Anderson E, Balushkin A et al (2008) Diversity, relative abundance, new locality records and population structure of Antarctic demersal fishes form the northern Scotia Arc islands and Bouvetoya. Polar Biol 31:1481–1497
Jørgensen C, Enberg K, Dunlop ES et al (2007) Managing evolving fish stocks. Science 318:124–125
Kellerman A (1990) Catalogue of early life stages of Antarctic notothenioid fishes. Ber Polarforsch 67:45–136
Kock KH (2005) Antarctic icefishes (Channichthyidae): a unique family of fishes. A review. Part I. Polar Biol 28:862–895
Kock KH, Purves M, Duhamel G (2005) Interactions between cetaceans and fish-eries in the Southern Ocean. Polar Biol 28:379–388
Kock KH, Reid K, Croxall J, Nicol S (2007) Fisheries in the Southern Ocean: an ecosystem approach. Phil Trans R Soc Lond B Biol Sci 29:2333–2349
Koubbi P, Moteki M, Duhamel G et al (2011) Ecoregionalization of mycto-phid fish in the Indian sector of the Southern Ocean: Results from generalized dissimilarity models. Deep-Sea Res II 58:170–180

Kuhn K, Gaffney P (2006) Preliminary assessment of population structure in the mackerel icefish (*Champsocephalus gunnari*). Polar Biol 29:927–935

Kuhn K, Gaffney P (2008) Population subdivision in the Antarctic toothfish (*Dissostichus mawsoni*) revealed by mitochondrial and nuclear single nucleotide polymorphisms (SNPs). Antarctic Sci 4:327–338

Kuhn K, Near T, Jones C, Eastman J (2009) Aspoects of the biology and population genetics of the Antarctic nototheniid fish *Trematomus nicolai*. Copeia 2:320–327

Laurel BJ, Bradbury IR (2006) "Big" concerns with high latitude marine protected areas (MPAs): trends in connectivity and MPA size. Can J Fish Aquat Sci 63:2603–2607

Lautrédou AC, Bonillo C, Denys G et al (2010) Molecular taxonomy and identification within the Antarctic genus *Trematomus* (Notothenioidei, Teleostei): how valuable is barcoding with COI? Polar Sci 4:333–352

Lawver LA, Gahagan LM (2003) Evolution of Cenozoic seaways in the circum-Antarctic region. Palaeogeogr Palaeoclim Palaeoecol 198:11–37

Lecointre G (2012) Phylogeny and systematics of Antarctic teleosts: methodological and evolutionary issues. In: di Prisco G, Verde C (eds) Adaptation and evolution in marine environments—The impacts of global change on biodiversity, vol 1. Series "From Pole to Pole". Springer, Berlin, pp 97–117

Linse K, Griffiths HJ, Barnes DKA, Clarke A (2006) Biodiversity and biogeography of Antarctic and sub-Antarctic mollusca. Deep-Sea Res II 53:885–1008

Lubchenko J, Palumubi SR, Gaines SD, Andelman S (2003) Plugging a hole in the ocean: the emerging science of marine reserves. Ecol Appl 13:S3–S7

Marshall C (2012) Aspects of protein cold adaptation in Antarctic fish. In: di Prisco G, Verde C (eds) Adaptation and evolution in marine environments—The impacts of global change on biodiversity, vol 1. Series "From Pole to Pole". Springer, Berlin, pp 143–155

Matschiner M, Hanel R, Salzburger W (2009) Gene flow by larval dispersal in the Antarctic notothenioid fish *Gobionotothen gibberifrons*. Mol Ecol 18:2574–2587

Meredith M, Brandon M, Murphy E et al (2005) Variability in hydrographic conditions to the east and northwest of South Georgia, 1996–2001. J Mar Syst 53:143–167

Nielsen EE, Hemmer-Hansen J, Poulsen NA et al (2009) Genomic signatures of local directional selection in a high gene flow marine organism; the Atlantic cod (*Gadus morhua*). BMC Evol Biol 9:276

North A (2001) Early life history strategies of notothenioids at South Georgia. J Fish Biol 58:496–505

North A, Murray A (1992) Abundance and diurnal vertical distribution of fish larvae in the early spring and summer in a fjord at South Georgia. Antarctic Sci 4:405–412

North A, White M (1987) Reproductive strategies of Antarctic fish. In: Kullander S, Fernholm B (eds). Proceedings 5th Congress European Ichthyol, pp 381–391. Swedish Museum of Natural History, Stockholm

Palumbi SR (2003) Population genetics, demographic connectivity, and the design of marine reserves. Ecol Appl 13(Suppl 1):S146–S158

Papetti C, Susana E, La Mesa M et al (2007) Microsatellite analysis reveals genetic differentiation between year-classes in the icefish *Chaenocephalus aceratus* at South Shetlands and Elephant Island. Polar Biol 30:1605–1613

Papetti C, Susana E, Patarnello T, Zane L (2009) Spatial and temporal boundaries to gene flow between *Chaenocephalus aceratus* populations at South Orkney and South Shetlands. Mar Ecol Prog Ser 376:269–281

Papetti C, Pujolar JM, Mezzavilla M et al (2012) Population genetic structure and gene flow patterns between populations of the Antarctic icefish *Chionodraco rastrospinosus*. J Biogeogr (in press)

Parker R, Paige K, DeVries A (2002) Genetic variation among populations of the Antarctic toothfish: evolutionary insights and implications for conservation. Polar Biol 25:256–261

Patarnello T, Marcato S, Zane L, Varotto V, Bargelloni L (2003) Phylo-geo-graphy of the *Chionodraco* genus (Perciformes, Channichthydae) in the Southern Ocean. Mol Phylog Evol 28:420–429

Patarnello T, Verde C, di Prisco G et al (2011) How will fish that evolved at constant sub-zero temperatures cope with global warming? Notothenioids as a case study. Bioessays 33:260–268

Peck LS, Clark MS, Clarke A et al (2005) Genomics: applications to Antarctic ecosystems. Polar Biol 28:351–365

Peck LS, Clark MS, Morley SA et al (2009) Animal temperature limits and ecological relevance: effects of size, activity and rates of change. Fund Ecol 23:248–256

Peck LS, Clark MS (2012) Understanding adaptations and responses to change in Antarctica: recent physiological and genomic advances in marine environments. In: di Prisco G, Verde C (eds) Adaptation and evolution in marine environments—The impacts of global change on biodiversity, vol 1. Series "From Pole to Pole". Springer, Berlin, pp 157–182

Petit JR, Jouzel J, Raynaud D, Barkov NI et al (1999) Climate and atmospheric history of the past 420,000 years from the Vostok ice core, Antarctica. Nature 399:429–436

Pettis RW, Forest A (1979) Chemical composition of ferromanganese nodules from the Southern Ocean. Aust J Mar Freshw Res 30:535–539

Pörtner HO (2006) Climate dependent evolution of Antarctic ectotherms: an integrative analysis (EASIZ, SCAR). Deep-Sea Res II 53:1071–1104

Reilly A, Ward RD (1999) Microsatellite loci to determine stock structure of the Patagonian toothfish *Dissostichus eleginoides*. Mol Ecol 8:1753–1756

Rock J, Costa FO, Walker DI et al (2008) DNA barcodes of fish of the Scotia Sea, Antarctica indicate priority groups for taxonomic and systematic focus. Antarct Sci 20:253–262

Rogers A, Morley S, Fitxcharles E et al (2006) Genetic structure of Patagonian toothfish (*Dissostichus eleginoides*) populations on the Patagonian Shelf and Atlantic and western Indian Ocean Sectors of the Southern Ocean. Mar Biol 149:915–924

Rogers AD (2007) Evolution and biodiversity of Antarctic organisms: a molecular perspective. Phil Trans R Soc B 362:2191–2214

Rusch DB, Halpern AL, Sutton G et al (2007) The *Sorcerer II* Global Ocean Sampling expedition: northwest Atlantic through eastern tropical Pacific. PLoS Biol 5:e77

Scheffer M, Carpenter S, Foley JA, Folkes C, Walker R (2001) Catastrophic shifts in ecosystems. Nature 413:591–596

Scheffer M, Carpenter S, de Young B (2005) Cascading effects of overfishing marine systems. Trends Ecol Evol 20:579–581

Schneppenheim R, Kock K, Duhamel G, Janssen G (1994) On the taxonomy of the *Lepidonotothen squamifrons* group (Pisces, Perciformes, Notothenioidei). Arch Fish Mar Res 42:137–148

Selkoe KA, Watson JR, White C et al (2010) Taking the chaos out of genetic patchiness: seascape genetics reveals ecological and oceanographic drivers of genetic patterns in three temperate reef species. Mol Ecol 19:3708–3726

Shaw P, Arkhipkin A, Al-Khairulla H (2004) Genetic structuring of Patagonian toothfish populations in the Southwest Atlantic Ocean: the effect of the Atlantic Polar Front and deep-water troughs as barriers to genetics exchange. Mol Ecol 13:3293–3303

Smith P, Gaffney P (2005) Low genetic diversity in the Antarctic toothfish (*Dissostichus mawsoni*) observed with mitochondrial and intron DNA markers. CCAMLR Science 12:43–51

Smith P, McVeagh M (2000) Allozyme and microsatellite DNA markers of toothfish population structure in the Southern Ocean. J Fish Biology 57:72–83

Susana E, Papetti C, Barbisan F et al (2007) Isolation and characterization of eight microsatellite loci in the icefish *Chaenocephalus aceratus* (Perciformes, Notothenioidei, Channichthyidae). Mol Ecol Notes 7:791–793

Tavares M, De Melo GAS (2004) Discovery of the first known benthic invasive species in the Southern Ocean: the North Atlantic spider crab *Hyas araneus* found in the Antarctic Peninsula. Antarctic Sci 16:129–131

Thatje S, Hillenbrand CD, Larter R (2005) On the origin of Antarctic marine benthic community structure. Trends Ecol Evol 20:534–540

Thomson PG, Davidson AT, van den Enden R et al (2010) Distribution and abundance of marine microbes in the Southern Ocean between 30 and 80 degrees E. Deep-Sea Res II 57:815–827
Bilyk KT, DeVries AL (2010) Heat tolerance and its plasticity in Antarctic fishes. Comp Biochem Physiol 158:382–390
Tin T, Fleming ZL, Hughes Ka et al (2009) Impacts of local human activities on the Antarctic environment. Antarctic Sci 21:3–33
Tittensor DP, Mora C, Jezt W et al (2010) Global patterns and predictors of marine biodiversity across taxa. Nature 466:1098–1103
Trivelpiece WZ, Hinken JT, Miller AK et al (2011) Variability in krill biomass links harvesting and climate warming to penguin population changes in Antarctica. Proc Natl Acad Sci USA 108:7625–7628
Van Houdt JKJ, Hellemans B, van de Putte A, Koubbi P, Volckaert FAM (2006) Isolation and multiplex analysis of six polymorphic microsatellites in the Antarctic notothenioid fish, *Trematomus newnesi*. Mol Ecol Notes 6:157–159
Van de Putte AP, Van Houdt JKJ, Maes GE et al (2009) Species identification in the trematomid family using nuclear genetic markers. Polar Biol 32:1731–1741
Van de Putte AP, Van Houdt JKJ, Hellemans B, Collins M, Volckaert FAM (2012) High genetic diversity and connectivity in a common mesopelagic fish of the Southern Ocean: the myctophid *Electrona antarctica*. Deep-Sea Res II. In press
Verde C, Giordano D, Russo R, di Prisco G (2012) The adaptive evolution of polar fishes. Lessons from the function of hemoproteins. In: di Prisco G, Verde C (eds) Adaptation and evolution in marine environments—The impacts of global change on biodiversity, vol 1. Series "From Pole to Pole". Springer, Berlin, pp 197–213
Vogiatzi E, Lagnel J, Pakaki V et al (2011) In silico mining and characterization of simple sequence repeats from gilthead sea bream (*Sparus aurata*) expressed sequence tags (EST-SSRs); PCR amplification, polymorphism evaluation and multiplexing and cross-species assays. Mar Genom 4:83–91
Walther GR, Post E, Convey P et al (2002) Ecological responses to recent climate change. Nature 416:389–395
Ward NL, Steven B, Penn K, Methé BA, Detrich WH III (2009) Characterization of the intestinal microbiota of two Antarctic notothenioid fish species. Extremophiles 13:679–685
Webb DJ, de Cuevas BA (2007) On the fast response of the Southern Ocean to changes in the zonal wind. Ocean Sci 3:417–427
White M, Veit R, North A, Robinson K (1996) Egg-shell morphology of the Antarctic fish, *Notothenia rossii* Richardson, and the distribution and abundance of pelagic eggs at South Georgia. Antarctic Sci 8:267–271
White M (1998) Development, dispersal and recruitment: a paradox for survival among Antarctic fish. In: di Prisco G, Pisano E, Clarke A (eds) Fishes of Antarctica. Springer, Milan, pp 53–62
White M, North A, Twelves E, Jones S (1982) Early development of *Notothenia neglecta* from the Scotia Sea, Antarctica. Cybium 6:43–51
Williams R, Smolenski A, White R (1994) Mitochondrial DNA variation of (*Champsocephalus gunnari*) Lonnberg (Pisces: Channichthyidae) stocks on the Kerguelen Plateau, southern Indian Ocean. Antarctic Sci 6:347–352
Williams ID, Walsh WJ, Schroeder RE et al (2008) Assessing the importance of fishing impacts on Hawaiian coral reef fish assemblages along regional-scale human population gradients. Env Cons 35:261–272
Worm B, Hilborn R, Baum JK et al (2009) Rebuilding global fisheries. Science 325:578–585
Zane L, Marcato S, Bargelloni L, Bortolotto E (2006) Demographic history and population structure of the Antarctic silverfish *Pleuragramma antarcticum*. Mol Ecol 15:4499–4511
Zdzitowiecki K, White M (1992) Acanthocephalan infection of inshore fish in two fjords at South Georgia. Antarctic Sci 4:197–203

Chapter 6
Phylogeny and Systematics of Antarctic Teleosts: Methodological and Evolutionary Issues

Guillaume Lecointre

Since the first publication of a molecular phylogeny of notothenioids by Bargelloni et al. (1994), the face of systematics has changed. Molecular phylogenies are no longer obtained from a single genetic marker and the increasing sequencing and computing powers allow better taxonomic samplings in molecular trees. The speed of resolution of the "tree of life" (Lecointre and Le Guyader 2001) is increasing and will continue to do so, through better coordination among teams and projects. Moreover, the International Polar Year between 2007 and 2009 has stimulated coordinated access to the Antarctic field for large population samplings. In parallel, barcoding approaches are modifying the work flow chart of taxonomy, separating routine identifications that can be efficiently performed using pre-identified sequence data bases from taxonomic research devoted to species delimitation. Taxonomic skills are promoted in the sense that authority is needed at the early step of validation of identifications given to specimens used in sequence databases, and in parallel the time of taxonomists is saved for their favourite research, leaving the burden of routine identification to the barcode tool (Ward et al. 2005, 2009). On the research side of taxonomy, and more specifically concerning species delimitations, taxonomy is facing deep changes in its interactions between traditional morphology-based taxonomic skills and DNA sequence-based approaches through "integrative taxonomy" (Dayrat 2005; Will et al. 2005; Ciprandi-Pires and Marinoni 2010).

Notothenioids are obviously not the only taxonomic component of the Antarctic fish diversity (Table 6.1) (Eastman 2005; Gon and Heemstra 1990; Miller 1993; Causse et al. 2011). In terms of number of species, they represent 134 of the 327 species recorded in the Southern Ocean, and 45.5% of the fish species of the shelf;

G. Lecointre (✉)
Département Systématique and Evolution, Muséum National d'Histoire Naturelle, 57 rue Cuvier, 75231 Paris cedex 05, France
e-mail: lecointr@mnhn.fr

G. di Prisco and C. Verde (eds.), *Adaptation and Evolution in Marine Environments, Volume 1*, From Pole to Pole, DOI: 10.1007/978-3-642-27352-0_6,

Table 6.1 Families of benthic fishes of the Antarctic shelf (from Eastman 2005)

Taxon of the shelf	Number of species	Fauna (%)
Myxinidae (hagfishes)	1	0.5
Petromyzontidae (lampreys)	1	0.5
Rajidae (skates)	8	3.6
Carapidae (pearlfishes)	1	0.5
Moridae (deepsea cods)	4	1.7
Muraenolepididae (eel cods)	4	1.7
Gadidae (cods)	1	0.5
Congiopodidae (horsefishes)	1	0.5
Bathylutichthyidae	1	0.5
Liparidae (snailfishes)	70	31.5
Zoarcidae (eelpouts)	24	10.8
Notothenioidei (six families)	101	45.5
Tripterygiidae (triplefins)	1	0.5
Achiropsettidae (southern flounders)	4	1.7
Total	222	100

31% are liparids (snailfishes) and 11% are zoarcids (eelpouts). However, they account for more than 90% of the fish biomass of the shelf (Eastman 2005) and are the most studied Antarctic fishes (Eastman 1993; di Prisco et al. 1998).

6.1 Taxonomy

Barcodes consist first in sequencing a gene of reference for well identified specimens deposited in collections. For most vertebrates the gene of reference is the mitochondrial gene encoding cytochrome oxydase I. The gene was chosen because its mutational constraints are such that intraspecific pairwise distances between individuals do not overlap interspecific pairwise distances (the "barcoding gap"). These sequences are therefore obtained for a collection of individuals of a given species and deposited in a database with an associated data analysis system, the Barcode of Life Data System (BOLD; Ratnasingham and Hebert 2007, http://www.boldsystems.org). The database obviously needs initial taxonomic expertise to be useful and reliable. Then this database can be used for the assignment of a new specimen sequence to a given species already recorded in the database. This is made to improve the efficiency of routine identifications, neither to replace taxonomists nor to be the only criterion for research in species delimitation. The project has been presented as a powerful tool for molecular taxonomy (Hebert et al. 2003) which provoked debates (De Salle et al. 2005; Will et al. 2005; Rubinoff 2006; De Salle 2006). Several studies showed that the barcoding gap is efficient for most "fishes" (Ward et al. 2005, 2009; Rock et al. 2008; Steinke et al. 2009; Lautrédou et al. 2010; Dettaï et al. 2011). The barcoding of Antarctic fishes has started recently (Rock et al. 2008; Lautrédou et al. 2010) and is rapidly increasing. The usefulness of the approach in Antarctic cruises is shown in Dettaï et al. (2011) and Grant et al. (2011). Notably, identifications of eggs and larvae of notothenioids are now available and increasingly reliable; in BOLD databases there are already 106 notothenioid species (out of 134) barcoded from 1,586 specimens (as of June 2011). Other markers are also developed for the same purposes (Van de Putte et al. 2009). Taxonomy of notothenioids is therefore alive and well. In their updated species list, Eastman and Eakin (2000) recorded 8 families, 43 genera and 122 species (96 Antarctic and 26 non-Antarctic). As of June 2011, the count is 134 species distributed in 8 families and 44 genera (Fig. 6.1), i.e. an increase in species of 10% within 10 years. Note that some genera should disappear to render their stem genus monophyletic, such as *Pagothenia*, *Cryothenia* (the corresponding species should become *Trematomus*) and *Paranotothenia*, which should be *Notothenia*. The number of valid species slightly differs in Fishbase (Froese and Pauly 2011), notably with regard to the number of valid species in the genera *Channichthys* and *Harpagifer*.

6.2 Phylogeny

The rise of phylogenetics as a visible science dates back to the early 1990s when gene amplification and sequencing technologies could be affordable enough to become available to systematists, at a time when computing power was already accessible to those labs (see for instance Hillis and Moritz 1990; Hillis 2004). However, we should keep in mind that the fundamental revolution had been conceptual long before becoming technological. Hennig (1950, 1966) found the methodological solution that was lacking to Charles Darwin (1859) to establish unambiguous correspondence between degree of kinship and classification, i.e. to only admit monophyletic groups in the classifications. As in any scientific investigation, modern phylogenies, either molecular or morphologically-based, can be hampered by artefacts. Since the beginnings of molecular systematics, there have been contributions in the literature about avoiding false trees due to non-random homoplasy (Hillis and Moritz 1990), calculating robustness of nodes, or about how to conduct phylogenetic analyses from different sources of data (Miyamoto and Fitch 1995), including the ongoing debates about articulating molecular and morphological data, homologies and phylogenies. Since the mid 1990s, literature also increased on misinterpreted trees, i.e. correctly reconstructed but reflecting the history of genes that is not the history of species (Doyle 1992, 1997; Maddison 1997). The ultimate criterion to know whether a clade of species (i.e. a monophyletic group of species) is reliable or not corresponds to its repeatability (Grande 1994; Chen et al. 2003; Dettaï and Lecointre 2004, 2005) across different type of data, different teams and even different examples to represent the same terminal entities. One of the reason is that there can be so many trees for a given set of terminal entities to classify, that the chance to get the same clade twice just by chance is extremely low (Page and Holmes 1998). Each dataset has its own source of potential errors with regard to species interrelationships. Recovering the same clade from independent data is a strong indication of its reliability because the clade is recovered in spite of all possible biases specific to each dataset.

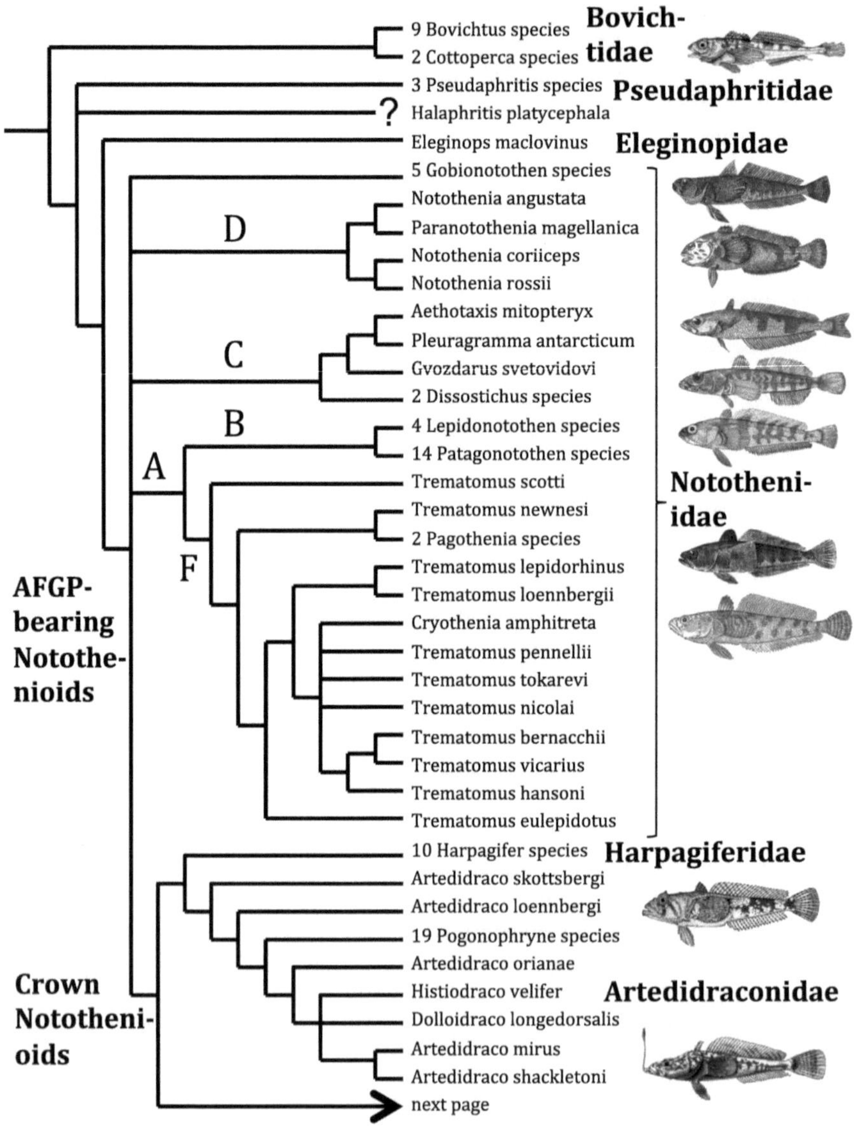

Fig. 6.1 Summary tree exhibiting the state of knowledge of notothenioid interrelationships. *Gvozdarus* and *Vomeridens* are placed according to Balushkin (2000). Position of *Halaphritis* is provisional (and not that of Last et al. 2002). The phylogeny within the Trematominae manages different areas of uncertainties and resolution of Kuhn and Near (2009) and Janko et al. (2011). *Notothenia microlepidota*, *Notothenia cyanobrancha*, and *Artedidraco glabeobarbatus* are lacking

The phylogeny of the notothenioids is now clear at the level of genera, and most of the times at the interspecific level, except one area concerning the family Nototheniidae. The picture can be drawn from a significant number of molecular studies from

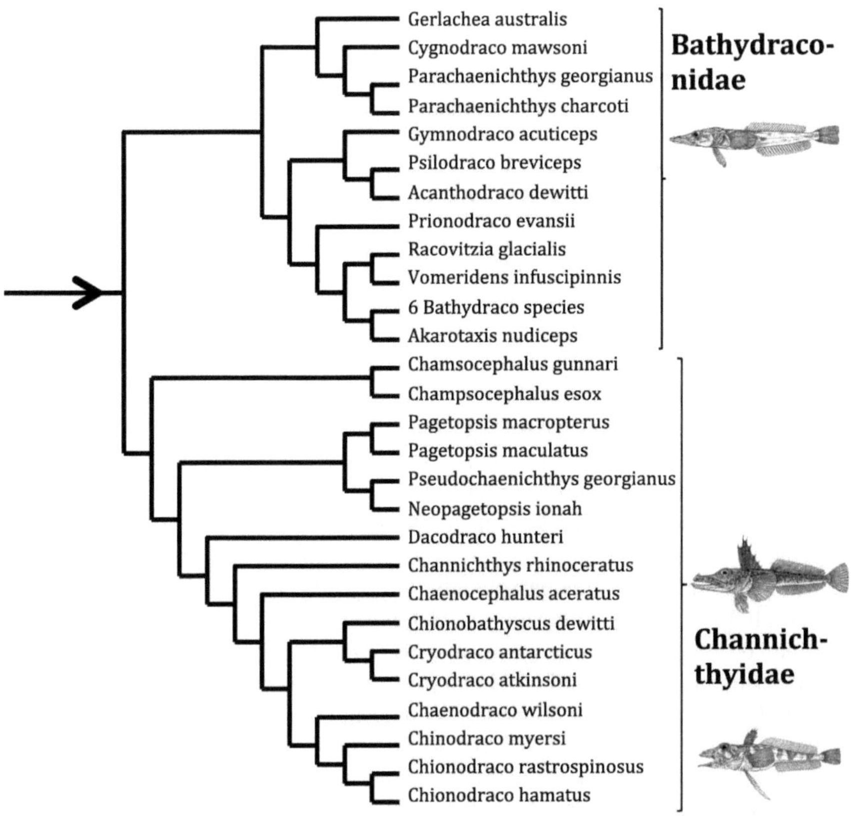

Fig. 6.1 (continued)

1994 onwards (Fig. 6.1). The number of phylogenetic studies using morphological characters is rather low in comparison (for instance Eakin 1981; Iwami and Abe 1984; Iwami 1985; Hastings 1993; Balushkin 2000; Voskoboynikova 2000). Only Hastings (1993) and Balushkin (2000) cover all notothenioids. Unfortunately, none of these studies are explicitly using coded characters into a data matrix followed by a standard parsimony approach. Such phylogenetic analyses using computerized standard parsimony have been performed later, separately, family by family, along with a parallel molecular comparison of the same species or groups (Chen et al. 1998; Derome et al. 2002; Near et al. 2003; Sanchez et al. 2007).

6.2.1 Bovichtidae and Pseudaphritidae

Bovichtidae classically contained three genera (Eastman 1993): *Bovichtus* (9 species known today), and monotypic *Cottoperca* and *Pseudaphritis* (Eastman 1993; Hastings 1993). Several species have been discovered since then. From 28S rDNA sequence data, Lecointre et al. (1997) found paraphyly of the bovichtidae, *Pseudaphritis* being more closely related to the rest of the non-bovichtid notothenioids than to the two other bovichtids. This was already suggested by Voskoboynikova (1993) in a non-cladistic framework and Lecointre et al. (1997) retained a number of her morphological characters common to *Pseudaphritis* and non-bovichtid notothenioids as synapomorphies, along with the major Hb 1 component (95%) in the hemoglobins (di Prisco et al. 1991; Verde et al. 2012 Chap. 11). This phylogenetic position of *Pseudaphritis* was confirmed by Ritchie et al. (1997) from mitochondrial 12S and 16S ribosomal DNA sequences in a wider taxomonic sample, and confirmed later by Near et al. (2004) from the complete 16S rDNA gene sequences, and by Near and Cheng (2008) from mitochondrial 16S rDNA, ND2 gene, three tRNAs sequences and sequences from the nuclear gene S7. The Pseudaphritidae are now considered as a separate family, an option chosen by Near et al. (2004), Eastman (2005, 2006) and BOLD. In that framework, Eastman (2006) found a distinct hypoglossal gland as a synapomorphy of the bovichtids sensu *stricto* (i.e. *Bovichtus* and *Cottoperca*). In 2002 a new genus, *Halaphritis*, was described from South eastern Australia (Last et al. 2002). The authors provisionally classified the genus in the Bovichtidae sensu *lato*, i.e. as in Eastman (1993) and Fishbase, using too loose a definition of this paraphyletic family, thus calling for more molecular and morphological data. Interestingly, *Halaphritis* shares some of the six synapomorphies proposed by Lecointre et al. (1997) to support the clade grouping the Pseudaphritidae and the rest of non-bovichtid notothenioids (Last et al. 2002). As the Bovichtidae sensu *lato* were defined on symplesiomorphies, this brings *Halaphritis* potentially closer to *Pseudaphritis* than to *Bovichtus* and *Cottoperca*. Here we will provisionally place *Halaphritis* in the Pseudaphritidae, but wait for more molecular and morphological data, as did Last et al. (2002), in order to clarify the relationships of *Halaphritis*.

6.2.2 Eleginopidae

Nototheniidae in the classical sense contained 50 species, among which *Eleginops maclovinus*. Lecointre et al. (1997) found that the nototheniid *Dissostichus* was more closely related to non-nototheniid notothenioids than to *Eleginops*, confirming Balushkin's (1992) point of view. This position of *Eleginops* was later confirmed by Bargelloni et al. (2000) with well supported nodes obtained in a tree from mitochondrial 12S and 16S rDNA sequences, and also by Near et al. (2004) and Near and Cheng (2008). The new family Eleginopidae is an option kept by Eastman and Eakin (2000), Near et al. (2004) and Eastman (2005, 2006).

6.2.3 Nototheniidae

Part of the interrelationships among Nototheniidae (without *Eleginops*) form a consensus, however, the monophyly of the family remains controversial. From mitochondrial 12S and 16 rDNA sequences, Bargelloni et al. (1994, 2000) found paraphyletic nototheniids, *Pleuragramma* being (in the second paper) more closely related to crown notothenioids than to the rest of the nototheniids. This relationship had a low support. Interestingly, from the complete 16S gene Near et al. (2004) found the monophyly of the family but with low support. Near and Cheng (2008) found the paraphyly with low support: when ND2 sequences, three tRNA sequences and the nuclear gene S7 are added to the 16S data, the clade *Notothenia-Paranotothenia* is more closely related to the crown notothenioids. From four markers (the nuclear genes of Rhodopsin, MLL, the mitochondrial gene encoding cytochrome b and the d-loop) Sanchez et al. (2007) did not resolve the interrelationships among nototheniid components. So in spite of numerous studies we don't know yet whether the Nototheniidae are monophyletic or not.

However, some clades are clearly repeated in various studies. Recoding Balushkin's (2000) characters, Sanchez et al. (2007) found the "pelagic clade C" (*Aethotaxis*, *Pleuragramma*, *Dissostichus*, *Gvozdarus*), clade "F" containing *Trematomus* and *Pagothenia*, clade "D" containing *Notothenia* and *Paranotothenia*, and a clade grouping *Lepidonotothen*, *Patagonotothen* and *Gobionotothen*. The two latter clades form the Nototheniinae *sensu* Balushkin (2000). Clades C, D and F are recovered from d-loop sequence data, and from these data *Gobionotothen* is outside *Lepidonotothen-Patagonotothen*. Clade A emerges, grouping *Lepidonotothen-Patagonotothen* with F. The Nototheniinae of Balushkin (2000) are therefore polyphyletic. Two of these clades were recovered by Bargelloni et al. (2000) from 12S to 16S data: F and *Lepidonotothen-Patagonotothen*. With a better taxonomic sampling on the complete 16 gene, Near et al. (2004) found the two latter clades, plus clades A, C and D. When the nuclear gene S7 is added along with ND2 and three tRNA gene sequences, those clades are confirmed except the "pelagic clade", *Pleuragramma* escaping from it but with low support. In all these analyses the position of *Gobionotothen* remains unresolved. To summarize, nototheniids are made of the following components:

- *Gobionotothen*
- *Notothenia* with *Paranotothenia* (clade D)
- *Trematomus* with *Pagothenia* (clade F)
- *Lepidonotothen* with *Patagonotothen*
- Clade A = the last two
- Clade C (the clade of pelagic nototheniids): *Aethotaxis*, *Dissostichus*, *Pleuragramma* and *Gvozdarus*.

Interrelationships among these components are not resolved yet. They are those that remain to be clarified in the near future.

In all analyses the genera *Lepidonotothen*, *Patagonotothen* and *Dissostichus* are each monophyletic, *Pagothenia* and *Cryothenia* are nested within *Trematomus* (so they should become *Trematomus*) and *Paranotothenia* is nested within *Notothenia* (so it should become a *Notothenia*). None of the molecular studies contain the monotypic *Gvozdarus*.

6.2.4 Harpagiferidae

Crown notothenioids are made of four families: Harpagiferidae, Artedidraconidae, Bathydraconidae and Channichthyidae. This crown group is found in all studies (Eakin 1981; Bargelloni et al. 1994, 2000; Derome et al. 2002, Near et al. 2004). Harpagiferidae and Artedidraconidae are monophyletic and are sister groups in all studies.

6.2.5 Artedidraconidae

Lecointre et al. (2011) published the phylogeny of the genus *Artedidraco* from sequences obtained from 77 specimens in three mitochondrial genes (cytochrome oxidase I, cytochrome *b* and d-loop) and the nuclear rhodopsin retro gene, and Eakin et al. (2009) obtained the phylogeny of 12 of the 19 *Pogonophryne* species from the ND2 mitochondrial gene. The genus *Artedidraco* is paraphyletic and contains the genera *Pogonophryne*, *Histiodraco* and *Dolloidraco*, a paraphyly that was already suggested by Derome et al. (2002) from an incomplete taxonomic sample. The *Pogonophryne* intrarelationships are partly resolved.

6.2.6 Bathydraconidae

The family Bathydraconidae was found paraphyletic by Bargelloni et al. (2000) from mitochondrial 12S–16D rDNA sequences, but the support was low and the number of species sampled was weak (3). From the mitochondrial d-loop region and cytochrome *b* gene sequences, Derome et al. (2002) found three highly supported subgroups: the group made of *Bathydraco*, *Racovitzia*, and *Prionodraco* provisionally called “Bathydraconinae”, the group containing *Gymnodraco*, *Acanthodraco* and *Psilodraco* called “Gymnodraconinae”, and the group assembling *Parachaenichthys*, *Cygnodraco* and *Gerlachea* called “Cygnodraconinae”. Those molecular data did not contain any signal for or against the monophyly of bathydraconids. When adding recoded morphological characters from the literature, bathydraconids became monophyletic in the tree based on molecules and morphology altogether. The only character that supported monophyly of the family was the loss of the anterior spinous dorsal fin. This showed that a morphological matrix can impose its topology when mixed with a molecular matrix.

These three components were confirmed by the complete 16S sequence data of Near et al. (2004), who added *Akarotaxis* to Bathydraconinae as the sister group of *Bathydraco* [in accordance with Balushkin (2000) who also added *Vomeridens* as the sister group of *Racovitzia*]. The family was paraphyletic, *Gymnodraco* being closer to Channichthyidae, however, with poor support. This was the same situation from the data of Near and Cheng (2008). To conclude, the three components are reliable but the monophyly of the family only holds on the loss of the anterior dorsal fin, among those morphological characters that have been coded to date.

6.2.7 Channichthyidae

The icefishes are among the most fascinating teleosts, mainly because of their large size associated with the loss of erythrocytes and hemoglobins, exhibiting a unique case of 16 "white-blooded" species among vertebrates. 15 of the 16 icefish species have lost the adult β-globin gene but retain a truncated α-globin pseudogene (Near et al. 2006). Interrelationships among genera were already proposed by Iwami (1985) from coded anatomical characters. However, the parsimony of the tree retained by Iwami (1985) was not formally controlled. Chen et al. (1998) controlled Iwami's matrix using computerized standard parsimony and found four equiparsimonious trees, among which there was the tree retained by Iwami. Moreover, from cytochrome *b* and d-loop mitochondrial sequences, Chen et al. (1998) found a tree that is one of the trees of Iwami, a tree that was confirmed later by Derome et al. (2002). Thus it appears that there is no conflict between Iwami's data and their mitochondrial data, just a lack of signal in the morphological data used. In other words, this exemplifies the fact that parsimony performed "by hand" can lead an author to miss equiparsimonious solutions. Near et al. (2003) tended to interpret these differences as incongruent, and associated ND2 and 16S mtDNA sequences with 58 recoded morphological data taken from several authors (Iwami 1985; Balushkin 2000; Voskoboynikova 2000). The tree obtained from the whole data differed from Derome et al. (2002) only for two species out of 16, *Channichthys rhinoceratus* and *Chaenocephalus aceratus*. The conflict was actually not strong because the relevant nodes in Derome et al. (2002) were not highly supported. Using the same tree, Near et al. (2006) explained the presence of a complete, but non-functional, adult alpha–beta-globin genetic cluster in *Neopagetopsis ionah* through incomplete lineage sorting of ancestral polymorphism. This cause is going to be increasingly invoked for recent flocks (see below) such as for instance the Trematominae (Janko et al. 2011; Lautrédou et al. 2011, in prep.).

6.2.8 Species Interrelationships

Most of the genera have less than five species, then a significant number of species found their interrelationships resolved in a sampling context where interrelationships at the wide scale were investigated (for instance it is clear for pseudaphritids, eleginopids, bathydraconids and channichthyids which genera are not species-rich). The most species-rich genera deserve specific studies: *Bovichtus* (9 species), *Patagonotothen* (14 species), *Trematomus* (14 species), *Artedidraco* (6 species) and *Pogonophryne* (19 species). The work has been carried out in detail for *Trematomus* (Ritchie et al. 1996; Sanchez et al. 2007; Kuhn and Near 2009; Lautrédou et al. 2010; Janko et al. 2011), *Pogonophryne* (Eakin et al. 2009) and *Artedidraco* (Lecointre et al. 2011). In *Trematomus* the picture drawn through mitochondrial markers is only partly resolved in spite of a comparatively high number of studies, and a noticeable result is that *T. lepidorhinus* and *T. loennbergii* are not distinguishable by the Cytochrome Oxidase I gene used for barcoding, similar to *T. bernacchii* and *T. vicarius*. These last two species have *T. hansoni* as a sister group. The most basal Trematominae species is *T. scotti*, then the clade emerges of cryopelagic species made of *T. newnesi* and the genus *Pagothenia*. In the sister group the picture becomes contradictory. Among nuclear markers, interrelationships are either contradictory or unresolved: only *T. scotti* remains as the sister group of the rest. However, the nuclear markers used by Kuhn and Near (2009) and Janko et al. (2011) are most probably too conserved and further investigations are needed in order to check for congruence between mitochondrial and nuclear markers, or to explain incongruence. Indeed it would not be surprising to find a lack of coalescence of nuclear alleles in the very recent *Trematomus* radiation (Janko et al. 2011; Lautrédou et al. 2011). From three mitochondrial genes (COI, Cyt b and d-loop) and the nuclear rhodopsin retrogene sequences from 77 specimens, interrelationships within *Artedidraco* are now resolved and the genus appears to be paraphyletic; it contains the species-rich *Pogonophryne,* and the monotypic genera *Dolloidraco* and *Histiodraco*. Interrelationships within the genus *Pogonophryne* have started to be investigated from the mitochondrial ND2 gene sequences of 12 species out of 19. For the moment resolution remains poor: five main groups of specimens have been identified, among which three are monospecific and two associate several species (Eakin et al. 2009).

6.2.9 Zoarcidae and Liparidae

These two important families have not been studied at the scale at which notothenioids have been studied. It is partly because these benthic animals are more difficult to catch. They are really minor components of the benthic fauna in terms of numbers of individuals, sizes and biomass: in the Southwest Ross Sea, notothenioids are 91.2% of the fish biomass caught and they dominate by 91.6% in terms of number of individuals (Eastman 2005). Liparids are 31.5% in number of species of the shelf, however, most of the 334 snailfish species known worldwide are uncommon or rare (Chernova et al. 2004) and most of them are small, fragile animals. This is the reason why it takes longer to get an equivalent picture, the time for cruises to be able to gather a significant amount of fresh tissue of the Antarctic components of the family. Molecular systematics of the family has already begun, however, focused on taxa from the northern hemisphere (Knudsen et al. 2007). The molecular work on Antarctic taxa is just starting (Dettaï et al. 2011; Duhamel et al. 2010; Lautrédou, 2009, in prep.). As a result, in June 2011 there is no public liparid sequences in BOLD databases and very few Antarctic zoarcids.

6.3 Evolutionary Issues

6.3.1 Species Flocks

A substantial part of the marine Antarctic species richness may be framed into species flocks. Indeed, the Antarctic continental shelf has been described as a giant generator of species flocks (Eastman and McCune 2000). Species flocks are bursts of closely related endemic species, where species are numerous relatively to surrounding areas, and ecologically diverse (Ribbink 1984). Physical similarities of the Antarctic shelf with ancient lakes where species flocks were found have been described by these authors in terms of isolation, depth and age. Indeed currents, deep water, sub-zero temperatures and distance isolate the waters of the Antarctic shelf from all other shelves of the Southern Ocean. The Antarctic shelf is ca. 500 m deep, eight times deeper than the world average, because of the weight of the ice sheet on the continent (Anderson 1999). The age of shelf isolation is at least 40 my, the ice cap began to form 38 mya and onwards, the unrestricted circumpolar surface current might have favoured thermal isolation of the Southern Ocean since 25 mya, and the shelf has existed under polar conditions for 14–12 my (Kennett 1982). Moreover, repeated advances and retreats of the ice sheet on the shelf itself probably caused benthic faunal extinctions, forcing some other benthic populations to reach refugia in sub-Antarctic islands or in isolated shelf areas or in the deep sea (Thatje et al. 2005). This probably stimulated speciation (Patarnello et al. 2011). These physical and historical factors are likely to have promoted species flocks at different times and taxonomic levels, in an area that includes the Southern Ocean (i.e. south of the Polar Front) and the sub-Antarctic islands.

According to Eastman and McCune (2000), who considered the points of view of Ribbink (1984) and Greenwood (1984), there are five criteria to recognize a species flock: monophyly: high speciosity, high level of endemism, morphological and ecological diversification, and habitat dominance. Some of these criteria seem easy to grasp (monophyly, endemism, speciosity) while others are more difficult to assess because more parameters seem to play. The first three criteria are historical-geographic ones, the two others are ecological ones. In spite of these potential difficulties, it is nevertheless possible to recognize taxonomic entities that clearly correspond to species flocks (for instance the Notothenioidei at the scale of the Southern Ocean, as convincingly proposed by Eastman and McCune 2000), and taxonomic entities that clearly do not correspond to flocks (for instance harpagiferids or liparids of the Antarctic shelf). Notothenioids are monophyletic at the scale of the Southern Ocean (including sub-Antarctic islands); in this area they have a level of endemicity of 97%, which is very high for a marine group. Teleosts in general are less species-rich in the Antarctic shelf than in other shelves (Eastman 1993; Aronson et al. 2007), however, the ichthyofauna is unique. The speciosity of notothenioids (134 species, Fishbase 2011) is very high with regard to other components of the ichthyofauna. Indeed notothenioids represent at least 50% of it, taken into account the recent increase of the number of known species of liparids to 93 (Andriashev 2003), an ecologically monotonous family which is a non-monophyletic component of the shelf (Lautrédou 2009, in prep.). A number of studies summarized by Eastman and McCune (2000) have confirmed that notothenioids clearly dominate the fish biomass above 90% (see also Eastman 2005). Morphological and ecological diversification of notothenioids is obvious. In spite of benthic origins, the group secondarily diversifies to niches in the water column, involving pelagic or partially pelagic zooplanktivory and piscivory (Eastman 1993). The group contains benthic, epibenthic, cryopelagic and pelagic species, with morphological diversification more centred on vertical motion in the water column and neutral buoyancy control than on diversification in trophic morphology (Eastman 1993). The notothenioids are therefore seen as a giant species flock at the scale of the Southern Ocean by Eastman and McCune (2000), providing a model used here as a point of reference for further comparisons.

It is possible to recognize subflocks and the best candidate for that are the Trematominae (made of the genera *Trematomus*, *Cryothenia* and *Pagothenia*). With 14 species all endemic to the coastal areas of the Southern Ocean, the Trematominae (Notothenioidei, Teleostei) correspond to a sudden burst of diversification, more exactly the sister group of *Trematomus scotti* (Janko et al. 2011; Lautrédou et al. 2011, in prep.). The group is monophyletic (Sanchez et al. 2007; Near et al. 2004; Near and Cheng 2008). There is a noticeable degree of ecological diversity which does not fit the phylogeny (Klingenberg and Ekau 1996; Ritchie et al. 1996; Sanchez et al. 2007). They represent an important part of the biomass of coastal ichtyofauna. The speciosity is the weakest criterion (13 species, i.e. 10% of the notothenioids) however, Trematominae can be considered as a small recent flock.

The icefishes (family Channichthyidae, Notothenioidei) are endemic, monophyletic (Derome et al. 2002; Near et al. 2004) and display some degree of speciosity (16 species) representing a very important part (more than 25%) of the fish biomass (Eastman 2005), and exhibiting a noticeable ecological diversity (Eastman 1993; Chen et al. 1998) linked to the ability to feed in the water column. Molecular data indicate that their phyletic diversification might not have been as quick as in the Trematominae.

Even if the Antarctic shelf has been described as a giant species flock generator, it would not be accurate to see flocks everywhere. A species flock is partly a historical concept and other patterns of fish diversification could be found, obtained from different historical patterns. For example, the Liparidae of the

Antarctic shelf cannot be a species flock because of the absence of monophyly of the components. Some components are more related to Arctic liparids than to other liparids of the Antarctic shelf (Lautrédou 2009, in prep.). Some recent diversifications fail to fall into the definition of a flock because ecological diversity is not found, or not yet studied enough. Harpagiferids are a good example of a monophyletic family not corresponding to a flock: they are 10 species of ecologically monotonous fishes (Eastman 2005). Another example is the 19 species of *Pogonophryne*, that could be interpreted as a recent subflock at a first glance. The apparently sudden burst of their diversification, as shown by the 12 species-tree of Eakin et al. (2009), would suggest so, however, we need more molecular markers to confirm that picture. Last but not least, it is difficult at the moment to assess the ecological diversity of the 19 species. The available ecological data would indicate a rather different picture: *Pogonophryne* would be a phyletic radiation without ecological or morphological diversification (Eastman 2005). The morphology is so characteristic and constant that the taxonomy of the group is the most difficult of all notothenioids. As *Pogonophryne* species are nested within the genus *Artedidraco* (Lecointre et al. 2011), they *are* actually members of this genus, which contains more ecologically diversified animals (Lombarte et al. 2003). Then the question is raised whether the entire Artedidraconidae could be a giant species flock, just as channichthyids, though with less spectacular ecological diversity.

6.3.2 The Controversial Dating of Notothenioids

Dates of divergence of non-bovichtid notothenioids have been proposed to be between 38 (Eastman 1993) and 10 my (Bargelloni et al. 1994). Bargelloni et al. (1994) may have found the correct maximum time of divergence for the Trematominae that constitute the sister group of *T. scotti* (7–15 my, Lautrédou et al. 2011, in prep.) by using an ectotherm rate of change in mitochondrial genes, but that age was transferred to the divergence time of non-bovichtid notothenioids because *Trematomus* had a basal position among them in their tree. However, we know today that Trematominae do not have such a basal position: they are the sister group of the clade made of *Patagonotothen* and *Lepidonotothen* (Near et al. 2004; Sanchez et al. 2007). So the age of approx. 10 my appears today not correctly argumented. Near (2004, Table 1, 2009) summarized the various estimates of the maximum divergence time for antifreeze-glycoprotein bearing notothenioids that have been proposed from different genes and studies: the range is from 5 to 21 my. The width of the range may be due to differences in estimation techniques. Using a calibration based on the presence of an Eocene fossil presumed to be a notothenioid (*Proeleginops grandeastmanorum*–40 mya) and sequences of the 12S–16S rDNA genes, Near (2004, 2009) found a divergence time of 24 my. However, doubts remain about the assignment of the fossil to notothenioids (Near 2004; Eastman, pers. comm.). The estimation was made using a single calibration point, and it is clear that it has to be checked using a second point with outgroup acanthomorph sequences, and strategies that account for heterogeneity of molecular evolutionary rates. For the moment the hypothesis holds for external reasons. It is provisionally accepted because there is no better supported challenging hypothesis and it is convergent with the onset of the unrestricted circum-Antarctic surface currents leading to thermal isolation of the Southern Ocean and the development of widespread seasonal freezing waters around the continent (Eastman 1993). There is also evidence for widespread continental glaciation and sharp drops in oceanic surface temperatures between 38 and 35 mya (Eastman 1993; Eastman and Clarke 1998). Antifreeze glycoproteins (AFGPs) could then have appeared at that time of cooling, far before the phyletic diversification that led to taxa known today. However, there is also evidence contradicting that fact. Antifreeze genes have been obtained independently in several group of teleosts, for instance zoarcoids (Cheng et al. 2006), herring, smelt, sea raven, sculpin, winter flounder, arctic cod through unrelated molecular processes (Logsdon and Doolittle 1997), thus such functional convergence is not rare in teleosts. Interestingly, AFGP genes do not have exactly the same structure in different notothenioids in which they have been studied. For instance, the AFGP genes of *Dissostichus mawsoni*, *Dissostichus eleginoides* and *Notothenia coriiceps* all differ from one another by the number of repeats (Logson and Doolittle 1997; Chen et al. 1997; Hsiao et al. 1990). This could be interpreted either as the result of changes that occurred after a single common origin of AFGP gene recruitment, or multiple recruitments of pancreatic trypsinogen protease genes in different phyla, after lineage separations. Interestingly, by using a substitution rate calculated from salmon mitochondrial DNA, Chen et al. (1997) estimated the origin of the *Dissostichus* AFGP gene at 5–14 mya, far later than the origin of the clade of AFGP-bearing notothenioids estimated (Near 2004) at 24 mya, though the calibration used by Chen et al. (1997) may not be correct because a mitochondrial genome rate was used to calculate divergence among nuclear genes (Near 2009). It is then not certain that the presence of AFGPs in the blood can be considered as a symapomorphy of the sister group of *Eleginops*, i.e. the AFGP-bearing notothenioids, even if they are monophyletic. These AFGPs might be subsequent multiple convergences, however, more data are needed to further test these hypotheses. Sequencing more genes encoding AFGPs in various lineages of notothenioids is an interesting future field of research. It is going to measure and document the nature and intensity of convergent molecular evolution when selective pressures imposed by cooling temperatures are (supposedly) strong.

Dating the *Trematomus* radiation has also been controversial in a similar way. The divergence of the sister group of *T. scotti* was estimated at 3.4 mya (Ritchie et al. 1996) using a rate of 0.14% of transversional changes per million year. Near (2004) recorded estimations across studies varying between 2.5 and 4.5 mya. Using a fossil-based calibration, Near (2004) found 7.4 mya and more recently from nuclear markers Lautrédou et al. (2011, in prep.) found 10 mya.

6.3.3 Origins

What are notothenioids? They have always been classified within the perciform teleosts (Nelson 1976, 2006), however, the Perciformes have never been defined. Not surprisingly, recent molecular systematics, all based on sequences from several genes, revealed that perciforms are polyphyletic (Chen et al. 2003; Holcroft and Wiley 2008; Li et al. 2009). In such a context, where are notothenioids? From a morphological and anatomical point of view, notothenioid families have been placed among percoids for a long time (Regan 1913; Norman 1937, 1938, 1966; Berg 1940; Andriashev 1964; Lindberg 1971). A number of authors suggested more precise hypotheses of kinship for the Notothenioidei, placing them together with some trachinoid components (weeverfish-like fishes: Berg 1947; Bertin and Arambourg 1958; Gosline 1968; Hastings 1993; Balushkin 2000), or with the Zoarcoidei (eelpouts: Anderson 1984, 1990). Among the polyphyletic trachinoids (Li et al. 2009), the Trichonotidae (sanddivers: Gosline 1968; Hastings 1993), Pinguipedidae (sandperches: Gosline 1968; Pietsch 1989; Hastings 1993; Balushkin 2000) and Cheimarrichthyidae (torrentfishes: Gosline 1968) have been proposed as sister groups of the notothenioids. Blennioidei have also been invoked in the literature (Eastman 1993, Balushkin 2000), as Gosline included those trachinoid components as well as notothenioids under a wider understanding of the Blennioidei. Most modern authors have left the Notothenioidei in the big perciform bush (Greenwood et al. 1966, Nelson 1976). Molecular studies rejected a number of the aforementioned morphology-based hypotheses. The torrent fish of New Zealand, Cheimarrichthys (generally considered as a pinguipedid, but sometimes put in its own family of Cheimarrichthyidae) is not the sister group of the sub-Antarctic and Antarctic Notothenioidei, but is instead closer to sand lances (Ammodytidae, Chen et al. 2003). Other trachinoids such as the Chiasmodontidae group with the Scombroidei (mackerels) and the Stromateoidei, zoarcoids (eelpouts) are close relatives of cottoids (sculpins). The blenny group with the Gobiesocoidei (clingfishes, Li et al. 2009) were found (Dettaï and Lecointre 2004, 2005) within a clade provisionally called "X" along with Trachinidae (weeverfishes), Scorpaenoidei (scorpion fishes), Zoarcoidei (eelpouts), Serranidae (groupers), Percidae (perches), Cottoidei (sea raven, sculpins, snailfishes), Gasterosteidae (sticklebacks), Triglidae (gurnards). Interestingly, that clade contains the zoarcids (suborder Zoarcoidei) and the liparids (suborder Cottoidei), and therefore it contains 88% of the teleost species found in the Southern Ocean. That clade was recovered by Miya et al. (2003) from a important length of mitochondrial sequences but with less taxa, and by Smith and Wheeler (2004) and Smith and Craig (2007, Fig. 2) with some other additional components [Bembridae (duckbills) and Platycephaloidei (flatheads), for instance] in a different taxonomic sampling context. Thus Li et al. (2009) decided to propose a name for the clade: Serraniformes. Some confusion remains about the extant sister group of Notothenioidei: Smith and Wheeler (2004) found the Congiopodidae (the Antarctic horse fish), Smith and Wheeler (2006) and Smith and Craig (2007) found the Bembridae using almost the same genes, Chen et al. (2003) and Dettaï and Lecointre (2004) found the Percidae, which are not far from the clade Bembridae + Notothenioidei in clade E of Smith and Craig (2007); Li et al. (2009), who did not sample the Bembridae, found the Trachinidae, which are not far from the clade Bembridae + Notothenioidei in Smith and Wheeler (2006). Further studies should provide the answer.

6.4 Systematic Issues

The general picture of notothenioid interrelationships can be considered now as well established (Fig. 6.1). Most of the large-scale interrelationships being reliable, it is time to name the groups that have been recurrently found by different teams and data. Only a few areas remain having unclear interrelationships:

- Position of *Halaphritis;*
- Interrelationships of nototheniid components and assessment of the monophyly of the family;
- Assessment of the monophyly of Bathydraconidae (they are most probably monophyletic: molecules provide no signal for or against monophyly while anatomical data provide the loss of the first anterior spiny dorsal fin as a synapomorphy);

The remaining fields of taxonomic work to be done concern interrelationships of species within some genera:

- interrelationships within *Bovichtus* (9 species);
- interrelationships within *Harpagifer* (10 species);
- interrelationships within *Patagonotothen* (14 species);
- interrelationships within *Pogonophryne* (19 species);
- evaluate effects of incomplete lineage sorting on interrelationships within *Trematomus*.

When relationships remain to be completed by better taxonomic samplings, which is the case with the acanthomorph example (Dettai and Lecointre 2005; Li et al. 2009), giving names to clades can only be approximate with regard to ranking. Indeed nobody has sampled yet the 314 families of acanthomorph teleosts in a single data matrix, and the fact is complicated by the possible polyphyly of some of them (the Serranidae, for instance). Then we will never know whether the clade is complete or not until we have sampled all the genera. But when relationships are fully known for all species, which is going to be the case very soon for the 134 species of notothenioids, naming clades could then follow the rules of the Phylocode, which is going to be the next step in notothenioid taxonomic research.

References

Anderson ME (1984) On the anatomy and phylogeny of the Zoarcidae. PhD thesis, College of William & Mary, Williamsburg, VA, p 253
Anderson ME (1990) The origin and evolution of the Antarctic ichthyofauna. In: Gon O, Heemstra PC (eds) Fishes of the southern ocean. JBL Smith Institute of Ichthyology, Grahamstown, pp 28–33
Anderson JB (1999) Antarctic marine geology. Cambridge University Press, Cambridge, p 289
Andriashev AP (1964) A review of Antarctic fishes. Issledovaniya fauny morei (Investigations of Marine Fauna) 2:355–386
Andriashev AP (2003) Liparid fishes (Liparidae, Scorpaeniformes) of the southern ocean and adjacent waters. Explorations of the fauna of the seas. In: Balushkin AV, Chernova NV (eds) Results of the Russian Antarctic expeditions, vol 9. Russian Academy of Sciences, Sankt-Petersburg, p 475 (In Russian)
Aronson RB, Thatje S, Clarke A, Peck LS, Blake DB, Wilga CD, Seibel BA (2007) Climate change and invasibility of the Antarctic benthos. Annu Rev Ecol Evol Syst 38:129–154
Balushkin AV (2000) Morphology, classification, and evolution of notothenioid fishes of the southern ocean (Notothenioidei, Perciformes). J Ichthyol 40:S74–S109
Baluskin AV (1992) Classification, phylogenetic relationships, and origins of the families of the suborder Notothenioidei (Perciforms). J Ichthyol 32:90–110
Bargelloni L, Ritchie PA, Battaglia B, Lambert DM, Meyer A (1994) Molecular evolution at subzero temperatures: mitochondrial and nuclear phylogenies of fishes from Antarctica (suborder Notothenioidei), and the evolution of antifreeze glycopeptides. Mol Biol Evol 11:854–886
Bargelloni L, Marcato S, Zane L, Patarnello T (2000) Mitochondrial phylogeny of notothenioids: a molecular approach to antarctic fish evolution and biogeography. Syst Biol 49(1):114–129
Berg LS (1940) Classification of fishes, both recent and fossil. Inst Zool Acad Sci USSR 5:87–517
Berg LS (1947) Classification of fishes, both recent and fossil. JW Edwards Brothers, Ann Arbor
Bertin L, Arambourg C (1958) Super-ordre des téléostéens (Teleostei). In: Grassé PP (ed) Traité de Zoologie Tome XIII: Agnathes et Poissons 3: 2204–2500. Masson et Cie, Paris
Causse R, Ozouf-Costaz C, Koubbi P, et al. (2011) Demersal ichtyofauna from the Dumont d'Urville Sea (East Antarctica) during the CEAMARC surveys in 2007–2008. Polar Sci (in press)
Chen L, DeVries AL, Cheng C-HC (1997) Evolution of antifreeze glycoprotein gene from a trypsinogen gene in Antarctic notothenioid fish. Proc Natl Acad Sci USA 94:3811–3816
Chen WJ, Bonillo C, Lecointre G (1998) Phylogeny of the channichthyidae (Notothenioidei, Teleostei) based on two mitochondrial genes. In: di Prisco G, Pisano E, Clarke A (eds) Fishes of Antarctica. A biological overview. Springer, Milano, pp 287–298
Chen WJ, Bonillo C, Lecointre G (2003) Repeatability of clades as a criterion of reliability: a case study for molecular phylogeny of Acanthomorpha (Teleostei) with larger number of taxa. Mol Phylogenet Evol 26:262–288
Cheng C-HC, Cziko PA, Evans CW (2006) Nonhepathic origin of notothenioid antifreeze reveals pancreatic synthesis as common mechanism in polar fish freezing avoidance. Proc Natl Acad Sci USA 103:10491–10496
Chernova NV, Stein DL, Andriashev AP (2004) Family liparidae scopoli 1777 - snailfishes. Calif Acad Sci Annoted Checklists of Fishes 31:72
Ciprandi-Pires A, Marinoni L (2010) DNA barcoding and traditional taxonomy unified through integrative taxonomy: a view that challenges the debate questioning both methodologies. Biota Neotrop 10(2):339–346
Darwin C (1859) The origin of species, 1st edn. John Murray, London (Reissue by Penguin Classics, Penguin Books, 1995)
Dayrat B (2005) Towards integrative taxonomy. Biol J Linn Soc 85:407–415

De Salle R (2006) Species discovery versus species identification in DNA barcoding efforts: response to Rubinoff. Conserv Biol 20:1545–1547

De Salle R, Egan MG, Siddall M (2005) The unholy trinity: taxonomy, species delimitation and DNA barcoding. Phil Trans Royal Soc B 360:1905–1916

Derome N, Chen WJ, Dettaï A, Bonillo C, Lecointre G (2002) Phylogeny of Antarctic dragonfishes (Bathydraconidae, Notothenioidei, Teleostei) and related families based on their anatomy and two mitochondrial genes. Mol Phylogenet Evol 24:139–152

Dettaï A, Lecointre G (2004) In search of nothothenioid (Teleostei) relatives. Antarctic Sci 16(1):71–85

Dettaï A, Lecointre G (2005) Further support for the clades obtained by multiple molecular phylogenies in the acanthomorph bush. CR Biol 328:674–689

Dettaï A, Lautrédou AC, Bonillo C et al (2011) The actinopterygian diversity of the CEAMARC cruises: barcoding and molecular taxonomy as a multi-level tool for new findings. Deep Sea Res II 58:250–263

di Prisco G, Maresca B, Tota B (1991) The biochemistry of oxygen transport in red-blooded Antarctic fish. In: di Prisco G, Maresca B, Tota B (eds) Biology of Antarctic fish. Springer, Berlin, pp 263–281

di Prisco G, Pisano E, Clarke A (1998) Fishes of Antarctica. A biological overview. Springer, Milano

Doyle JJ (1992) Gene trees and species trees: molecular systematics as one-character taxonomy. Syst Bot 17(1):144–163

Doyle JJ (1997) Trees within trees: genes and species, molecules and morphology. Syst Biol 46:537–553

Duhamel G, Hautecoeur M, Dettaï A, Causse R, Pruvost P, Busson F, Couloux A, Koubbi P, Williams R, Ozouf-Costaz C (2010) Liparids from the eastern sector of southern ocean and first information from molecular studies. Cybium 34(4):319–343

Eakin RR (1981) Osteology and relationships of the fishes of the Antarctic family Harpagiferidae (Pisces, Notothenioidei). In: Kornicker LS (ed) Biology of the Antarctic seas IX. Antarctic Res Series 31: 81–147. American Geophys Union, Washington

Eakin RR, Eastman JT, Near TJ (2009) A new species and a molecular phylogenetic analysis of the Antarctic fish genus *Pogonophryne* (Notothenioidei: Artedidraconidae). Copeia 4:705–713

Eastman JT (1993) Antarctic fish biology. Academic, San Diego

Eastman JT (2005) The nature of the diversity of Antarctic fishes. Polar Biol 28:3–107

Eastman JT (2006) Aspects of the morphology of phyletically basal bovichtid fishes of the Antarctic suborder Notothenioidei (Perciformes). Polar Biol 29:54–763

Eastman JT, Clarke A (1998) Radiations of Antarctic and non-Antarctic fish. In: di Prisco G, Pisano E, Clarke A (eds) Fishes of Antarctica, a biological overview. Springer, Milano, pp 3–26

Eastman JT, Eakin RR (2000) An updated species list for notothenioid fish (Perciformes; Notothenioidei), with comments on Antarctic species. Arch Fish Mar Res 48(1):11–20

Eastman JT, McCune AR (2000) Fishes on the Antarctic continental shelf: evolution of a marine species flock? J Fish Biol 57:84–102

Froese R, Pauly D (eds) (2011) FishBase. World Wide Web electronic publication./ www.fishbase.org. (version 06/11)

Gon O, Heemstra PC (1990) Fishes of the southern ocean. JLB Smith Institute of Ichthyology, Grahamstown

Gosline WA (1968) The suborders of perciform fishes. Proc US Natl Museum 124:1–77

Grande L (1994) Repeating patterns in nature, predictability, and "impact" in science. In: Grande L, Rieppel O (eds) Interpreting the hierarchy of nature. Academic, New York, pp 61–84

Grant RA, Griffiths HJ, Steinke D, Wadley V, Linse K (2011) Antarctic DNA barcoding; a drop in the ocean? Polar Biol 34:775–780

Greenwood PH (1984) What is a species flock? In: Echelle AA, Kornfield I (eds) Evolution of fish species flocks. Orono Press, Maine, pp 13–19

Greenwood PH, Rosen DE, Weitzman SH, Myers GS (1966) Phyletic study of teleostean fishes, with a provisional classification of living forms. Bull Am Museum Natl Hist 131:339–456

Hastings PA (1993) Relationships of the fishes of the perciform suborder Notothenioidei. In: Miller RG (ed) A history and atlas of the fishes of the Antarctic ocean. Foresta Inst for Ocean and Mountain Studies, Carson City, pp 99–107

Hebert PDN, Cywinska A, Ball SL, de Waard JR (2003) Biological identifications through DNA barcodes. Proc Roy Soc Lond B 270:313–321

Hennig W (1950) Grundzüge einer Theorie der phylogenetischen Systematik. Deutscher Zentralverlag, Berlin

Hennig W (1966) Phylogenetic systematics. University of Illinois Press, Urbana

Hillis DM (2004) The tree of life and the grand synthesis of biology. In: Cracraft J, Donoghue MJ (eds) Assembling the tree of life. Oxford University Press, New York, pp 545–547

Hillis DM, Moritz C (1990) Molecular systematics. Sinauer Associates, Sunderland

Holcroft N, Wiley EO (2008) Acanthuroid relationships revisited: a new nuclear gene-based analysis that incorporates tetraodontiform representatives. Ichthyol Res 55(3):274–283

Hsiao KC, Cheng C-HC, Fernandes IE, Detrich HW, DeVries A (1990) An antifreeze glycopeptide gene from the Antarctic cod *Notothenia coriiceps neglecta* encodes a polyprotein of high peptide copy number. Proc Natl Acad Sci USA 87:9265–9269

Iwami T (1985) Osteology and relationships of the family Channichthyidae. Mem Natl Inst Polar Res, Tokyo, Ser E 36:1–69

Iwami T, Abe T (1984) Gill arches of fishes of the suborder Notothenioidei Pisces, Perciformes. Mem Natl Inst Polar Res Tokyo: 32

Janko K, Marshall C, Musilová Z, Van Houdt J, Couloux C, Cruaud C, Lecointre G (2011) Multilocus analyses of an Antarctic fish species flock (Teleostei, Notothenioidei, Trematominae): phylogenetic approach and test of the early-radiation event. Mol Phylogenet Evol 60:305–316

Kennett JP (1982) Marine geology. Prentice-Hall, Englewood Cliffs

Klingenberg CP, Ekau W (1996) A combined morphometric and phylogenetic analysis of an ecomorphological trend: pelagization in Antarctic fishes (Perciformes: Nototheniidae). Biol J Linn Soc 59:143–177

Knudsen SW, Møller PR, Gravlund P (2007) Phylogeny of the snailfishes (Teleostei: Liparidae) based on molecular and morphological data. Mol Phylogenet Evol 44(2):649–666

Kuhn K, Near TJ (2009) Phylogeny of *Trematomus* (Notothenioidei: Nototheniidae) inferred from mitochondrial and nuclear gene sequences. Antarctic Sci 2009:1–6

Last PR, Balushkin AV, Hutchins JB, Schaefer SA (2002) *Halaphritis platycephala* (Notothenioidei: Bovichtidae): a new genus and species of temperate icefish from southeastern Australia. Copeia 2:433–440

Lautrédou AC (2009) Relations de parenté et phylogéographie au sein des Liparidae (Gill, 1861), Mémoire de Master 2

Lautrédou AC, Bonillo C, Denys C, Cruaud C, Ozouf-Costaz C, Lecointre G, Dettaï A (2010) Molecular taxonomy and identification within the Antarctic genus *Trematomus* (Notothenioidei, Teleostei): how valuable is barcoding with COI? Polar Sci 4(2):333–352

Lautrédou AC, Hinsinger D, Gallut C, Cruaud C, Ozouf-Costaz C, Lecointre G, Dettaï A (2011) Estimating incomplete lineage sorting and divergence time of the trematominae species flock from four nuclear markers. Ichtyol Herpetol Joint Meet, Minneapolis

Lecointre G, Le Guyader H (2001) Classification Phylogénétique du Vivant, 3edn. Berlin, Paris. English edition: Lecointre G, Le Guyader H (2006) The tree of life. Harvard University Press, Cambridge, MA

Lecointre G, Bonillo C, Ozouf-Costaz C, Hureau JC (1997) Molecular phylogeny of the antarctic fishes: paraphyly of the Bovichtidae and no indication for the monophyly of the Notothenioidei (Teleostei). Polar Biol 18(3):193–208

Lecointre G, Gallut C, Bonillo C, Couloux A, Ozouf-Costaz A, Dettaï A (2011) The Antarctic fish genus *Artedidraco* is paraphyletic (Teleostei, Notothenioidei, Artedidraconidae). Polar Biol (in press)

Li B, Dettaï A, Cruaud C, Couloux A, Desoutter-Meniger M, Lecointre G (2009) RNF213, a new nuclear marker for acanthomorph phylogeny. Mol Phylogenet Evol 50(2):345–363
Lindberg GU (1971) Opredelitel'i kharakteristika semeistv ryb mirovoi fauny [identification guide and characteristics of fish families of the world]. Nauka, Leningrad, p 469
Logsdon JM, Doolittle FW (1997) Origin of antifreeze protein genes: a cool tale in molecular evolution. Proc Natl Acad Sci USA 94:3485–3487
Lombarte A, Olaso I, Bozzano A (2003) Ecomorphological trends in the Artedidraconidae (Pisces: Perciformes: Notothenioidei) of the Weddell sea. Antarctic Sci 15:211–218
Maddison WP (1997) Gene trees in species trees. Syst Biol 46(3):523–536
Miller RG (1993) A history and atlas of the fishes of the Antarctic ocean. Foresta Institute, Nevada, p 792
Miya M, Takeshima H, Endo H et al (2003) Major patterns of higher teleostean phylogenies: a new perspective based on 100 complete mitochondrial DNA sequences. Mol Phylogenet Evol 26:121–138
Miyamoto MM, Fitch WM (1995) Testing species phylogenies and phylogenetic methods with congruence. Syst Biol 44:64–76
Near TJ (2004) Estimating divergence times of notothenioid fishes using a fossil-calibrated molecular clock. Antactic Sci 16:37–44
Near TJ (2009) Notothenioid fishes (Notothenioidei). In: SB Hedges, Kumar S (eds) The timetree of life. Oxford University Press, New York, pp 339–343
Near TJ, Cheng C-HC (2008) Phylogenetics of notothenioid fishes (Teleostei: Acanthomorpha): inferences from mitochondrial and nuclear gene sequences. Mol Phylogenet Evol 47:832–840
Near TJ, Pesavento JJ, Cheng CC (2003) Mitochondrial DNA, morphology and the phylogenetic relationships of Antarctic icefishes (Notothenioidei: Channichthyidae). Mol Phylogenet Evol 28:87–98
Near TJ, Pesavento JJ, Cheng C-HC (2004) Phylogenetic investigations of Antarctic notothenioid fishes (Perciformes: Notothenioidei) using complete gene sequences of the mitochondrial encoded 16S rRNA. Mol Phylogenet Evol 32:81–891
Near TJ, Parker SK, Detrich HW III (2006) A genomic fossil reveals steps in hemoglobin loss by the Antarctic icefishes. Mol Biol Evol 23(11):2008–2016
Nelson JS (1976) Fishes of the world, 1st edn. John Wiley-Interscience, New York, p 416
Nelson JS (2006) Fishes of the world, 4th edn. Wiley, New York, p 601
Norman JR (1937) Coast fishes, II: the Patagonian region. Discovery Rep 16:1–150
Norman JR (1938) Coast Fishes, III: the Antarctic Zone. Discovery Rep 18:1–104
Norman JR (1966) A draft synopsis of the orders, families, and genera of recent fishes and fish-like vertebrates. Trustees of the British Museum (Natural History), London, p 649
Page RDM, Holmes EC (1998) Molecular evolution: a phylogenetic approach. Blackwell Science, Abingdon
Patarnello T, Verde C, di Prisco G, Bargelloni L, Zane L (2011) How will fish that evolved at constant sub-zero temperatures cope with global warming? Notothenioids as a case study. Bioessays 33:260–268
Pietsch TW (1989) Phylogenetic relationships of trachinoid fishes of the family Uranoscopidae. Copeia 1989:253–303
Ratnasingham S, Hebert PDN (2007) BOLD: the barcode of life data system. Mol Ecol Notes 7:355–364
Regan CT (1913) Classification of the Percoid fishes. Ann Mag Natl Hist 8:111–145
Ribbink AJ (1984) Is the species flock concept tenable? In: Echelle AA, Kornfield I (eds) Evolution of fish species flocks. Orono Press, Maine, pp 21–25
Ritchie PA, Bargelloni L, Meyer A, Taylor JA, Macdonald JA, Lambert DM (1996) Mitochondrial phylogeny of trematomid fishes (Nototheniidae, Perciformes) and the evolution of Antarctic fish. Mol Phylogenet Evol 5:383–390
Ritchie PA, Lavoué S, Lecointre G (1997) Molecular phylogenics and evolution of Antarctic notothenioid fishes. Comp Biochem Physiol A 118:1009–1027

Rock J, Costa FO, Walker DI, North AW, Hutchinson WF, Carvalho GR (2008) DNA barcodes of fish of the Scotia sea, Antarctica, indicate priority groups for taxonomic and systematics focus. Antarctic Sci 20:253–262

Rubinoff D (2006) Utility of mitochondrial DNA barcodes in species conservation. Conserv Biol 20:1026–1033

Sanchez S, Dettaï A, Bonillo C, Ozouf-Costaz C, Detrich WH, Lecointre G (2007) Molecular and morphological phylogenies of the Nototheniidae, with taxonomic focus on the Trematominae. Polar Biol 30:155–166

Smith L, Craig M (2007) Casting the percomorph net widely: the importance of broad taxonomic sampling in the search for the placement of serranid and percid fishes. Copeia 2007:35–55

Smith L, Wheeler W (2004) Polyphyly of the mail-cheeked fishes (Teleostei: Scorpaeniformes): evidence from mitochondrial and nuclear sequence data. Mol Phylogenet Evol 32:627–646

Smith L, Wheeler W (2006) Venom evolution widespread in fishes: a phylogenetic road map for the bioprospecting of piscine venoms. J Hered 97:206–217

Steinke D, Zemlak TS, Boutillier JA, Hebert PD (2009) DNA barcoding of Pacific Canada's fishes. Mar Biol 156:2641–2647

Thatje S, Hillenbrand CD, Larter R (2005) On the origin of Antarctic marine benthic community structure. Trends Ecol Evol 20:534–540

Van de Putte AP, Van Houdt JKJ, Maes GE, Janko K, Koubbi P, Rock J, Volckaert FAM (2009) Species identification in the trematomid family using nuclear genetic markers. Polar Biol 32:1731–1741

Verde C, Giordano D, Russo R, di Prisco G (2012) The adaptive evolution of polar fishes. Lessons from the function of hemoproteins. In: di Prisco G, Verde C (eds) Adaptation and evolution in marine environments—The impacts of global change on biodiversity, vol 1. Series "From Pole to Pole". Springer, Berlin, pp 197–213

Voskoboynikova OS (2000) Comparative osteology of *Dacodraco hunteri* and its position within the family Channichthyidae (Notothenioidei). Zool Zh 79:321–332

Voskoboynikova OS (1993) Evolution of the visceral skeleton and phylogeny of Notothenioidei. J Ichthyol 33:23–47

Ward RD, Zemlak TS, Innes BH, Last PR, Hebert PD (2005) DNA barcoding Australia's fish species. Phil Trans Royal Soc Lond B Biol Sci 360:1847–1857

Ward RD, Hanner R, Hebert PD (2009) The campaign to DNA barcode all fishes. FISH-BOL. J Fish Biol 74:329–356

Will KW, Mishler BD, Wheeler QD (2005) The perils of DNA barcoding and the need for integrative taxonomy. Syst Biol 54:844–851

Part III
Theme 2: Response to Stress—Adaptations

Chapter 7
Evolutionary Adaptation and Disaptation in the Cold: the Icefish Paradigm

Bruno Tota, Daniela Amelio, Filippo Garofalo and Daniela Pellegrino

7.1 Introduction

The aim of this chapter is to illustrate in the context of cold adaptation of the Antarctic teleosts the icefish as a unique case-study of physiological responses to genetic changes, i.e. loss of hemoglobin (Hb) and myoglobin (Mb), without apparent immediate compensatory mutations. This offers the opportunity to study the effects of epigenetic compensations and how these have been integrated at different hierarchic levels in the emergent new phenotype. However, the available evidence does not allow to clarify whether the disaptive icefish phenotype, despite its exposure to stably high environmental pO_2, may have been predisposed to increased sensitivity to hypoxic disturbance (hypoxemic and intracellular hypoxia); nor to which extent this organism has been able to reprogram gene expression within aerobic tissues (including the heart), recruiting silent, alternative, or redundant pathways for correcting a deleterious, but non-lethal O_2-transport phenotype. Therefore, in updating the pertinent literature, we will emphasize our view (Garofalo et al. 2009) that an inherent morpho-functional plasticity of the basic

B. Tota (✉) · D. Amelio · F. Garofalo · D. Pellegrino
Department of Cellular Biology, University of Calabria,
Arcavacata di Rende, 87030 Cosenza, Italy
e-mail: tota@unical.it

D. Amelio
e-mail: daniela.amelio@unical.it

F. Garofalo
e-mail: filigaro@libero.it

D. Pellegrino
e-mail: danielapellegrino@unical.it

G. di Prisco and C. Verde (eds.), *Adaptation and Evolution in Marine Environments, Volume 1*, From Pole to Pole, DOI: 10.1007/978-3-642-27352-0_7,

teleost cardio-circulatory system was sufficient to allow structural and functional expansion of an alternative (Hb-free blood and Mb-free cardiac muscle) design to cope with new demands. Conceivably, this disaptive condition, followed by adaptive recovery, was facilitated by the lack of competition from the comparatively sparse non-notothenioid ichthyofauna (Montgomery and Clements 2000).

7.2 Cold-Adaptive Radiation in an "Evolutionary Caldron"

The remarkable paleogeographic and palaeoclimatic changes occurring during the Cenozoic Era in Antarctica resulted in the opening of the Drake Passage and in the formation of the Antarctic Polar Front (APF), which persists to the present (Eastman 1993). The cold APF "wall" prevents mixing of the waters of the Southern Ocean with those of the Indian, Pacific and Atlantic Oceans and inhibits mixing of the Antarctic cold-adapted fauna with the fauna of the cold temperate oceans (Petricorena and Somero 2007). The resulting frigid, glaciated water environment, with a history of loss and formation of near shore and coastal habitats, is characterized by relatively constant temperature of $-1.8 + 1°C$, and stably high content of O_2. Such very stable "ice bath" environment has acted as an "evolutionary caldron" promoting phylogenetic and phenotypic specializations of cold adaptation of the endemic teleosts. Their species diversity has been severely restricted to such extent that the modern Southern Ocean ichthyofauna appears dominated by the single perciform suborder, the Notothenioidei (Eastman 1993, 2005). This negatively buoyant, bottom-dwelling ancestral perciform stock arose $\sim$40–60 Mya and underwent a process of diversification which enabled it to occupy the ecological niches left by the shallow-water, cosmopolitan, temperate competing taxa that were eliminated during the establishment of the freezing habitat (Eastman 2005). Approximately 35% of all Antarctic fish belong to the Notothenioidei, this figure increasing to 55% in the subzero shelf and continental slope waters, often constituting in excess of 90% of the species collected (Eastman 1993, 1995; Montgomery and Clements 2000). Consequently to their high degree of cold adaptation, the endemic species exhibit stenothermia (and corresponding low tolerance to heat) which in turn has restricted their biogeographic distribution (Pörtner et al. 2007, and references therein).

The extant Notothenioidei comprise 8 families, 43 genera and 122 species, of which 6 families are south of the Antarctic Polar Front (Balushkin 2000; Eastman and Eakin 2000). Their ecological diversifications have been enhanced by morphological traits, including changes in body shape and the capacity of being almost neutrally buoyant through accumulation of lipid (Eastman 1995), suited for pelagic or semi-pelagic lifestyle, i.e. secondary pelagicism. To a significant extent, secondary pelagicism, which arose independently several times in different clades (Eastman 1993, 1997), was attained through paedomorphism, i.e. retention of ancestrally larval characteristics by adults of a descendent taxon (Eastman 1997), probably related to temperature-elicited changes of developmental processes (heterochrony). Examples are represented by the long pelagic larval stage of many Antarctic species, the attainment of sexual maturity at a large maximum size, the persistence of the notochord in adult stages of some species, like *Pleuragramma antarcticum* (Eastman 1997), incomplete canal formation in the lateral line, the decreased bone mineralization and other skeletal modifications (Detrich and Amemiya 2010), lack of scales (facilitating cutaneous respiration) in channichthyids and many bathydraconids (Montgomery and Clements 2000), and reliance of embryonic and postembryonic muscle growth entirely supported by stratified hyperplasia (Johnston et al. 2002). Above all, the success of these teleosts was especially tailored by the evolution of adaptive molecular and cellular changes towards cold stability and efficient function, such as the expansion of the tubulin gene families for efficient microtubule assembly in the cold (Detrich et al. 2000) and effective protein translocation in the cold (Römisch et al. 2003). However, a key novelty preventing the animal from freezing was the evolution of antifreeze glycoproteins, AFGP, that, present in the blood and tissue fluids, recognize ice nuclei, bind to them and arrest ice growth below 1.2–1.5°C, the equilibrium (colligative) freezing point (Fletcher et al. 2001). The encoding gene for AFGP has resulted from innovative recruitment of portions of a structurally and functionally distinct pancreatic trypsinogen-like protease gene, together with de novo amplification of a tripeptide (ThrAlaAla) coding element which generated a new AFGP coding region (Cheng and Chen 1999; Cheng and Detrich 2007 and references therein).

7.3 Disaptation Under Relaxed Selection Pressure: the Icefish Paradigm

In addition to specialized traits, notothenioids exhibit a variety of regressive changes or losses of function, i.e. *disaptation*, according to the nomenclature of Baum and Larson (1991). It has been speculated that, in comparison with fish flocks of other habitats, the relaxed selection pressure of the stably frigid, well oxygenated waters and the low competition of the Antarctic habitat has allowed this higher tolerance of disaptation, eventually compensated by subsequent adaptive recovery (Clarke and Johnston 1996; Johnston et al. 2003; Sidell and O'Brien 2006; Cheng and Detrich 2007). Examples of disaptation include:

- loss of the heat-shock response (HSR), as documented in *Trematomus bernacchii* and similar species, in which, despite constitutive expression of heat-inducible genes, no type of heat shock proteins (HSP) could be induced by thermal stress (Place et al. 2004; Petricorena and Somero 2007);
- lack of Mb expression in the skeletal muscle of notothenioids, which occurred early in their radiation some 7–15 Mya (Sidell et al. 1997);
- presence of large-diameter myotomal fibres, reaching 600 μm (Egginton et al. 2002), progressive reduction in number and loss in body size-specific maximum

number of fast fibres in the more derived species, i.e. Channichthyidae and Harpagiferidae; consequently, the decreased surface-to-volume ratio, reducing membrane leak pathways, results in lower basal energy costs due to a fewer energy-utilising pumps necessary for keeping ionic equilibria (Johnston et al. 2003).

Channichthyidae provide the most dramatic example of disaptation, being unique among adult vertebrates for their loss of respiratory pigments Hb (Ruud 1954) and, in some species, Mb (Hamoir and Gerardin-Otthiers 1980; Kock 2005a, b; Sidell and O'Brien 2006). Their monophyletic evolution appears a relatively recent event, but the period of their origin, oscillating between 20–15 Mya and 8, 5 Mya remains to be defined (see Kock 2005a, b). While the Hb loss is caused by a single mutational event that deleted the entire β-globin gene and the 5$'$ end of the linked α-globin gene (Cocca et al. 1995; Near et al. 2006 and references therein; Chap. 11 Verde et al. 2012), Mb, which is not expressed in the heart ventricle of 6 of the 16 species of icefishes (Moylan and Sidell 2000), is thought to have been lost as result of at least three independent mechanisms. That is, a 5-nucleotide insertion leading to premature termination in *Champsocephalus gunnari*, an aberrant polyadenylation signal in *Pagetopsis macropterus* (Vayda et al. 1997), and a duplicated TATAAAA sequence that interferes with transcription in *Chaenocephalus aceratus* (Small et al. 2003). On the basis of phylogenetic evidence showing different mutational mechanisms among Mb non-expresser (Mb^-) species and the disparate positions of Mb^- within Channichthydae, it is conceivable that multiple independent mutational events led to loss of Mb expression during icefish evolution.

In an attempt to clarify the functional significance of the successful evolution of the icefish, despite the loss of, what were once considered to be, essential-life O_2-binding chromoproteins, a number of studies identified important multilevel (from biochemical and ultrastructural to tissue and organismal) compensatory changes (reviewed by Garofalo et al. 2009). These are illustrated below.

7.4 Major Icefish Compensatory Adjustments

In the last two decades a conceptual homeostatic network for the physiology of hypoxia tolerance has emerged (Hochachka and Somero 2002) in which tissue-specific O_2 sensing systems, e.g. heme oxygenases, induction of transcriptional factors (HIF, VEGF, erythropoietin, HSPs, cardiac natriuretic peptides, etc.), O_2-sensitive genes and consequent hypoxia-induced reprogramming gene expression may interact to set the stage for downstream phenotypic changes. Accordingly, Garofalo et al. (2009) in their speculative synoptic diagram (Fig. 7.1) have hypothesized that in the icefish, the major compensations to Hb and Mb losses may be achieved by homeostatic circuits activated against hypoxemic and intracellular hypoxia. The multilevel signal transduction network appears adjusted to ensure an efficient transcellular movement of O_2 from capillaries to

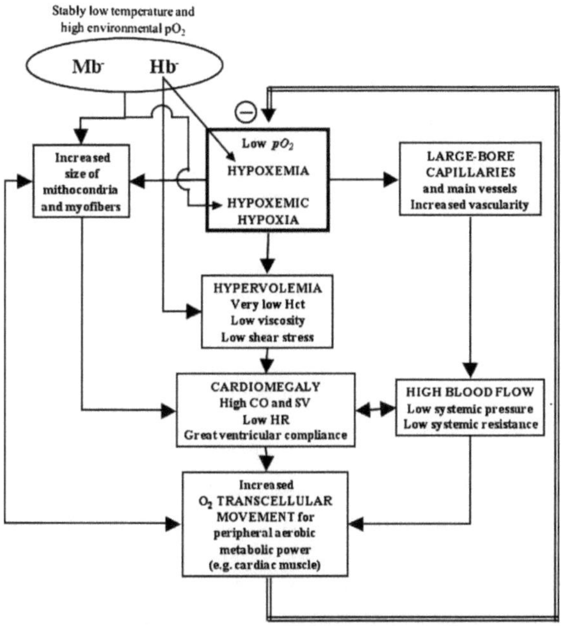

Fig. 7.1 Putative hierarchic organization of homeostatic circuits and signal-transduction processes involved in the response to the loss of respiratory pigments. The diagram, based on several studies of the literature, shows complex loops activated at different levels to provide an efficient cellular oxygenation. This multilevel homeostatic cascade includes, (1) tissue specific oxygen/redox sensing systems (e.g. heme oxygenases, Lushchak and Bagnyukova 2007), (2) signal transduction pathways (e.g. hypoxia-response systems, Hochachka et al. 1999; Heise et al. 2007), (3) hypoxia/redox responsive genes (e.g. VEGF-, eNOS-, HIF-1-genes, Meeson et al. 2001), (4) phenotypic compensatory responses (circulatory, vascular, and cardiac modifications). All the above results downstream adaptative cardio-circulatory readjustments, avoid the risk of hypoxemic and intracellular hypoxia. For references and details, see text

mitochondria, keeping in balance the icefish redox biome. It includes extensive circulatory (larger blood volume, low blood viscosity and hematocrit), vascular (large-bore capillaries, increased vascularization, low-resistance/high capacitance vascular tree) and cardiac (heart enlargement and specialized volume pump performance) adjustments (see below) for regulating the systemic O_2 transport capacities and for maximizing the driving pO_2 gradient across the exchange surfaces of the aerobic tissues.

7.4.1 Circulatory Adjustments

In the icefish, the arterial O_2-carrying capacity is one tenth that of the closely related red-blooded species, O_2 being carried in physical solution (Ruud 1954; Holeton 1970). This is compensated by an impressive increase of mass-specific blood volume 2–4 times greater than in red-blooded teleosts (Holeton 1970; Acierno et al. 1995), which, in turn, requires remarkably increased cardiac output (CO). Although the specific signal transduction processes responsible for this compensatory hypervolemia remain to be elucidated, the fact that it is present both in the (Hb^-)/(Mb^-) (*C. aceratus*) and in the (Hb^-)/(Mb^+) (*Chionodraco hamatus*) phenotypes points to the Hb^- condition *per se*, i.e. the hypoxemic status, as the principal stimulus triggering the homeostatic responses. These might be achieved via the orchestration of major fluid regulating hormones (renin-angiotensin system, aldosterone, cardiac natriuretic peptides ANP and BNP, and arginine vasopressin), as epitomized by myocardiocytes (and other tissues) in which chronic hypoxia activates a protective program steered by HIF (hypoxia inducible transcription factor), with consequent promotion of ANP and BNP expression (Weidemann et al. 2008) and enhanced concerted influence of NO and BNP (Luciano et al. 2008). Both NO and the cardiac natriuretic peptides exert relevant vasodilatory and hypotensive actions, while HIF-1 α synthesis is regulated by tissue-specific O_2 sensing mechanisms (Hochachka and Somero 2002; Meeson et al. 2001).

Cold strikingly increases the viscosity of aqueous solutions (blood, interstitial fluid, aqueous cytoplasm: Sidell 1998). Cold-induced increase of whole blood viscosity depends on (a) the physical properties of fluids; (b) an increase in the stiffness of the red blood cells (RBCs) which lowers their deformability, and (c) blood velocity alterations (Macdonald and Wells 1991).

Blood viscosity at 0°C is about 40% higher in red-blooded *T. bernacchii* than at 10°C and the viscosity of red-blooded species is 25% higher compared to white-blooded *C. hamatus* when compared at high shear 25 rate. The low resistance of the icefish vascular tree counterbalances the augmented plasma viscosity caused by the high plasma AFGP concentrations (10–35 mg mL^{-1}). Antarctic fishes can reduce RBCs rigidity (membrane fluidity) through homeoviscous adaptation (membrane phospholipid remodelling: Egginton 1996; Storelli et al. 1998). Furthermore, many notothenioid species show reduction of intracellular viscosity and increased RBC deformability due to their lower mean corpuscular Hb concentrations (MCHC) compared to temperate counterparts.

A well known hematological trait concerns the hematocrit (Hct) values which are lower in notothenioids than in temperate fish (Egginton and Davison 1998). It has been argued that this common trend may have adaptational significance because of the consequent reduction of the cardiovascular workload (vascular resistance changes in direct proportion to changes in viscosity). Pioneering studies of Detrich and Yergeau (2004) and Yergeau et al. (2005) addressed the question on how Hb loss may influence the program of terminal erythroid differentiation from the proerythroblast progenitor during icefish development. Analysing comparatively the genetic pathway of erythrocyte formation in the Antarctic red-blooded notothenioid *N. coriiceps* and in zebrafish (*Danio rerio*), they identified candidate genes for erythropoiesis, such as bloodthirsty (bty) and found that disruption of bty synthesis in zebrafish suppresses both erythrocyte and Hb production.

All the above-mentioned factors affect the vascular shear stress, which is proportional to blood viscosity and flow velocity and inversely proportional to the vessel radius. Interestingly, the presence of large-bore capillaries (below) contributes to lower both blood viscosity and shear stress. Changes in capillary diameter are also important considering the resistance to a laminar flow, which in a cylinder varies directly with viscosity and inversely with the fourth power of the radius.

7.4.2 Vascular Adjustments

The finding that the very large fibre diameter of the skeletal musculature is common to both the Antarctic endemic and the sub-Antarctic species living at much higher temperature (Beagle Channel) prompts the view that it may be considered a phyletic notothenioid trait (Johnston et al. 2003). Consequently, the occurrence of wide-bore capillaries appears a structural prerequisite for maintaining adequate tissue oxygenation, particularly in the bulk of aerobic skeletal myotomal musculature in which longer diffusional distances aggravate the problem of O_2 and metabolite transport in the cold. In fact, Egginton et al. (2002) showed that at the appropriate permeability and capillary radius values, mean pO_2 in channichthyid slow muscle fibres is modest, but it would be dangerously low (0.52 ± 1.22 kPa) if their capillary circumference were similar to that of the nototheniids. For the minimum pO_2 to be above zero would require a 50% greater capillary radius, approaching 6 μm (in which case mean O_2 tension would be 1.99 ± 1.46 kPa), while if capillary radius were half that detected, a 5-fold reduction in pO_2 (to 0.23 ± 0.87 kPa) is predicted. In conclusion, it is likely that channichthyids utilize capillaries of the minimum radius needed for adequate O_2 delivery, underscoring, at the same time, the mechanical cost (upper size limitations) imposed by a microcirculation with larger-circumference capillaries. Therefore, the remodelling of the icefish cardiovascular system must obey to the major determinants of intracellular pO_2, namely, the capillary supply and the intracellular diffusion distance.

To accommodate in the circulation the larger blood volume, channichthyids have also increased in most tissues the diameter and densities (number and length) of the main vessels (arteries and veins), even more than the increases detected in the red-blooded notothenioids, whose main vessels are somewhat larger than the norm for vertebrates (Egginton et al. 2002). This angio-morphometric pattern is typically exemplified by the vasculature associated with the retina of the eye, which is extremely dense in the Hb^- compared to the Hb^+ counterparts (Eastman and Lannoo 2004; Wujcik et al. 2007). Recently Beers and colleagues (2010) hypothesized that higher levels of circulating NO in icefishes, due to a slower rates of NO degradation associated with the loss of Hb (see below), may induce this extraordinary vascularization since the early developmental stages.

Increased vascularity has been detected in the scale-less and thin skin and in the fins (Walvig 1960; Jakubowski 1982), as well as in the gills (Vogel and Kock 1981) in which the vessel arrangement appears well suited to supply the considerably larger respiratory surface area of the icefish (Jakubowski 1982).

The *bulbus arteriosus*, the fourth chamber of the fish heart, is much larger in the icefish than the general size of teleostean *bulbus* and endowed with NP receptors (Cerra et al. 1997). Its morphology may allow to accommodate the very high stroke volume ejected from the icefish ventricle (Icardo et al. 1999), contributing to the uncommonly low pressure drops from the ventral aorta to the dorsal aorta (Hemmingsen et al. 1972; Hemmingsen 1991).

7.4.3 Cardiac Adjustments

In comparison to Hb^+ species, icefish heart has dramatically increased its CO [i.e. the product of stroke volume(SV) and heart rate (HR)], pumping remarkably large SV (6–15 times greater than in other teleosts, for references see Table 7.1) at relatively low HR (Hemmingsen et al. 1972; Tota et al. 1991b). Since the ventral aorta pressure is also relatively low (Hemmingsen et al. 1972), we emphasized the typical volume pump performance of the icefish heart, i.e. it moves the relatively large blood volumes related to the compensatory hypervolemia against modest afterloads, perfusing a high flow/low pressure/low resistance vascular circuit (Tota and Gattuso 1996 and references therein). The elevated blood flow maximizes the driving pO_2 gradient across the length of the exchange area ensuring adequate O_2 delivery to the aerobic tissues, including the heart (Hemmingsen et al. 1972).

In response to growth stimuli such as the permanent volume overload, the icefish heart has undergone remarkable cardiac enlargement (physiological cardiomegaly). After Twelves's early observation on the large sized heart of *C. aceratus* (Twelves 1972), a morphometric analysis based on a vast number of fish species showed that most notothenioids possess relative heart weights similar to those of the majority of teleosts, including temperate counterparts, while the icefish relative heart weight is remarkably higher than that of other fish, 3–4 times the typical values found in poikilotherm vertebrates (Johnston et al. 1983; Tota et al. 1991a, b). This cardiac enlargement, which reflects the remarkable phenotypic plasticity of the fish heart, has been attained maintaining the basic phyletic myoarchitecture of the ancestral group from which Notothenioidei have radiated. The ventricle is made up of a fully trabeculated myocardium (*spongiosa*) supplied from the luminal venous blood through the intertrabecular spaces (*lacunae*). Because of cold-limited diffusional processes and in absence of respiratory pigments, the upper size limits of the cardiac enlargement are imposed by the maintenance of an efficient intramyocardial pO_2. This means that the relationship between a minimum distance for O_2 diffusion from the lacunae and maximal myocardial cross-sectional area (*trabecula* diameter) adequate for stroke work (SW) generation needs tight morphogenetic regulation from the early developmental stage to, and throughout, the adult icefish life. Accordingly, a higher degree of trabeculation and a shorter average diffusion

Table 7.1 Values of mitochondrial volume in oxidative muscle cells. Modified from Tota et al. (1997)

Tissue type	Animals	Mitochondrial volume (% of cell volume)	References
	Insects		
Tymbal muscle	Cicada	33	Josephson and Young, 1985
Flight muscle	Insect	44	Elder 1975
	Birds		
Pectoral muscle	Columba livia	29	James and Meek 1979
Heart ventricle muscle	Finch	34	Bossen et al. 1978
	Fishes		
	Temperate		
Skeletal muscle			
	Salvelinus fontinalis	31	Johnston and Moon 1981
	Scomber scomber	36	Bone et al. 1978
	Engraulis encrasicolus	45	Johnston and Moon 1981
	Antarctic		
Skeletal muscle			
	Gobionotothen gibberifrons (Hb^+/Mb^+)	25	O'Brien et al. 2003
	Notothenia coriiceps (Hb^+/Mb^+)	27	Egginton et al. 2002
	Trematomus newnesi (Hb^+/Mb^+)	31	Egginton et al. 2002
	Chionodraco rastrospinosus (Hb^-/Mb^+)	39	O'Brien et al. 2003
	Chionodraco hamatus (Hb^-/Mb^+)	30	Tota et al. 1991b
	Chaenocephalus aceratus (Hb^-/Mb^-)	53	Egginton et al. 2002
Heart ventricle muscle			
	Gobionotothen gibberifrons (Hb^+/Mb^+)	16	O'Brien and Sidell 2000
	Notothenia coriiceps (Hb^+/Mb^+)	18	Urschel and O'Brien 2008
	Chionodraco rastrospinosus (Hb^-/Mb^+)	20	O'Brien and Sidell 2000
	Chionodraco hamatus (Hb^-/Mb^+)	44	Tota et al. 1991b
	Chaenocephalus aceratus (Hb^-/Mb^-)	37	O'Brien and Sidell 2000

distance to a mitochondrion were detected in the *spongiosa* of the icefish as compared with that of the red-blooded notothenioid (O'Brien et al. 2000). These structural differences in the ventricular trabeculation suggest that stimuli related to specific characteristics of the icefish (hypervolemic condition and/or cardiac pO_2 levels) exert morphogenetic influence on its heart remodelling. Ultrastructural studies on the myocardial trabeculae (below) provide an important insight on the subcellular basis of their enlargement.

7.4.4 Myocardiocyte Hypertrophy with Mitochondrial Prolife-Ration

Under enhanced mechanical demands (chronic strenuous exercise, prolonged increases of blood pressure or peripheral vascular resistance or sympatho-adrenergic overstimulation, etc.) the cardiomyocytes become hypertrophic increasing particularly the myofibrils, the force generating compartment. The ventricular myocytes of both red-blooded and Hb-less notothenioid species have higher diameters (ranging from 4 to 6 μm) than those usually detected in teleosts. However, in parallel with the mitochondrial volume density of 30–60% detected in notothenioid red skeletal muscle (Johnston 1989; Egginton and Sidell 1989), the icefish cardiomyocyte enlargement is determined by an extraordinary increase in mitochondrial volume densities, the highest reported in any fish and among the largest in vertebrate hearts (Table 7.2). As illustrated in Fig. 7.2, the mitochondrial proliferation is at the expense of the myofibrils, whose volume densities is towards the lower end of the range reported in fish hearts (*C. aceratus*: Johnston et al. 1983; Harrison et al. 1991; *C. hamatus*: Tota et al. 1991a; Zummo et al. 1995). Therefore, being the mitochondrial and the myofibril compartments associated in the cardiomyocyte space, representing each other's boundary conditions, there is a severe topological limitation for increasing myofibril number in the icefish myocardium.

It is known that the decrease in acclimation temperature of various teleosts increases the mitochondrial density of their skeletal muscles (Johnston 1989; Egginton and Sidell 1989) and ventricular cardiomyocytes (see for references, Tota et al. 1991b). This phenomenon is considered as a typical cold adaptation response (Johnston and Maitland 1980) in line with the correlation between mitochondrial up-regulation and enhanced aerobic capacity (Hoppeler and Lindstedt 1985) which boosts the rate of ATP synthesis, inversely related to temperature, at the same time compensating for the cold-limited O_2, ion and metabolite diffusion rates (Egginton and Sidell 1989; Johnston et al. 1994). Under cold-limited diffusion conditions, the proliferation of mitochondria reduces the average path lengths between them and the myofibrils, increasing the area of exchange surface (Londraville and Sidell 1990). Interestingly, however, recent studies by Sidell and coworkers uncovered a selective pattern of mitochondrial

Table 7.2 Resting cardiac power output and myocardial oxygen consumption in fish

	Temp (°C)	Power output of the heart (mW g^{-1} heart wt)	Power output of the heart (mWg^{-1} body wt)	Myocardial VO_2 μl min^{-1} g^{-1}	Myocardial VO_2 μl s^{-1} mW^{-1}	Animal VO_2 μl min^{-1} kg^{-1}	O_2 cost of cardiac pump (%)	Ref
Tropical fish								
Katsuwomis pelamis	25	6.31	29.2	–	–	18,200	2.3*	*a*
Thunnus albacares	25	5.65	27.2	-	-	10,500	-	*a*
Temperate fish								
Anguilla anguilla	10	0.84	0.55	-	-	-	-	*b*
Hemitripterus americanus	10	1.08	0.82	18.96	0.29	-	0.6	*c*
Oncorhyncus mykiss	10	1.27–1.56	1.23–1.52	19.9	0.23	582–645	2.7–3.6	*d–f*
Oncorhyncus mykiss	5	0.86	0.94	13.74	0.27	-	-	*g*
Oncorhyncus mykiss	15	1.91	1.62	28.27	0.25	-	-	*g*
Squalus acanthias	15	2.27	0.92	31.2	0.23	-	-	*h*
Sub-antarctic fish								
Eleginops maclovinus	10	1.37	0.84	24.1	0.29	-		*i*
Paranotothenia magellanica	10	0.67	0.64	16.32	0.41	-	-	*i*
Patagonothenia tessellata	10	0.76	0.71	25.59	0.56	-	-	*i*
Antarctic fish								
Trematomus bernacchii	–0.5	1.98	0.89	-	-	-	-	*j*
Chionodraco rastrospinosus	0.5	0.84	3.94	20.0	0.40	-	-	*k*
Chaenocephalus aceratus	1	1.54	4.62	37.8	0.41	333-533	21-34	*k-m*

Modified from: Axelsson et al. (1998). [a] Bushnell and Brill (1992), [b] Imbrogno and Tota (unpublished), [c] Farrell et al. (1985) (the % cost of cardiac pump has been estimated on the basis of the VO_2 of the lingcod); [d] Kiceniuk and Jones (1977), [e] Houlihan et al. (1988), [f] Wood et al. (1979), [g] Graham and Farrell (1990) (animals acclimated at the reported temperature); [h] Davie and Franklin (1992), [i] Agnisola et al. (1997), [j] Pellegrino, Garofalo and Tota (unpublished), [k] Acierno et al. (1997), [l] Holeton (1970), [m] Hemmingsen et al. (1972). *As estimated by Farrell and Jones (1992)

proliferation in the icefish skeletal and myocardial muscle cells. O'Brien and Sidell (2000) in their ultrastructural and aerobic capacity analysis of the heart ventricles of three notothenioid species with variable expression of respiratory pigments, i.e. the (Hb^+/Mb^+) *Gobionotothen gibberifrons*, the (Hb^-Mb^+) *C. rastrospinosus* and the (Hb^-/Mb^-) *C. aceratus*, showed that the cell volume percentage occupied by mitochondria, Vv (mit,f), was highest in *C. aceratus* with a dramatic value of 36.53 ± 2.07, intermediate in *C. rastrospinosus* (20.10 ± 0.74) and lowest in red-blooded *G. gibberifrons* (15.87 ± 0.74). In all three species, the surface area of inner mitochondrial membrane per volume of mitochondria was inversely related to mitochondrial volume density; however, the surface area of mitochondrial cristae per gram of heart was greater in *C. aceratus* than in the other two notothenioids, which showed similar surface areas. No differences among the three species were detected in the maximal activities (per gram wet mass of tissue) of several aerobically poised enzymes, including citrate synthase and cytochrome oxidase.

Therefore, O'Brien and Sidell (2000) concluded that rather than increasing aerobic metabolic capacity, such high Vv (mit,f) values in the icefish cardiac and skeletal muscle can be interpreted as a response to reduce the path lengths for both the O_2 transcellular movement from lacunae to mitochondria, and its intracellular diffusion throughout the lipid conduits of the membranes. The cold-constrained ion and metabolite diffusion rate is also counterbalanced by the decreased average distance between mitochondria and myofibrils and the increased exchange surface area between cytoplasm and mitochondria (Londraville and Sidell 1990).

Of note, the enlargement of the myocardial and myotomal fibres decreases the surface-to volume ratio, hence lowering basal energy requests and number of energy-utilising pumps needed for maintaining ionic equilibria (the estimated cost being up to 40% of basal energy requirements: Jobling 1994). In a more recent study, Urschel and O'Brien (2008) convincingly showed that this ultrastructural mitochondrial response is not a genetically fixed trait of the icefish but, instead, is directly influenced by Hb and Mb expression. In fact, the increased mitochondrial density in *C. aceratus* (Hb^-/Mb^-) heart is not the result of a canonical biogenetic pathway but is caused by an increased organelle size through proliferation of the outer mitochondrial membrane without corresponding increases in inner membrane surface density, protein synthesis or mtDNA replication. In contrast, the synthesis of phospholipids is increased, being targeted to the outer mitochondrial membrane with consequent higher protein-to-lipid ratio than in red blooded counterparts. Since the pectoral adductor muscles of the notothenioids examined so far do not express Mb (Sidell et al. 1997), it is of interest that mitochondria in the pectoral adductor muscle of *C. rastrospinosus* lacking both Hb and Mb are enlarged compared with those in the ventricular myocardiocytes expressing Mb (O'Brien and Sidell, 2000; O'Brien et al. 2003). In fact, mitochondria in *C. rastrospinosus* pectoral muscles appear strikingly similar to those reported in the pectoral adductor and heart ventricle of *C. aceratus* (Hb^-/Mb^-) (O'Brien et al. 2003), suggesting an informational feedback between mitochondrial membrane

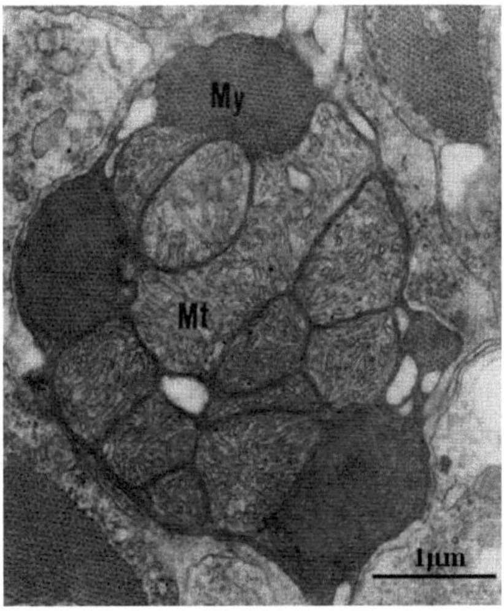

Fig. 7.2 Transverse section through a ventricular myocyte of *Chaenocephalus aceratus*. Note the proliferation of mithochondria (Mt) and low myofibrillar (My) packing at the cell periphery. (Modified from: Harrison et al. 1991)

biosynthesis and the presence of O_2-binding proteins. Very recently, O'Brien and Mueller (2010) provided a detailed account on the form and function of the icefish mitochondria.

7.4.5 Cardiac NOS/NO System and Redox Biome Homeostasis in the Icefish

One of the oldest universal functions of the NOS/NO system is to integrate cell biochemistry and energetics. NO synthesis is achieved through several enzymatic pathways (NOS-dependent NO generation from the guanidine group of L-Arg), as well as non-enzymatic processes, which can coexist within the same cell or tissue. In vertebrates, three different genes encode the NOS isoforms, i.e. NOS1 or neuronal NOS (nNOS), NOS2 or inducible NOS (iNOS) and NOS3 or endothelial NOS (eNOS), which can be found in a variety of tissues and cell types (Andreakis et al. 2010). Beside the NOS-dependent NO generation, NO can also derive from nitrite, a major biologic reservoir of NO, through nitrite reductase processes catalyzed by a number of hemoxygenases, including deoxyHb and deoxyMb. Thus, these evolutionarily old proteins, in addition to their classical respiratory role and apart from being the major NO scavengers, can contribute to NO-dependent hypoxic vasodilatation, cellular respiration and signalling, particularly in skeletal and cardiac muscles (Gladwin and Kim-Shapiro 2008).

Following our preliminary observations (Pellegrino et al. 2003) and Morlà et al. (2003), who firstly documented the presence of NOS/NO system in icefish, we further showed extensively NOS expression (Amelio et al. 2006) and NO-induced cardiac modulation (Pellegrino et al. 2004; Garofalo et al. 2009) in both Antarctic red-blooded and icefish species. The fact that myocardial NO activity depends from local environmental conditions, particularly tissue hypoxia, acidosis, presence of Mb and/or superoxide anions (Thomas et al. 2001), is of importance in the context of the redox balance of the icefish myocardium. In fact, its cardiac highly restricted subcellular NOS localization is suitable to provide an appropriate cold "NO microenvironment" for the working *trabeculae* of the *spongiosa*. The dynamic NO behaviour can largely contribute to ensure adequate O_2 transcellular transport from the lacunae to the mitochondria, in line with the computer modelling analysis of Lancaster's group (Thomas et al. 2001). Namely, "reversible NO inhibition of cellular mitochondrial O_2 consumption substantially extends the zone of adequate tissue cellular oxygenation away from the blood vessel, with an especially dramatic effect during conditions of increased tissue work". Likely, in the icefish *spongiosa*, the venous blood is retained for a longer period in the myriads of the lacunary spaces, hence exposing myocardiocytes to relatively large pO_2 fluctuations with the risk of hypoxia under increased metabolic demands. In this context, NO can react with membrane lipids. These can function, especially in mitochondria, as NO carriers and/or NO scavengers; at the same time, they represent very effective conduits for diffusive flux of O_2 and NO. This may allow O_2 flow compensation to individual cells or mitochondria (Londraville and Sidell 1990). NO solubility in hydrophobic solvents is approximately 9-times greater than that in water (Shaw and Vosper 1977), while that of O_2 is 3-fold greater; therefore, NO can be 9-times more concentrated in cell membranes, including mitochondria.

On the basis of this recent knowledge, we and Sidell's team have argued that the NOS/NO system may contribute to the major channichthyid circulatory and myocardial compensations. On the other hand, the Antarctic red-blooded and Hb-less notothenioids provide exclusive opportunities to investigate, in naturally occurring genetic knockouts for Hb/Mb, whether these disaptive losses may have elicited modifications of the intracardiac NOS enzymatic pattern and/or susceptibility to NO. For example, Sidell and O'Brien (2006) suggested that high levels of NO eventually occurring in absence of the Hb and Mb scavengers have promoted major adjustments in the icefish. Accordingly, Urschel and O'Brien (2008) and O'Brien and Mueller (2010), discussed the possible influence exerted by NO on mitochondrial morphology and function in the icefish myocardium.

7.4.6 Icefish Heart Vulnerability

The icefish heart acts as a typical volume pump (see above). This hemodynamic profile entails a very high sensitivity of the Starling's response, i.e. these hearts are capable to achieve significant changes of SV and SW in a limited range of preloads values (Tota et al. 1991). While SV and SW are unaffected by temperature changes (from 0.5 to 5.8°C), icefish HR is highly sensitive to temperature changes, being unaffected by both preload and afterload changes (Tota et al. 1991). The icefish cardiac work requires a remarkably high energetic cost, being approximately twice the weight specific power expenditure of a red-blooded notothenioid, i.e. 22% of resting metabolic energy production (Hemmingsen 1991; Sidell and O'Brien 2006). Therefore, according to the theory of O_2-limited thermal tolerance in Antarctic fish (Mark et al. 2002), even small temperature increases will further augment the energetic cost of the cardiac work. To which extent such thermal cardiac sensitivity may have predictive significance regarding the potential effects of climate change on icefish remains to be clarified (see below).

A hallmark of the maladaptive icefish phenotype is represented by its heart incapability to pump against even modest increases in ventral aorta pressure (from 1.76 to 2.65 kPa), as detailed in an in vitro analysis on the isolated and perfused working heart of *C. hamatus* (Tota et al. 1991) which confirmed the pioneering in vivo study on unrestrained *C. aceratus* by Hemmingsen et al. (1972).

According to Hemmingsen and Douglas (1977), "without a low flow resistance, cardiac work probably would be intolerable for the Chaenichthyids". This limitation, we believe, may result from the above-mentioned structural (myocyte diameter) and ultrastructural (mitochondria/myofibrils ratio) myocardial remodelling imposed by the compensation to the loss of respiratory pigments (Tota and Gattuso 1996). As emphasized in the myotomal muscle (Egginton et al. 2002; Johnston et al. 2003), also in the heart basic geometry (myocardiocyte space economy) imposes strict topological constraints in the arrangement of organelle logistics and metabolic demand. Namely, in the icefish cardiomyocytes, an enlarged mitochondrial compartment is necessary to allow a sufficient transcellular O_2 flux, but this makes impossible, at the same time, any ulterior myofibrillar increment that would determine consequent deleterious increase of fiber diameter (Garofalo et al. 2009).

This striking case of non-optimal adaptive recovery in response to evolutionary regressive changes or losses of function clearly calls into question the accustomed view of all species evolving to the same adaptionist pinnacle.

7.4.7 Perspectives: Global Change and Icefish Survival

Knowledge from fish species routinely exposed to dramatic environmental challenges identified a variety of adaptive response strategies, including the "resistance" adaptation which enhances resistance to the potentially lethal effects of environmental extremes. By highlighting fundamental aspects of phenotypic plasticity, these studies indicate that while adjustments performed during juvenile or adult life are usually reversible, exposure to environmental fluctuations during developmental stages may have life-long impact (Johnston et al. 2003; Pörtner et al. 2007, and references therein). The latter cannot, therefore, be extrapolated from short- and medium-term physiological experiments which may be informative on adult phenotype plasticity, but cannot be used for predicting the effects of environmental changes over evolutionary time. This lesson is particularly appropriate in the context of the Southern Ocean's climate change, currently characterized by a continuous rise in temperature (Clarke et al. 2007, and references therein). Conceivably, this may hamper survival of extremely stenotherm fish, like most ice fish species. However, carefully designed studies are necessary for providing prognostic evaluation (including putatively different inter-species vulnerability) on the channichthyids fate in the context of the global warming scenario. Such intriguing issue needs also to be reconsidered in relation to the two, not necessarily alternative, interpretations regarding the physiological significance that has permitted the ice fish to successfully evolve and live despite the loss of, what were once believed to be, essential-life O_2-binding chromo proteins.

According to one hypothesis, the adjustments for the lower O_2 transport capacity of the icefish blood (hypoxemia) and the consequent dangerous reduction of peripheral pO_2 necessary for adequate tissue oxygenation (hypoxemic hypoxia) need to take into account the increased O_2 solubility in body fluids and cold-induced dense mitochondrial membrane networks of lipid-rich tissues (Egginton et al. 2002; Sidell and O'Brien 2006, O'Brien and Mueller 2010, and references therein). Conceivably, at the severely cold body temperature of the nototheniod stenotherms, the Hb and Mb losses of genetic information may not be of crucial functional significance, thus relaxing all selective pressure on the retention of chromoprotein expression and/or structure. This may represent a "loss without-penalty" phenomenon, i.e. losses potentially reflecting traits no longer exerting negative selective effects for the animal fitness under the stably frigid conditions of the Southern Ocean (Somero et al. 1998). According to the other hypothesis (Pörtner et al. 2007), the scenario is dominated by excessive O_2 supply at low cellular cost and O_2 demands in the cold. It is supported by presence of large cell sizes, low mitochondrial capacity, concomitant reduction of metabolic energy and thus O_2 requirements (Johnston et al. 2003). Rather than reflecting O_2 limitation, these features may point to structural overcompensation of O_2 supply reductions at cellular/tissue levels, thereby lowering the selective pressure on Hb or Mb conservation. Likely, this allowed the development of an alternative physiological design, Hb-free blood and Mb-free muscle cells (Cheng and Detrich 2007) that could function similarly well in the cold Antarctic environment.

In conclusion, both interpretations suggest that the icefish thermal resistance and survival in the warming Antarctic habitat will depend on how, and to which extent, highly aerobic tissues like the myocardium will be able to cope with reactive O_2 species (ROS), reactive nitrogen intermediates (RNI) and NO levels in absence of two major ferrous heme deoxygenases known to tightly regulate cell oxygenation and redox state. Hopefully, this article discusses some major homeostatic loops that we suggested to underpin the phenotypic plasticity/vulnerability of the environmentally relevant icefish phenotype shall provoke new research on this wonderful fish.

References

Acierno R, Agnisola C, Tota B, Sidell BD (1997) Myoglobin enhances cardiac performance in antarctic ice fish species that express the protein. Am J Physiol 273:100–106

Acierno R, MacDonald JA, Agnisola C, Tota B (1995) Blood volume in the hemoglobinless Antarctic teleost *Chionodraco hamatus* (Lönnberg). J Exp Zool 272:407–409

Agnisola C, Acierno R, Calvo J, Farina F, Tota B (1997) In vitro cardiac performance in the sub-Antarctic notothenioids *Eleginops maclovinus* (subfamily Eleginopinae), *Paranotothenia magellanica*, and *Patagonotothen tessellata* (subfamily Nototheniinae). Comp Biochem Physiol A 118:1437–1445

Amelio D, Garofalo F, Pellegrino D, Giordano F, Tota B, Cerra MC (2006) Cardiac expression and distribution of nitric oxide synthases in the ventricle of the coldadapted Antarctic teleosts, the hemoglobinless *Chionodraco hamatus* and the red-blooded *Trematomus bernacchii*. Nitric Oxide 15:190–198

Andreakis N, D'Aniello S, Albalat R, Patti FP, Garcia-Fernàndez J, Procaccini G, Sordino P, Palumbo A (2010) Evolution of the nitric oxide synthase family in Metazoans. Mol Biol Evol 28(1):163–179

Axelsson M, Agnisola C, Nilsson S, Tota B (1998) Fish cardio-circulatory function in the cold. In: Pörtner HO, Playle R (eds) Cold ocean physiology. Cambridge University Press, Cambridge, pp 327–364

Balushkin AV (2000) Morphology, classification and evolution of notothenioid fishes of the Southern Ocean. J Ichthyol 40:74–109

Baum DA, Larson A (1991) Adaptation reviewed: a phylogenetic methodology for studying character macroevolution. Syst Zool 40:1–18

Beers JM, Borley KA, Sidell BD (2010) Relationship among circulating hemoglobin, nitric oxide synthase activities and angiogenic poise in red- and white-blooded Antarctic notothenioid fishes. Comp Biochem Physiol A 156:422–429

Bone Q (1978) Myotomal muscle fiber types in Scomber and Katsuwonnus. In: Sharp GD, Dizon AE (eds) The physiological ecology of tunas. Academic Press, New York, pp 183–204

Bossen EH, Sommer JR,Waugh RA (1978) Comparative stereology of the mouse and finch left ventricle. Tissue Cell 10:773–784

Bushnell PG, Brill RW (1992) Oxygen transport and cardiovascular responses in Skipjack tuna (*Katsuwonus pelamis*) and yellow fin tuna (*Thunnus albacares*) exposed to acute hypoxia. J Comp Physiol 162B:131–143

Cerra MC, Canonaco M, Acierno R, Tota B (1997) Different binding activity of A- and B-type natriuretic hormones in the heart of two antarctic teleosts, the red-blooded *Trematomus bernacchii* and the hemoglobinless *Chionodraco hamatus*. Comp Biochem Physiol A 118: 993–999

Cheng C-HC, Chen L (1999) Evolution of an antifreeze glycoprotein. Nature 40:443–444

Cheng C-HC, Detrich HW III (2007) Molecular ecophysiology of Antarctic notothenioid fishes. Philos Trans R Soc London, Ser B 362:2215–2232

Clarke A, Johnston IA (1996) Evolution and adaptive radiation of Antarctic fishes. Trends Ecol Evol 11:212–218

Clarke A, Murphy EJ, Meredith MP, King JC, Peck LS, Barnes DK, Smith RC (2007) Climate change and the marine ecosystem of the western Antarctic Peninsula. Philos Trans R Soc London B Biol Sci 362(1477):149–166

Cocca E, Ratnayake-Lecamwasam M, Parker SK, Camardella L, Ciaramella M, di Prisco G, Detrich WH III (1995) Genomic remnants of α-globin genes in the haemoglobin less Antarctic icefishes. Proc Nat Acad Sci U S A 92:1817–1821

Davie PS, Franklin CE (1992) Myocardial oxygen consumption and mechanical efficiency of a perfused dogfish heat preparation. J Comp Physiol 162:256–262

Detrich WH III, Amemiya CT (2010) Antarctic notothenioid fishes: genomic resources and strategies for analyzing an adaptive radiation. Integr Comp Biol 50(6):1009–1017

Detrich WH III, Yergeau DA (2004) Comparative genomics in erythropoietic gene discovery: synergisms between the Antarctic icefishes and the zebrafish. Methods Cell Biol 77:475–503
Detrich III WH, Parker SK, Williams RC Jr, Nogales E, Downing KH (2000) Cold adaptation of microtubule assembly and dynamics. Structural interpretation of primary sequence changes present in the alpha- and beta-tubulins of Antarctic fishes. J Biol Chem 275(47):37038–37047
Eastman JT (1993) Antarctic Fish Biology. Evolution in a unique environment. Academic Press, San Diego, pp 1–322
Eastman JT (1995) The evolution of Antarctic fishes: questions for consideration and avenues for research. Cybium 19:371–389
Eastman JT (1997) Phyletic divergence and specialization for pelagic life in the Antarctic notothenioid fish *Pleuragramma antarcticum*. Comp Biochem Physiol A 118:1095–1101
Eastman JT, Eakin RR (2000) An updated species list for notothenioid fish (Perciformes; Notothenioidei), with comments on Antarctic species. Arch Fish Mar Res 48:11–20
Eastman JT (2005) The nature of the diversity of Antarctic fishes. Polar Biol 28:94–107
Eastman JT, Lannoo MJ (2004) Brain and sense organ anatomy and histology in hemoglobinless Antarctic icefishes (Perciformes: Notothenioidei: Channichthyidae). J Morphol 260:117–140
Egginton S (1996) Blood rheology of Antarctic fishes: viscosity adaptations at very low temperatures. J Fish Biol 48(3):513–521
Egginton S, Davison W (1998) Effects of environmental and experimental stress on Antarctic fishes. In: Pörtner HO, Playle R (eds) Cold ocean physiology. Cambridge University Press, Cambridge, pp 299–326
Egginton S, Sidell BD (1989) Thermal acclimation induces adaptive changes in subcellular structure of fish skeletal muscle. Am J Physiol 256:R1–R9
Egginton S, Stilbeck C, Hoofd L, Calvo J, Johnston IA (2002) Peripheral oxygen transport in skeletal muscle of Antarctic and sub-Antarctic notothenioid fish. J Exp Biol 205:769–779
Elder HY (1975) Muscle structure. In: Usherwood PR (ed) Insect muscle. Academic Press, London, pp 1–74
Farrell AP, Jones DR (1992) The heart. Fish Physiol 12A:1–88
Farrell AP, Wood S, Hart T, Driedzic WR (1985) Mycordial oxygen consumption in the sea raven, *Hemitripterus americanus*: the effects of volume loading, pressure loading and progressive hypoxia. J Exp Biol 117:237–250
Fletcher GL, Hew CL, Davies PL (2001) Antifreeze proteins of teleost fishes. Annu Rev Physiol 63:359–390
Garofalo F, Amelio D, Cerra MC, Tota B, Sidell BD, Pellegrino D (2009) Morphological and physiological study of the cardiac NOS/NO system in the Antarctic (Hb^-/Mb^-) icefish *Chaenocephalus aceratus* and in the red-blooded *Trematomus bernacchii*. Nitric Oxide 20(2): 69–78
Gladwin MT, Kim-Shapiro DB (2008) The functional nitrite reductase activity of the heme-globins. Blood 112:2626–2647
Graham MS, Farrell AP (1990) Myocardial oxygen consumption in trout acclimated to 5 and 15°C. Physiol Zool 63:536–554
Hamoir G, Gerardin-Otthiers N (1980) Differentiation of the sarcoplasmic proteins of white, yellowish and cardiac muscle of an antarctic hemoglobin-free fish, *Champsocephalus gunnari*. Comp Biochem Physiol B 65:199–206
Harrison P, Zummo G, Farina F, Tota B, Johnston IA (1991) Gross anatomy, myoarchitecture, and ultrastructure of the heart ventricle in the haemoglobinless icefish *Chaenocephalus aceratus*. Can J Zool 69:1339–1347
Heise K, Estevez MS, Puntarulo S, Galleano M, Nikinmaa M, Pörtner HO, Abele D (2007) Effects of seasonal and latitudinal cold on oxidative stress parameters and activation of hypoxia inducible factor (HIF-1) in zoarcid fish. J Comp Physiol B 177:765–777
Hemmingsen EA (1991) Respiratory and cardiovascular adaptations in hemoglobinfree fishes: resolved and unresolved problems. In: di Prisco G, Maresca B, Tota B (eds) Biology of antarctic fish. Springer, Berlin, pp 191–203

Hemmingsen EA, Douglas EL (1977) Respiratory and circulatory adaptations to the absence of hemoglobin in chaenichthyid fishes. In: Llano GA (ed) Adaptations within Antarctic ecosystems. In: Proceedings of 3rd SCAR symposium on Antarctic biology. Gulf Publishing Co, Houston, Texas, pp 479–487

Hemmingsen EA, Douglas EL, Johansen K, Millard RW (1972) Aortic blood flow and cardiac output in the hemoglobin-free fish *Chaenocephalus aceratus*. Comp Biochem Physiol A 43:1045–1051

Hochachka PW, Rupert JL, Monge C (1999) Adaptation and conservation of physiological systems in the evolution of human hypoxia tolerance. Comp Biochem Physiol A: Mol Integr Physiol 120:1–17

Hochachka PW, Somero GN (2002) Biochemical adaptation: mechanism and process in physiological evolution. Oxford University Press, New York

Holeton GF (1970) Oxygen uptake and circulation by a haemoglobinless fish (*Chaenocephalus aceratus* Lönnberg) compared with three red-blooded Antarctic fish. Comp Biochem Physiol 34:457–471

Hoppeler H, Lindstedt SL (1985) Malleability of skeletal muscle in overcoming limitations: structural elements. J Exp Biol 115:355–364

Houlihan DF, Agnisola C, Lyndon AR, Gray C, Hamilton NM (1988) Protein synthesis in a fish heart: responses to increased power output. J Exp Biol 137:565–587

Icardo JM, Colvee E, Cerra MC, Tota B (1999) Bulbus arteriosus of the Antarctic teleosts I. The white-blooded chionodraco hamatus. Anat Rec 254:396–407

Jakubowski M (1982) Dimensions of respiratory of the gills and skin in the Antarctic white-blooded fish, *Chaenocephalus aceratus* Lönnberg (Chaenichthyidae). Z mikrosk-anatom Forsch Leipzig 96:145–156

James NT, Meek GA (1979) Stereological analyses of the structure of mitochondria in pigeon skeletal muscle. Cell Tissue Res 202:493–503

Jobling M (1994) Fish bioenergetics. Chapman and Hall, London, pp 1–309

Johnston IA (1989) Antarctic fish muscles: structure, function and physiology. Antarct Sci 1:97–108

Johnston IA, Maitland B (1980) Temperature acclimation in crucian carp (*Carassius carassius* L.) morphometric analyses of muscle fibre ultrastructure. J Fish Biol 17:113–125

Johnston IA, Moon TW (1981) Fine structure and metabolism of multiple innervated fast muscle-fibers in teleost fish. Cell Tissue Res 219:93–109

Johnston IA, Fernandez DA, Calvo J, Vieira VLA, North AW, Abercromby M, Garland T (2003) Reduction in muscle fibre number during adaptive radiation in notothenioid fishes: a phylogenetic perspective. J Exp Biol 206:2595–2609

Johnston IA, Fitch N, Zummo G, Wood RE, Harrison P, Tota B (1983) Morphometric and ultrastructural features of the ventricular myocardium of the haemoglobinless icefish *Chaenocephalus aceratus*. Comp Biochem Physiol A 76:475–480

Johnston IA, Guderley H, Franklin CE, Crockford T, Kamunde C (1994) Are mitochondria subject to evolutionary temperature adaptation? J Exp Biol 195:293–306

Johnston IA, Vieira VLA, Fernández DA, Abercromby M, Brodeur JC, Peck L, Calvo J (2002) Muscle growth in polar fish: a study of Harpagifer species with sub-Antarctic and Antarctic distributions. Fish Sci 68(Suppl II):1023–1028

Josephson R, Young D (1985) A synchronous insect muscle with an operating frequency greater than 500 Hz. J Exp Biol 118:185–208

Kiceniuk JW, Jones DR (1977) The oxygen transport system in trout (*Salmo gairdneri*) during sustained exercise. J Exp Biol 69:247–260

Kock KH (2005a) Antarctic icefishes (channichthyidae): a unique family of fishes. A review (Part I). Polar Biol 28:862–895

Kock KH (2005b) Antarctic icefishes (channichthyidae): a unique family of fishes. A review (Part II). Polar Biol 28:895–909

Londraville RL, Sidell BD (1990) Ultrastructure of aerobic muscle in Antarctic fishes may contribute to diffusive fluxes. J Exp Biol 150:205–220

Luciano JA, Tan T, Zhang Q, Huang E, Scholz P, Weiss HR (2008) Hypoxia inducible factor-1 improves the actions of nitric oxide and natriuretic peptides after simulated ischemia-reperfusion. Cell Physiol Biochem 21:421–428

Lushchak VI, Bagnyukova TV (2007) Hypoxia induces oxidative stress in tissues of a goby, the rotan *Percottus glenii*. Comp Biochem Physiol B: Biochem Mol Biol 148(4):390–397

Macdonald JA, Wells RMG (1991) Viscosity of body fluids from Antarctic notothenioid fish. In: di Prisco G, Maresca B, Tota B (eds) Biology of Antarctic fish. Springer-Verlag, Berlin, pp 163–178

Mark FC, Bock C, Pörtner HO (2002) Oxygen-limited thermal tolerance in Antarctic fish investigated by MRI and 31P-MRS. Am J Physiol Regul Integr Comp Physiol 283:R1254–R1262

Meeson AP, Radford N, Shelton JM, Mammen PP, DiMaio JM, Hutcheson K, Kong Y, Elterman J, Williams RS, Garry DJ (2001) Adaptive mechanisms that preserve cardiac function in mice without myoglobin. Circ Res 88:713–720

Montgomery J, Clements K (2000) Disaptation and recovery in the evolution of Antarctic fishes. Trends Ecol Evol 15:267–271

Morlà M, Alvar GN, Rahman I, Motterlini R, Saus C, Morales-Nin B, Company JB, Busquets X (2003) Nitric oxide synthase type I (nNOS), vascular endothelial growth factor (VEGF) and myoglobin-like expression in skeletal muscle of Antarctic ice fishes (Notothenioidei: Channichthyidae). Polar Biol 26:458–462

Moylan TJ, Sidell BD (2000) Concentrations of myoglobin mRNA in heart ventricle from Antarctic fishes. J Exp Biol 203:1277–1286

Near TJ, Parker SK, Detrich HW III (2006) A genomic fossil reveals key steps in hemoglobin loss by the antarctic ice fishes. Mol Biol Evol 23(11):2008–2016

O'Brien K, Mueller IA (2010) The unique mitochondrial form and function of Antarctic channichthyid ice fishes. Int Comp Biol 50:993–1008

O'Brien KM, Skilbeck C, Sidell BD, Egginton S (2003) Muscle fine structure may maintain the function of oxidative fibres in haemoglobin less Antarctic fishes. J Exp Biol 206:411–421

O'Brien KM, Sidell BD (2000) The interplay among cardiac ultrastructure, metabolism and the expression of oxygen-binding proteins in Antarctic fishes. J Exp Biol 203:1287–1297

O'Brien KM, Xue H, Sidell BD (2000) Quantification of diffusion distance within the spongy myocardium of hearts from antarctic fishes. Respir Physiol 122:71–80

Pellegrino D, Acierno R, Tota B (2003) Control of cardiovascular function in the icefish *Chionodraco hamatus*: involvement of serotonin and nitric oxide. Comp Biochem Physiol A: Mol Integr Physiol 134(2):471–480

Pellegrino D, Palmerini CA, Tota B (2004) No haemoglobin but NO: the ice fish (*Chionodraco hamatus*) heart as a paradigm. J Exp Biol 207:3855–3864

Petricorena ZL, Somero GN (2007) Biochemical adaptations of notothenioid fishes: comparisons between cold temperate South American and New Zealand species and Antarctic species. Comp Biochem Physiol A: Mol Integr Physiol 147:799–807

Place SP, Zippay ML, Hofmann GE (2004) Constitutive roles for inducible genes: evidence for the alteration in expression of the inducible hsp70 gene in Antarctic notothenioid fishes. Am J Physiol 287:R429–R436

Pörtner HO, Peck LS, Somero G (2007) Thermal limits and adaptation in marine ectotherm: an integrative view. Philos Trans R Soc B 362:2233–2258

Römisch K, Collie N, Soto N, Logue J, Lindsay M, Scheper W, Cheng C-HC (2003) Protein translocation across the endoplasmic reticulum membrane in cold-adapted organisms. J Biol Chem 278:37998–38003

Ruud JT (1954) Vertebrates without erythrocytes and blood pigment. Nature 173:848

Shaw AW, Vosper AJ (1977) Solubility of nitric oxide in aqueous and nonaqueous solvents. J Chem Soc Faraday Trans 8:1239–1244

Sidell BD (1998) Intracellular oxygen diffusion: the roles of myoglobin and lipid at cold body temperature. J Exp Biol 201:1118–1127

Sidell BD, O'Brien KM (2006) When bad things happen to good fish: the loss of haemoglobin and myoglobin expression in Antarctic ice fishes. J Exp Biol 209:1791–1802

Sidell BD, Vayda ME, Small DJ, Moylan TJ, Londraville RL, Yuan ML, Rodnick KJ, Eppley ZA, Costello L (1997) Variable expression of myoglobin among the hemoglobinless Antarctic icefishes. Proc Nat Acad Sci U S A 94:3420–3424

Small DJ, Moylan T, Vayda ME, Sidell BD (2003) The myoglobin gene of the Antarctic ice fish, *Chaenocephalus aceratus*, contains a duplicated TATAAAA sequence that interferes with transcription. J Exp Biol 206:131–139

Somero GN, Fields PA, Hofmann GE, Weinstein RB, Kawall H (1998) Cold adaptation and stenothermy in Antarctic notothenioid fishes: what has been gained and what has been lost? In: di Prisco G, Pisano E, Clarke A (eds) Fishes of Antarctica. A biological overview. Springer-Verlag, Italy, pp 97–109

Storelli C, Acierno R, Maffia M (1998) Membrane lipid and protein adaptations in Antarctic fish. In: Pörtner HO, Playle R (eds) Cold ocean physiology. Cambridge University Press, Cambridge, pp 166–189

Thomas DD, Liu X, Kantrow SP, Lancaster JP Jr (2001) The biological lifetime of nitric oxide: implications for the perivascular dynamics of NO and O_2. Proc Nat Acad Sci U S A 98(1): 355–360

Tota B, Cerra MC, Mazza R, Pellegrino D, Icardo J (1997) The heart of the Antarctic icefish as paradigm of cold adaptation. J Therm Biol 22:409–417

Tota B, Gattuso A (1996) Heart ventricle pumps in teleosts and elasmobranchs: a morphodynamic approach. J Exp Zool 275:162–171

Tota B, Acierno R, Agnisola C (1991a) Mechanical performance of the isolated and perfused heart of the haemoglobinless Antarctic icefish *Chionodraco hamatus* (Loönnberg): effects of loading conditions and temperature. Philos Trans R Soc London B 332:191–198

Tota B, Agnisola C, Schioppa M, Acierno R, Harrison P, Zummo G (1991b) Structural and mechanical characteristics of the heart of the ice fish *Chionodraco hamatus* (Lönnberg). In: di Prisco G, Maresca B, Tota B (eds) Biology of antarctic fish. Springer, Berlin, pp 204–219

Twelves EL (1972) Blood volume of two Antarctic fishes. Br Antarct Surv Bull 31:85–92

Urschel M, O'Brien KM (2008) High mitochondrial densities in the hearts of Antarctic ice fishes are maintained by an increase in mitochondrial size rather than mitochondrial biogenesis. J Exp Biol 211:2638–2646

Vayda ME, Small DJ, Yuan M, Costello L, Sidell BD (1997) Conservation of the myoglobin gene among Antarctic notothenioid fishes. Mol Mar Biol Biotechnol 6:207–216

Verde C, Giordano D, Russo R, di Prisco G (2012) The adaptive evolution of polar fishes. Lessons from the function of hemoproteins. In: di Prisco G, Verde C (eds) Adaptation and evolution in marine environments—The impacts of global change on biodiversity, vol 1. Series "From Pole to Pole". Springer, Berlin, pp 197–213

Vogel W, Kock KH (1981) Morphology of gill vessels in icefish. Arch Fisch Wiss 31:139–150

Walvig F (1960) The integument of the ice fish *Chaenocephalus aceratus* (Loönnberg). Nytt Mag Zool Oslo 6:111–120

Weidemann A, Klanke B, Wagner M, Volk T, Willam C, Wiesener MS, Eckardt KU, Warnecke C (2008) Hypoxia, via stabilization of the hypoxia-inducible factor HIF-1alpha, is a direct and sufficient stimulus for brain-type natriuretic peptide induction. Biochem J 409:233–242

Wood CM, Pieprzak P, Trott JN (1979) The influence of temperature and anaemia on the adrenergic and cholinergic mechanisms controlling heart rate in the rainbow trout. Can J Zool 57:2440–2447

Wujcik JM, Wang G, Eastman JT, Sidell BD (2007) Morphometry of retinal vasculature in Antarctic fishes is dependent upon the level of haemoglobin in circulation. J Exp Biol 210:815–824

Yergeau DA, Cornell CN, Parker SK, Zhou Y, Detrich HW III (2005) Bloodthirsty, an RBCC/TRIM gene required for erythropoiesis in zebrafish. Dev Biol 283:97–112

Zummo G, Acierno R, Aginisola C, Tota B (1995) The heart of the ice fish: bio construction and adaptation. Braz J Med Biol Res 28:1265–1276

Chapter 8
Aspects of Protein Cold Adaptation in Antarctic Fish

Craig Marshall

8.1 Introduction

With the transition of Antarctica from seasonally cold to a permanently frozen state came a very significant transition in the surrounding waters (Zachos et al. 2001; Peters et al. 2010). During the seasonal cold period the formation of sea ice and reduction of water temperature created a set of problems for marine organisms. Seasonal changes that could previously be avoided by behavioural adaptations became permanent and had a profound effect on the marine fauna and flora (Eastman 1993).

Among the changes brought about by freezing water is the almost complete elimination of the intertidal zone that is now occupied by ice. This is compounded by the spontaneous formation of ice on substrates down to about 10 m caused by the presence of super-cooled water. Ice is less dense than water and the formation of ice crystals on the bottom causes mud and smaller objects to float and to be incorporated into the surface layer of ice where they are eventually released upon melting: the whole cycle resulting in a 'tilling' of the sub-tidal zone. This clearly has significance from an ecological point of view as it substantially alters sub-tidal and estuarine environments.

A reduction of water temperature itself also has consequences. Water viscosity and gas solubility are approximately double at −1.86°C (the freezing point of seawater) than at 20°C. Such an increase in viscosity has significant effects at macroscopic scales (such as locomotion) but much more profound effects at the scale of cilia and villi, and on diffusion. Increased gas solubility makes oxygen more available and is thought to be one of the underlying causes of polar gigantism

C. Marshall (✉)
Department of Biochemistry and Genetics Otago,
University of Otago, PO Box 56, Dunedin, New Zealand
e-mail: craig.marshall@otago.ac.nz

G. di Prisco and C. Verde (eds.), *Adaptation and Evolution in Marine Environments, Volume 1*, From Pole to Pole, DOI: 10.1007/978-3-642-27352-0_8,

(Chapelle and Peck 1999; Woods et al. 2009), and may also increase oxidative stress (Abele and Puntarulo 2004; Kassahn et al. 2009).

Whereas gas solubility increases at lower temperatures, salt solubility declines. This decrease probably has little direct effect although the activity of dissolved ions decreases at lowered temperatures and may alter some aspects of biological chemistry. Another similar affect is an increase in pH at a given temperature. This may be simply related to a reduction of H^+ activity at reduced temperatures, but complicating this simple analysis is an increase in the solubility of CO_2 at reduced temperatures which gives rise to an increase in $[H^+]$ (Andersson et al. 2008). Some combination of these effects can give rise to increased salt concentrations in body fluids at low temperature that seems to be affected by an increase in temperature (Gonzalez-Cabrera et al. 1995).

Temperature also has profound effects on the rates of reaction. In general, a 10°C change in temperature is proportional to a two-fold change in reaction rate (a Q_{10} of 2) (Hochachka and Somero 2002). For Antarctic organisms this corresponds to about a four-fold reduction at -1.86°C compared with 20°C. This generalized reduction affects all chemistry including enzyme-mediated reactions (Somero 1995). In some cases there is evidence for cold-adaptation in some enzymes whereby activity is greater that expected if $Q_{10} \approx 2$, something particularly important for enzymes involved in state change events such as neural conduction and muscle contraction where a minimum bound of activity exists. These observations are consistent with the idea of 'metabolic cold adaptation' first advocated by Krogh (Krogh 1914) and subsequently widely discussed from an Antarctic perspective (Sokolova and Portner 2003; Kawall et al. 2002; Clarke and Johnston 1999). Finding a structural basis for such adaptation has proved difficult with most studies finding only subtle structural evidence for metabolic cold adaptation (Georlette et al. 2004; Fields and Houseman 2004; Marshall et al. 2003; Feller 2003; Sharpe et al. 2001; Fields and Somero 1998).

Perhaps the most pressing problem faced by marine organisms in freezing seawater is the problem of freezing. Many invertebrate and plant species maintain internal fluids at about the same osmotic strength as seawater and thus avoid the problem. Fish however, have body fluids significantly hyposmotic to seawater and instead either live at a depth where they can avoid nucleation centres and remain supercooled, or produce antifreeze molecules: a family of highly varied proteins (and glycopeptides) that inhibit the growth of ice crystals even in supercooled solutions (DeVries and Cheng 2005). In the Antarctic, the notothenioid fish dominate shallow, and therefore freezing-susceptible water, both in biomass and in species number (Eastman 2000). Antifreeze glycopeptides in this family of fish seem to have been a crucial adaptation that allowed the exploitation of niches now unavailable to other fish. Almost all Antarctic notothenioid fish have AFGP and related proteins (DeVries and Cheng 2005) and the few high latitude notothenioid fish that do not probably reflect secondary losses. It seems likely that the presence of this key adaptation is responsible for the dominance of notothenioid fish in the Antarctic (Eastman and Clarke 1998; Near and Cheng 2008).

Notwithstanding these comments, notothenioid fish do not comprise the entire Antarctic fish fauna and other groups, particularly zoarcid and liparid fish are common (Eastman 2000; Gon and Heemstra 1990). These groups are less well-studied but those findings that are available suggest the general problems faced by these fish are the same although the specific responses may be different (Sharpe et al. 2001; Kelley et al. 2010).

8.2 Reduction of Reaction Rate: Scalable Effects

Perhaps the most studied biochemical problem of low temperatures is that of reduction of reaction rate with temperature. For enzymes, reductions in reaction rate are accompanied by decreases in kinetic parameters such as k_{cat} and K_m, and in structural properties such as enzyme flexibility. Reductions in rate of reaction may be readily accommodated by the synthesis of more protein, although clearly this strategy is limited by the total amount of protein possible in a cell. Furthermore, there is considerable evidence that the synthesis and folding of protein may have significant costs (Todgham et al. 2007).

Perhaps the best-studied enzyme in this context is lactate dehydrogenase (LDH). A number of studies have demonstrated conservation of both k_{cat} and K_m and a pattern of reduced thermal stability of the protein in LDH from enzymes from cold fish (Kawall et al. 2002; Fields and Houseman 2004; Sharpe et al. 2001; Fields and Somero 1998). Many of these findings are consistent with the notion of metabolic cold adaptation discussed above. Typically, enzymes from cold fish have larger k_{cat} and K_m at any given temperature than the homologous enzyme from fish from a warmer niche. However, at the ecologically significant temperatures, these parameters tend to be about the same suggesting that these properties are conserved. It has been suggested that this conservation reflects the underlying concentration of substrates and 'tuning' of enzyme kinetic parameters to these values relates to intrinsic Michaelis–Menten enzymatic control (Somero 1995; Lin and Somero 1995).

Although modelling studies suggest a structural basis for these changes in function typically associated with changes in the regions associated with ligand binding and catalytic movement (and not the catalytic residues themselves) (Georlette et al. 2004; Fields and Houseman 2004; Sharpe et al. 2001; Fields and Somero 1998; Holland et al. 1997) there are few structural studies that provide direct evidence of this assertion. Our own crystallographic studies show that any such changes are subtle and not obvious in structures of notothenioid fish LDH solved thus far (Kumar and Marshall, unpublished). Nonetheless, it is clear that such parameters as k_{cat} and K_m are conserved and it is possible that the effects are mediated by changes in the dynamics and structural changes associated with ligand binding and catalysis.

Other studies have looked at a variety of molecules and attempted to define a set of protein 'cold adaptations', but the results have proved equivocal

(Georlette et al. 2004; Feller 2003; Marx et al. 2007; Collins et al. 2008). In general, it seems that proteins from cold organisms have less well-ordered hydrophobic cores and generally show more flexibility at any given temperature than equivalent proteins from warmer organisms. It is common for cold-adapted proteins to denature at lower temperatures than their mesophilic counterparts, but the temperature at which this occurs is seldom of any physiological or ecological relevance.

In contrast, cold denaturation is potentially a problem (Temussi 2011; Marshall 1997). At low temperatures water become less ‘hydrophilic’ and the hydrophobic driving forces responsible for much of the structured folding characteristic of protein is diminished. Most cold denaturation is associated with the loss of quaternary interactions where these are primarily hydrophobic in nature. Less is known about loss of tertiary structure but the reduction in hydrophobic forces that stabilize protein may make protein folding more difficult (see below) and make the hydrophobic folded protein core more flexible (Collins et al. 2008).

In many cases, the effect of temperature is ‘scalable’: that is although it is reduced by decreased temperature, there is no point at which the protein function fails. Reductions in reaction rate associated with reduced temperatures are associated with reductions in the costs of living at such low temperatures (Eastman 1993) and a scaled-down metabolism may suffice although there is significant evidence that metabolic pathways as a whole show evidence of metabolic cold adaptation (Kawall et al. 2002; Clarke and Johnston 1999; Crockett and Sidell 1990; Johnston et al. 1998). The question of how metabolic cold adaptation manifests in Antarctic fish remains open.

8.3 State Changes in Protein Function

8.3.1 Membranes and Proteins

Some biological systems do exhibit ‘state changes’ with changes of temperatures: a non-linear response to some change in temperature. Perhaps the most common of such state changes occurs in cell membranes. Typically, membranes are considered to be optimized to a particular temperature and to be at risk of phase changes if that temperature changes (Williams 1990). To some extent, this is an anthropomorphic approach since most organisms do not maintain a constant body temperature the way that mammals do. However, most Antarctic fish live in a stenothermal environment and increases in temperature do have a significant affect on membrane structure (Cossins and Macdonald 1989; Becker et al. 1995).

It is generally considered that changes in the proportion of saturated and unsaturated fatty acids that comprise the phospholipids of membranes alter the properties of the membrane: something termed homeoviscous adaptation. Little is known of the membrane composition of Antarctic fish but there is some evidence for loss of function in some membrane proteins with an increase in temperature (Weinstein and Somero 1998). This is a profitable area for future study.

It is worth noting that most Antarctic notothenioid fish do not tolerate temperatures much above 5°C and at least one component of this limit may be related to membrane-protein interactions (Somero and DeVries 1967; Haschemeyer 1980). However, recent work suggests that at least some aspects of cardiovascular function show adaptation at elevated temperatures (Seebacher et al. 2005) and the upper limit of about 6°C may be less stringent than previously thought.

8.3.2 Microtubules

The best-studied 'state change' in protein function at low temperatures are the microtubule proteins. Microtubules are an important part of intracellular transport mechanisms and comprise a set of proteins, the tubulins, which polymerise to form elongated structures, the microtubules. These form part of the cytoskeleton important in cell division and vesicular transport in conjunction with cellular motors such as dynein and kinesin. They are particularly important in neural tissue such as brain and spinal cord where cellular components may be transported several millimetres or further (Nogales 2000).

Microtubule formation is characterized by a balance between assembly and disassembly of tubulin dimers. Tubulin monomers dimerize in a complex series of reactions that involve cofactors and chaperons, and the dimers aggregate to form long tubulin rods. This occurs in the presence of GTP but also contain a GTPase that slowly hydrolyses the GTP. The microtubule has plus and minus end and growth occurs most often at the minus end (although this depends on the circumstance), whereas the plus is end is relatively stable (Nogales and Downing 2008). Depolymerization is associated with this slow hydrolysis and tubulin dimers are released into the free pool.

Microtubule assembly is temperature sensitive. Mammalian microtubules polymerize slowly at 37°C and depolymerize by spontaneous 'catastrophes' at the plus end (Nogales and Downing 2008). At temperatures much below 37°C, the dissociated form of mammalian tubulins is favoured and the intracellular transport that depends on microtubule formation is not possible. However, microtubules from Antarctic notothenioid fish are stable at −1.9°C suggesting some significant changes from the mammalian forms of the molecule. Careful examination of the amino acid sequence (and modelling based on mammalian microtubule structures) suggests that a handful of amino acid changes at the C-terminus of the tubulin monomer increase the stability of the tubulin dimer by affecting the lateral interactions between tubulin monomers. These changes act to stabilize tubulin dimers by increasing the number of hydrophobic interactions and altering the flexibility of some of the amino acids in the contact region (Detrich et al. 2000).

8.4 Incidental Losses and Adaptations

8.4.1 Haemoglobin and Myoglobin

In addition to specific changes in proteins, there are several changes that may be only indirectly considered as cold adaptations but which are very interesting and throw light on protein structure and function. Perhaps the most interesting of these are ice fish that contain no haemoglobin. The ice fish (Channichthyidae) are a group of 16 species of notothenioid fish thought to comprise a species flock (Chen et al. 1998; Near et al. 2003). They are the only vertebrates that lack haemoglobin and consequently have 'white' blood rather than red in consequence of lipid within the plasma and an absence of red blood cells (Eastman 1993). In addition to the absence of haemoglobin, icefish have a number of secondary modifications that include very high cardiac output, large blood volume, increased vascular density, and alterations in the structure of mitochondrial membranes (Sidell 1991; Egginton 1997; Acierno et al. 1997; Rodnick and Sidell 1997; Feller and Gerday 1997; Sidell and O'Brien 2006).

What is the cause of the loss of haemoglobin? In all but one icefish there was no β-globin gene and an α-globin pseudogene: *Neopagetopsis ionah* contains a β-globin pseudogene as well (Near et al. 2006). These data suggest some globin gene rearrangements during the radiation of the Channichthyidae, but this does not explain how the loss of haemoglobin was not lethal. Studies on other notothenioid fish that do contain haemoglobin slow low loss can be tolerated, at least at rest. Carbon monoxide poisoning of many notothenioid fish (where the haemoglobin: CO complex prevents the transport of oxygen) is not lethal suggesting a significant haemoglobin-independent reserve capacity for oxygen carriage (di Prisco et al. 1992).

However, no other vertebrate has lost circulating haemoglobin, so is there some selective advantage in this unusual case? The removal of red cells reduces the haematocrit and this is particularly pronounced at subzero temperatures where water viscosity is increased. Many notothenioid fish sequester most of their red cells within the spleen except when they are needed (Axelsson et al. 1992; Davison et al. 1995) and there may be some selective advantage in the permanent loss of red cells at low temperatures (Sidell and O'Brien 2006).

In addition to the loss of haemoglobin in icefish, several lineages have also lost the capacity to synthesize myoglobin (Sidell and O'Brien 2006; Grove et al. 2004). Whereas the loss of haemoglobin in Channichthyidae seems to have been a single set of events, myoglobin loss appears to have occurred several times (Sidell and O'Brien 2006). Why should this secondary loss have occurred?

At least two explanations have been put forward to account for the loss of myoglobin. The first relates to the effectiveness of myoglobin as an intracellular transport molecule assisting in the diffusion of oxygen from the cell membrane to the inner mitochondrial membrane where it acts as a terminal electron acceptor in the electron transport chain. At low temperatures, the structural movement required to allow the entry and exit of an oxygen molecule from myoglobin is somewhat attenuated and myoglobin become less effective (Wittenberg and Wittenberg 1990; Sidell 1998; Petricorena and Somero 2007). At the same time, lipid bilayers become more soluble to oxygen and allow diffusion along the leaflets that are more extensive in the mitochondria of myoglobin-less icefish hearts (O'Brien and Sidell 2000; Urschel and O'Brien 2008; O'Brien 2011).

The second explanation relates to a secondary function of haemoglobin and myoglobin: that of nitric oxide oxygenase (Pellegrino et al. 2004; Tota et al. 2011). Nitric oxide (NO) has been implicated in angiogenesis and the development of capillary beds and in mitochondrial biogenesis (Sidell and O'Brien 2006; O'Brien 2011) and NO amounts are significantly increased in icefish in comparison with other notothenioid fish (Beers et al. 2010). It is easy to see how the loss of haemoglobin (and myoglobin) may have led to an increase in NO that in turn led to the very adaptations (see above) that assisted in compensating for reduced oxygen transport (Sidell and O'Brien 2006).

The loss of these two loci has the potential to severely limit the range of Channichthyid fish. However, at least one species, *Champsocephalus esox* is found in sub-Antarctic waters around Patagonia and the Falkland Islands/Malvinas at temperatures of up to 14°C (Nakamura et al. 1986). It is possible that the secondary adaptations to the loss of haemoglobin (and myoglobin) provide sufficient compensation for the transport of oxygen and may mean that icefish have the potential to adapt to warmer water.

8.4.2 Metallothioneins

In addition to the loss of haemoglobins in Channichthyids, the metal-binding protein metallothionein is also absent in this group of fish, but not in other notothenioids (Scudiero et al. 1997). Although the protein does not seem to be present, abundant mRNA is present in the liver suggesting some sort of translational block. The significance of this finding is unclear, but may be related to changes in iron metabolism in icefish.

8.4.3 Heat Shock Proteins

Heat shock proteins are a general class of proteins that are typically expressed in response to some environmental stress such as a change in temperature (Feder and Hofmann 1999). Many heat shock proteins act as molecular chaperones that assist in folding or refolding of proteins damaged by the stress and these chaperones are mostly conserved and found in a wide range of organisms. Heat shock proteins are typically found in both constitutive and inducible forms. Antarctic notothenioid fish appear to have lost the capacity to induce heat shock proteins in response to an environmental stress (Hofmann et al. 2000; Buckley et al. 2004). In these studies using isolated hepatocytes, exposure to neither temperature nor cadmium resulted in a stimulation of Hsp70 synthesis, but a high rate of constitutive expression of these proteins was observed under all conditions. Similar results were found in notothenioid fish found in temperate waters around southern New Zealand (Carpenter and Hofmann 2000).

This loss of inducible expression was not general however. A micro-array study of changes in mRNA expression in the Antarctic notothenioid *Trematomus bernacchii* in response to a heat shock found a significant set of changes (Buckley and Somero 2009). These data suggest a very specific loss of the inducible response in at least the Hsp70 family of chaperones.

Why might this have occurred? The elevation of constitutive expression of Hsp70 suggests a requirement for chaperones. Temperature is known to have a significant effect on protein folding (Dill et al. 2008) and there is evidence that folding success at low temperatures is somewhat reduced. Indeed, protein synthesis and folding is a balance between the rescue of folding intermediates by chaperones and their ubiquitylation and degradation by the proteasome (Glickman and Adir 2004). If the balance of folding is affected and fully-functional protein become harder to make, the overall cost of protein synthesis may increase: something for which there is evidence in Antarctic notothenioid fish (Todgham et al. 2007).

It is possible that selection of cold-adapted folding pathways may be more significant for protein cold adaptation than adaptations that affect protein binding and catalytic function (Lindquist 2010). This is another potentially fruitful area of research.

8.5 Discussion

Identifying the key adaptations that allow Antarctic fish to inhabit cold Southern waters is complex. Undoubtedly one most important adaptation is the ability to avoid freezing in water that is below the freezing point of fish body fluids. This is particularly important for fish living in shallow water or around ice sheets and shelves where supercooled water results in showers of small ice crystals (Leonard et al. 2011). Other adaptations are less clear-cut. The general problem of reduced enzyme activity at lower temperatures remains and presumably there are many ways to mitigate this effect. Cold denaturation is likely also to be problematic and again, experience suggests there are many ways to resolve this problem.

One of the questions that is asked about Antarctic fish is how they would cope in the event that the seas in which they live were to warm. This is another question that is difficult to answer (Somero 2011; Patarnello et al. 2011) but enough data exist to suggest that an increase in temperature itself is survivable. What is likely to be more important are changes in ecological niche associated with any increase in temperature. Such changes may alter habitat and affect interactions with other species and are likely to be essentially imponderable.

However, studies of cold adaptations in Antarctic fish have been very instructive in understanding the biochemistry and physiology of protein function, and work in this area should be encouraged.

References

Abele D, Puntarulo S (2004) Formation of reactive species and induction of antioxidant defence systems in polar and temperate marine invertebrates and fish. Comp Biochem Physiol 138:405–415. doi:10.1016/j.cbpb.2004.05.013

Acierno R, Maffia M, Rollo M, Storelli C (1997) Buffer capacity in the blood of the hemoglobinless antarctic fish *Chionodraco hamatus*. Comp Biochem Physiol A 118:989–992. doi:10.1016/S0300-9629(97)86787-4

Andersson AJ, Mackenzie FT, Bates NR (2008) Life on the margin: implications of ocean acidification on Mg-calcite, high latitude and cold-water marine calcifiers. Mar Ecol Progr Ser 373:265–273. doi:10.3354/meps07639

Axelsson M, Davison W, Forster ME, Farrell AP (1992) Cardiovascular responses of the red-blooded antarctic fishes *Pagothenia bernacchii* and *P. borchgrevinki*. J Exp Biol 167:179–201

Becker K, Wöhrmann A, Rahmann H (1995) Brain gangliosides and cold-adaptation in high-antarctic fish. Biochem Syst Ecol 23:695–707. doi:10.1016/0305-1978(95)00086-0

Beers JM, Borley KA, Sidell BD (2010) Relationship among circulating hemoglobin, nitric oxide synthase activities and angiogenic poise in red- and white-blooded Antarctic notothenioid fishes. Comp Biochem Physiol A Mol Integr Physiol 156:422–429. doi:10.1016/j.cbpa.2010.03.027

Buckley BA, Somero GN (2009) cDNA microarray analysis reveals the capacity of the cold-adapted Antarctic fish *Trematomus bernacchii* to alter gene expression in response to heat stress. Polar Biol 32:403–415. doi:10.1007/s00300-008-0533-x

Buckley BA, Place SP, Hofmann GE (2004) Regulation of heat shock genes in isolated hepatocytes from an Antarctic fish, *Trematomus bernacchii*. J Exp Biol 207:3649–3656

Carpenter CM, Hofmann GE (2000) Expression of 70 kDa heat shock proteins in antarctic and New Zealand notothenioid fish. Comp Biochem Physiol A Mol Integr Physiol 125:229–238

Chapelle G, Peck LS (1999) Polar gigantism dictated by oxygen availability. Nature 399: 114–115. doi:10.1038/20099

Chen W-J, Bonillo C, Lecointre G (1998) Phylogeny of the channichthyidae (Notothenioidei, Teleostei) based on two mitochondrial genes. In: di Prisco D, Pisano E, Clarke A (eds) Fishes of Antarctica: a biological overview. Springer, Milan, pp 287–298

Clarke A, Johnston NM (1999) Scaling of metabolic rate with body mass and temperature in teleost fish. J Animal Ecol 68:893–905. doi:10.1046/j.1365-2656.1999.00337.x

Collins T, Roulling F, Piette F, Marx J-C, Feller G, Gerday C, D'Amico S (2008) Fundamentals of cold-adapted enzymes. In: Margesin R, Schinner F, Marx J-C, Gerday C (eds) Psychrophiles: from biodiversity to biotechnology. Springer, Berlin, pp 211–227. doi:10.1007/978-3-540-74335-4

Cossins AR, Macdonald AG (1989) The adaptation of biological membranes to temperature and pressure: fish from the deep and cold. J Bioenerg Biomembranes 21:115–135

Crockett EL, Sidell BD (1990) Some pathways of energy metabolism are cold adapted in antarctic fishes. Physiol Zool 63:472–488

Davison W, Axelsson M, Forster ME, Nilsson S (1995) Cardiovascular responses to acute handling stress in the Antarctic fish Trematomus bernacchii are not mediated by circulatory catecholamines. Fish Physiol Biochem 14:253–257

Detrich HW, Parker SK, Williams RC, Nogales E, Downing KH (2000) Cold adaptation of microtubule assembly and dynamics—structural interpretation of primary sequence changes present in the alpha- and beta-tubulins of antarctic fishes. J Biol Chem 275:37038–37047. doi:10.1074/jbc.M005699200

DeVries AL, Cheng C-HC (2005) Antifreeze proteins and organismal freezing avoidance in polar fishes. In: Anthony PF, Steffensen JF (eds) Physiology of polar fishes, 22nd edn. Academic, New York, pp 155–201. doi:10.1016/S1546-5098(04)22004-0

di Prisco G, Macdonald JA, Brunori M (1992) Antarctic fishes survive exposure to carbon monoxide. Experientia 48:473–485. doi:10.1007/BF01928166

Dill KA, Ozkan SB, Shell MS, Weikl TR (2008) The protein folding problem. Annu Rev Biophys 37:289–316. doi:10.1146/annurev.biophys.37.092707.153558

Eastman JT (1993) Antarctic fish biology: evolution in a unique environment. Academic, San Diego

Eastman JT (2000) Antarctic notothenioid fishes as subjects for research in evolutionary biology. Antarct Sci 12:276–287

Eastman JT, Clarke A (1998) Radiations of Antarctic and non-Antarctic fish. In: di Prisco G, Pisano E, Clarke A (eds) Fishes of Antarctica: a biological overview. Springer, Milan, pp 3–26

Egginton S (1997) A comparison of the response to induced exercise in red- and white-blooded antarctic fishes. Comp Biochem Physiol B 167:129–134

Feder ME, Hofmann GE (1999) Heat-shock proteins, molecular chaperones, and the stress response: evolutionary and ecological physiology. Annu Rev Physiol 61:243–282. doi:10.1146/annurev.physiol.61.1.243

Feller G (2003) Molecular adaptations to cold in psychrophilic enyzmes. Cell Mol Life Sci 60:648–662

Feller G, Gerday C (1997) Adaptations of the hemoglobinless antarctic icefish (Channichthyidae) to hypoxia tolerance. Comp Biochem Physiol 118A:981–987

Fields PA, Houseman DE (2004) Decreases in activation energy and substrate affinity in cold-adapted A4-lactate dehydrogenase: evidence from the Antarctic notothenioid fish *Chaenocephalus aceratus*. Mol Biol Evol 21:2246–2255

Fields PA, Somero GN (1998) Hot spots in cold adaptation: localized increases in conformational flexibility in lactate dehydrogenase A4 orthologs of antarctic notothenioid fishes. Proc Natl Acad Sci USA 95:11476–11481. doi:10.1073/pnas.95.19.11476

Georlette D, Blaise V, Collins T, D'Amico S, Gratia E, Hoyoux A, Marx JC, Sonan G, Feller G, Gerday C (2004) Some like it cold: biocatalysis at low temperatures. FEMS Microbiol Rev 28:25–42. doi:10.1016/j.femsre.2003.07.003

Glickman MH, Adir N (2004) The proteasome and the delicate balance between destruction and rescue. PLoS Biol 2:E13. doi:10.1371/journal.pbio.0020013

Gon O, Heemstra PC (1990) Fishes of the southern ocean. JLB Smith Institute of Ichthyology, Grahamstown

Gonzalez-Cabrera PJ, Dowd F, Pedibhotla VK, Rosario R, Stanley-Samuelson D, Petzel D (1995) Enhanced hypo-osmoregulation induced by warm-acclimation in antarctic fish is mediated by increased gill and kidney Na^{+}/K^{+}-ATPase activities. J Exp Biol 198:2279–2291

Grove TJ, Hendrickson JW, Sidell BD (2004) Two species of Antarctic icefishes (genus *Champsocephalus*) share a common genetic lesion leading to the loss of myoglobin expression. Polar Biol 27:579–585. doi:10.1007/s00300-004-0634-0

Haschemeyer AEV (1980) Temperature effects on protein metabolism in cold-adapted fishes. Antarctic J US 15:147–149

Hochachka PW, Somero GN (2002) Biochemical adaptation: mechanism and process in physiological evolution. Oxford University Press, Oxford

Hofmann GE, Buckley BA, Airaksinen S, Keen JE, Somero GN (2000) Heat-shock protein expression is absent in the Antarctic fish *Trematomus bernacchii* (Family Nototheniidae). J Exp Biol 203:2331–2339

Holland LZ, McFall-Ngai M, Somero GN (1997) Evolution of lactate dehydrogenase-A homologs of barracuda fishes (genus *Sphyraena*) from different thermal environments: differences in kinetic properties and thermal stability are due to amino acid substitutions outside the active site. Biochemistry 36:3207–3215. doi:10.1021/bi962664k

Johnston IA, Calvo J, Guderley H, Fernandez D, Palmer L (1998) Latitudinal variation in the abundance and oxidative capacities of muscle mitochondria in perciform fishes. J Exp Biol 201:1–12

Kassahn KS, Crozier RH, Portner H-O, Caley MJ (2009) Animal performance and stress: responses and tolerance limits at different levels of biological organisation. Biol Rev Camb Philos Soc 84:277–292. doi:10.1111/j.1469-185X.2008.00073.x

Kawall HG, Torres JJ, Sidell BD, Somero GN (2002) Metabolic cold adaptation in Antarctic fishes: evidence from enzymatic activities of brain. Mar Biol 140:279–286. doi:10.1007/s002270100695

Kelley JL, Aagaard JE, MacCoss MJ, Swanson WJ (2010) Functional diversification and evolution of antifreeze proteins in the antarctic fish *Lycodichthys dearborni*. J Mol Evol 71:111–118. doi:10.1007/s00239-010-9367-6

Krogh A (1914) The quantitative relation between temperature and standard metabolism in animals. Int Z Phys-Chem Biol 1:491–498

Leonard GH, Langhorne PJ, Williams MJM, Vennell R, Purdie CR, Dempsey DE, Haskell TG, Frew RD (2011) Evolution of supercooling under coastal Antarctic sea ice during winter. Antarctic Sci 1:1–11. doi:10.1017/S0954102011000265

Lin J-J, Somero GN (1995) Temperature-dependent changes in expression of thermostable and thermolabile isozymes of cytosolic malate dehydrogenase in eurythermal goby fish *Gillicthys mirabilis*. J Exp Biol 198:551–560

Lindquist S (2010) Protein folding sculpting evolutionary change. Cold Spring Harb Symp Quant Biol 74:doi: 10.1101/sqb.2009.74.043

Marshall CJ (1997) Cold-adapted enzymes. Trends Biotechnol 15:359–364

Marshall CJ, Johnston NM, Murray BW, Brown PM, Verghese AI (2003) Does the enzyme citrate synthase from several Antarctic fish show evidence of cold adaptation? In: Huiskes AL, Gieskes WWC, Rozema J, Schorno RML, van der Vies SM, Wolff WJ (eds) Symposium of the VIII SCAR biology meeting. Backhuys Publishers, Leiden, pp 102–106

Marx JC, Collins T, D'Amico S, Feller G, Gerday C (2007) Cold-adapted enzymes from marine Antarctic microorganisms. Mar Biotechnol (NY) 9:293–304. doi:10.1007/s10126-006-6103-8

Nakamura I, Inada T, Takeda M, Hatanaka H (1986) Important fishes trawled off Patagonia. Japan Marine Fishery Resource, Tokyo

Near TJ, Cheng C-HC (2008) Phylogenetics of notothenioid fishes (Teleostei: Acanthomorpha): inferences from mitochondrial and nuclear gene sequences. Mol Phylogenet Evol 47: 832–840. doi:10.1016/j.ympev.2007.11.027

Near TJ, Pesavento JJ, Cheng C-HC (2003) Mitochondrial DNA, morphology, and the phylogenetic relationships of Antarctic icefishes (Notothenioidei: Channichthyidae). Mol Phylogen Evol 28:87–98. doi:10.1016/S1055-7903(03),00029-0

Near TJ, Parker SK, Detrich III HW (2006) A genomic fossil reveals key steps in hemoglobin loss by the antarctic icefishes. Mol Biol Evol 23:2008–2016. doi:10.1093/molbev/msl071

Nogales E (2000) Structural insights into microtubule function. Annu Rev Biochem 69:277–302. doi:10.1146/annurev.biochem.69.1.277

Nogales E, Downing KH (2008) Tubulin and microtubule structures. In: Fojo T (ed) The role of microtubules in cell biology, neurobiology, and oncology. Humana Press, Totowa, pp 211–225. doi:10.1007/978-1-59745-336-3

O'Brien KM (2011) Mitochondrial biogenesis in cold-bodied fishes. J Exp Biol 214:275–285. doi:10.1242/jeb.046854

O'Brien KM, Sidell BD (2000) The interplay among cardiac ultrastructure, metabolism and the expression of oxygen-binding proteins in Antarctic fishes. J Exp Biol 203:1287

Patarnello T, Verde C, di Prisco G, Bargelloni L, Zane L (2011) How will fish that evolved at constant sub-zero temperatures cope with global warming? Notothenioids as a case study. BioEssays 33:260–268. doi:10.1002/bies.201000124

Pellegrino D, Palmerini CA, Tota B (2004) No hemoglobin but NO: the icefish (*Chionodraco hamatus*) heart as a paradigm. J Exp Biol 207:3855–3864. doi:10.1242/jeb.01180

Peters SE, Carlson AE, Kelly DC, Gingerich PD (2010) Large-scale glaciation and deglaciation of Antarctica during the late eocene. Geology 38:723–726. doi:10.1130/G31068.1

Petricorena ZL, Somero GN (2007) Biochemical adaptations of notothenioid fishes: comparisons between cold temperate South American and New Zealand species and Antarctic species. Comp Biochem Physiol A Mol Integr Physiol 147:799–807. doi:10.1016/j.cbpa.2006.09.028

Rodnick KJ, Sidell BD (1997) Structural and biochemical analyses of cardiac ventricular enlargement in cold-acclimated striped bass. Am J Physiol 273R:252–258

Scudiero R, Carginale V, Riggio M, Capasso C, Capasso A, Kille P, di Prisco G, Parisi E (1997) Difference in hepatic metallothionein content in Antarctic red-blooded and haemoglobinless fish: undetectable metallothionein levels in haemoglobinless fish is accompanied by accumulation of untranslated metallothionein mRNA. Biochem J 322:207–211

Seebacher F, Davison W, Lowe CJ, Franklin CE (2005) A falsification of the thermal specialization paradigm: compensation for elevated temperatures in Antarctic fishes. Biol Lett 1:151–154. doi:10.1098/rsbl.2004.0280

Sharpe M, Love C, Marshall C (2001) Lactate dehydrogenase from the Antarctic eelpout, *Lycodichthys dearborni*. Polar Biol 24:258–269. doi:10.1007/s003000000206

Sidell BD (1991) The physiological roles of high lipid content in tissues of Antarctic fish species. In: di Prisco G, Maresca B, Tota B (eds) Biology of Antarctic fish. Springer, Berlin, pp 220–231

Sidell BD (1998) Intracellular oxygen diffusion: the roles of myoglobin and lipid at cold body temperature. J Exp Biol 201:1119–1128

Sidell BD, O'Brien KM (2006) When bad things happen to good fish: the loss of hemoglobin and myoglobin expression in Antarctic icefishes. J Exp Biol 209:1791–1802. doi:10.1242/jeb.02091

Sokolova IM, Portner H-O (2003) Metabolic plasticity and critical temperatures for aerobic scope in a eurythermal marine invertebrate (*Littorina saxatilis*, Gastropoda: Littorinidae) from different latitudes. J Exp Biol 206:195–207

Somero GN (1995) Proteins and temperature. Annu Rev Physiol 57:43–68. doi:10.1146/annurev.ph.57.030195.000355

Somero GN (2011) Invited review: comparative physiology: a "crystal ball" for predicting consequences of global change. Am J Physiol Regul Integr Comp Physiol. doi:10.1152/ajpregu.00719.2010

Somero GN, DeVries AL (1967) Temperature tolerance of some antarctic fishes. Science 156:257–258. doi:10.1126/science.156.3772.257

Temussi PA (2011) Cold denaturation and protein stability. In: Brnjas-Kraljević J, Pifat-Mrzljak G (eds) Supramolecular structure and function 10. Springer Netherlands, Dordrecht, pp 75–85. doi:10.1007/978-94-007-0893-8

Todgham AE, Hoaglund EA, Hofmann GE (2007) Is cold the new hot? Elevated ubiquitin-conjugated protein levels in tissues of Antarctic fish as evidence for cold-denaturation of proteins in vivo. J Comp Physiol B 177:857–866. doi:10.1007/s00360-007-0183-2

Tota B, Angelone T, Mancardi D, Cerra MC (2011) Hypoxia and anoxia tolerance of vertebrate hearts: an evolutionary perspective. Antioxid Redox Signal 14:851–862. doi:10.1089/ars.2010.3310

Urschel MR, O'Brien KM (2008) High mitochondrial densities in the hearts of Antarctic icefishes are maintained by an increase in mitochondrial size rather than mitochondrial biogenesis. J Exp Biol 211:2638–2646. doi:10.1242/jeb.018598

Weinstein RB, Somero GN (1998) Effects of temperature on mitochondrial function in the Antarctic fish *Trematomus bernacchii*. J Comp Physiol B: Biochem, Systemic Environ Physiol 168:190–196. doi:10.1007/s003600050136

Williams WP (1990) Cold-induced lipid phase transitions. Phil Trans R Soc London Ser B 326:555–570. doi:10.1098/rstb.1990.0031
Wittenberg JB, Wittenberg BA (1990) Mechanisms of cytoplasmic hemoglobin and myoglobin function. Annu Rev Biophys Biophys Chem 19:217–241
Woods HA, Moran AL, Arango CP, Mullen L, Shields C (2009) Oxygen hypothesis of polar gigantism not supported by performance of Antarctic pycnogonids in hypoxia. Proc Biol Sci 276:1069–1075. doi:10.1098/rspb.2008.1489
Zachos J, Pagani M, Sloan L, Thomas E, Billups K (2001) Trends, rhythms, and aberrations in global climate 65 Ma to present. Science 292:686–693. doi:10.1126/science.1059412

Chapter 9
Understanding Adaptations and Responses to Change in Antarctica: Recent Physiological and Genomic Advances in Marine Environments

Lloyd S. Peck and Melody S. Clark

Antarctic marine environments are amongst the most extreme on Earth in several characteristics. They combine the globally lowest and most stable temperatures (Fig. 9.1) with the highest oxygen content and the greatest variability in other variables such as light intensity, ice cover and phytoplankton productivity (Peck et al. 2006). The extremely long time over which these conditions have existed (Zachos et al. 2001; Billups et al. 2008), at least in some parts of Antarctica, has allowed the evolution of adaptations in the marine fauna not seen elsewhere. These factors have elicited strong interest from biologists for over 100 years.

The recent rapid climate change has produced a great focus on Antarctica because of its pivotal role in the Earth system, and also because temperature rises in both the sea and air around the Antarctic Peninsula are amongst the fastest on Earth (e.g. Meredith and King 2005). There have been few documented cases of biological responses associated with climate change in Antarctica. Those that have occurred have been almost exclusively along the Antarctic Peninsula and include the second largest natural global negative feedback in new coastal areas of biological productivity (Peck et al. 2010a), southern shifts in penguin distributions (Forcada and Trathan 2009), and southerly shifts in phytoplankton biomass and size (Schofield et al. 2010). Interest has also been enhanced by the International Polar Year, and this has included a markedly increased effort into quantifying and understanding biodiversity through bodies such as the Census of Antarctic Marine Life (CAML: http://www.caml.aq/barcoding/index.html) and the Marine Barcode of Life (http://www.marinebarcoding.org/). These initiatives and imperatives have

L. S. Peck (✉) · M. S. Clark
British Antarctic Survey, Natural Environment Research Council,
High Cross, Madingley Road, Cambridge, CB3 0ET, UK
e-mail: lspe@bas.ac.uk

M. S. Clark
e-mail: mscl@bas.ac.uk

G. di Prisco and C. Verde (eds.), *Adaptation and Evolution in Marine Environments, Volume 1*, From Pole to Pole, DOI: 10.1007/978-3-642-27352-0_9,

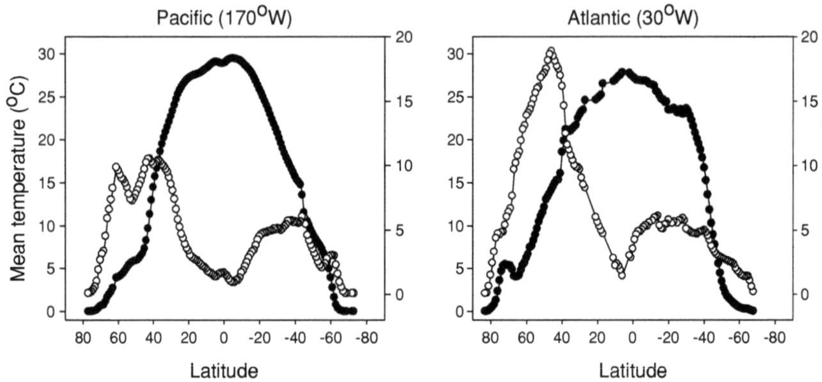

Fig. 9.1 Mean annual temperature (●, °C) and annual range (maximum–minimum experienced temperatures, ○, °C) for transects through the Pacific and Atlantic oceans from the high Arctic to Antarctica. The polar regions have the lowest and most stable temperatures. The temperate regions, with most human inhabitation are the most variable. Figure modified from Clarke and Gaston (2006)

also lead to large efforts being made to improve understanding of physiological adaptations to Antarctic marine systems. The rapid technological advances in genomic technologies have begun to be applied in Antarctica and the first 454 pyrosequencing efforts have been published on the infaunal clam *Laternula elliptica* (Clark et al. 2010) and krill, *Euphausia superba* (Clark et al. 2011). These technological advances are providing further opportunities to make a step change across several biological fields, as molecular biology is gradually being integrated as a standard tool in a wide array of applications from understanding basic adaptations, through responses to change in sea temperature and acidity, and then on to providing enhanced data for understanding gene flow between populations and evolution. These integrated approaches have supported a dramatic increase in efforts to identify novel genes and proteins, essential for life in this extreme environment. The molecular investigations are embedded within well characterised physiological systems, hence we provide an up-dated overview of physiological adaptations to life in the cold, with enhanced detail on gene level studies where known.

9.1 Slowed Growth and Development

Work in the last 3 decades showed that Antarctic marine species generally exhibit slow rates of growth, development and metabolic rate (Peck 2002; Peck et al. 2006, 2007). Barnes et al. (2007) investigated 6 species of Antarctic bryozoa and reported the slowest growth rates found anywhere on Earth for this taxon. The rates reported were very variable between species and genera, but were an order of magnitude slower than tropical species. A similar general pattern of slowed growth at higher latitude has been demonstrated in the past for fish (e.g. Kock and Everson 1998). Recent studies, however, have found relatively fast growth in the Spiny Icefish *Chaenodraco wilsoni* (La Mesa et al. 2009), showing it lives only to 4 or 5 years of age. Another icefish, *Chionobathyscus dewitti* on the other hand lives to 11 years and grows at a more moderate rate (Sutton et al. 2008), and *Chionodraco rastrospinosus* lives to at least 8 years (La Mesa and Ashford 2008).

Embryonic and larval development in Antarctic marine ectotherms is slow or very slow compared to temperate and tropical species, as demonstrated by work primarily on broadcast spawning echinoderms and molluscs (Bosch et al. 1987; Stanwell-Smith and Peck 1998; Peck et al. 2006). More recent studies have shown development is markedly slowed in brooding gastropods, with some species taking over 2 years to complete development. Brooding periods in Antarctic species to date range from 20 to 110 weeks compared to <1–26 weeks for tropical, temperate and cool temperate species (Peck et al. 2006). Recent investigations have documented the brooding habit of deep sea corals in Antarctica (Waller et al. 2010), the importance of egg lipids in developmental success in echinoderms (Moore and Manahan 2007), and higher levels of DNA damage in Antarctic echinoderm embryos on exposure to UV than New Zealand species (Lamare et al. 2007). Following on from the latter study, gene expression approaches were taken to identify activity of the DNA repair photolyase enzyme in the Antarctic sea urchin and correlate this with environmental conditions. Whilst normal repair mechanisms operate in this species, the rate at which they operate can be insufficient given increased UV-B exposure in shallow waters from the ozone hole (Isley et al. 2009; Lister et al. 2010). At the translational level, fast rates of protein synthesis in larvae of the echinoid *Sterechinus neumayeri* were reported by Pace and Manahan (2007). Subsequently Pace et al. (2010) demonstrated rates of protein elongation by ribosomes from these larvae were as fast as those from temperate species at temperatures 15°C warmer. Calculations suggested that protein synthesis in these larvae is very efficient, and the lowest cost of production for protein synthesis on record (Pace and Manahan 2007), although some care may be needed when interpreting these results as the reported costs appear to be below the theoretical minimum required for making proteins (Bowgen et al. 2007; Fraser and Rogers 2007). Previously, studies on protein ubiquitination and degradation have shown that some Antarctic fish species have unexpected high levels of damaged proteins (Place et al. 2004), hence this whole issue of translational rates, protein production and folding efficiency in the cold is an area ripe for further research.

9.2 Seasonality and Metabolism

Seasonality has been viewed as a major driver of the biology, and especially metabolism, of Antarctic marine species for several decades (e.g. Clarke 1988; Clarke and Peck 1991). Recent work has quantified levels of seasonality in the

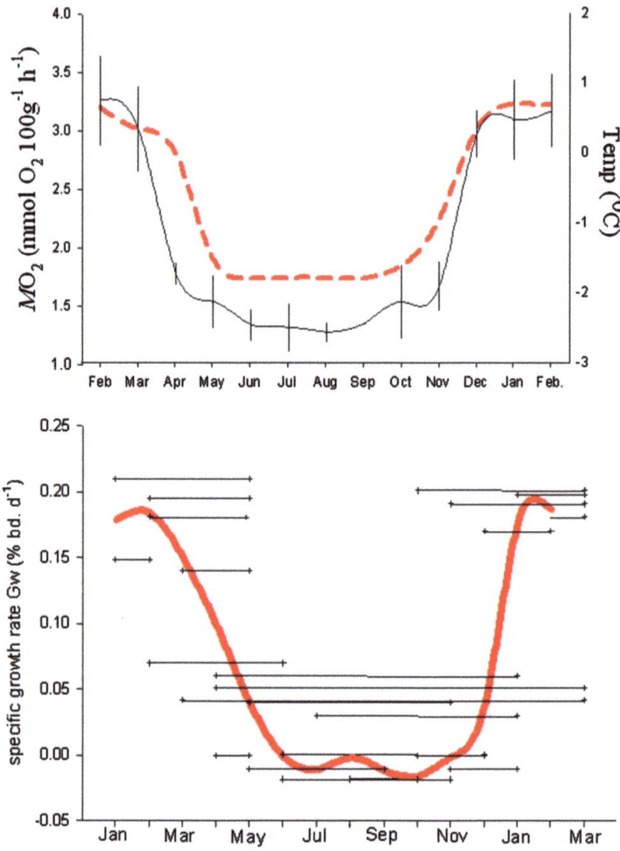

Fig. 9.2 *Upper graph*: free-ranging metabolic rate of *N. coriiceps*. The mean monthly MO_2 of wild *N. coriiceps* (black line, n = 6) was extrapolated from continual field recordings. Water temperature was measured by an onboard temperature sensor (*red*). *Lower graph*: mass specific growth rate (Gw) of free-ranging and sea-caged *N. coriiceps*. A total of 21 immature adult *N. coriiceps* (4 recaptured twice) were tagged, released and then recaptured between 2nd Jan 2004 and 12th March 2005. The black crosses indicate the date of first and subsequent capture and the average Gw for an individual fish during its days at liberty (connecting *black* line). The red line connects calculations of Gw (n = 6) measured from sea-caged fish every 8 weeks. For clarity the standard error is not shown on the graph, which from May to Oct was 0.003% bd. wt. d21, and from Nov to Apr 0.031% bd. wt. d21. Figures reproduced from Campbell et al. (2008)

metabolism of 5 species of shallow water scavengers and carnivores (Obermüller et al. 2010). Seasonal patterns of metabolism were less extreme than those previously reported for herbivores, but all of the species studied, except the amphipod *Paraceradocus miersi*, ceased feeding for a period of weeks to months in winter. Metabolic ratios of oxygen consumed to nitrogen excreted (O:N) also varied seasonally, and indicated changes in substrate used to fuel metabolism. Overall the starfish *Odontaster validus* had an O:N ratio of over 230, indicating that lipid or carbohydrate is the predominant substrate used, whereas the nemertean *Parborlasia corrugatus* had a ratio of 9, indicating protein is almost the sole substrate used to fuel metabolic requirements.

Extreme seasonality in activity and metabolism has also been demonstrated in Antarctic fish. In a year-round field study of the benthic *Notothenia coriiceps* Campbell et al. (2009) showed very large seasonality of the biology, similar to changes seen in hibernation or aestivation in mammals and insects. Metabolism varied from 1.3 to 3.3 mmol O_2 per 100 g of fish per hour from winter to summer and specific growth rate ranged from zero in winter to around 0.2% body weight per day in summer (Fig. 9.2).

This intense seasonality of the biology of a benthic predator is somewhat unexpected, and may reflect the balance of success of a visual predator when light levels are very low. The changes in metabolism and growth of *N. coriiceps* are similar to those only seen previously in herbivorous invertebrates such as the bryozoans *Isoseculiflustra tenuis* and *Kymella polaris* (Barnes and Peck 2005) and the urchin *Sterechinus neumayeri* (Brockington and Peck 2001).

In recent studies seasonality of activity has been attributed to be an important factor in governing levels of arginine metabolism in the bivalve *L. elliptica* (Rodrigues et al. 2009). *L. elliptica* has very strong seasonal variation in activity. When not pumping water through the mantle cavity this species retracts its siphons below the sediment surface. In summer it spends 86% of its time with siphons open and pumping, but in winter this is only 32% (Morley et al. 2007a). Significant seasonality in metabolism, growth, and activity have also been reported recently in marine bacteria (Pearce et al. 2007).

9.3 Temperature Compensation

Recent cold adaptation mechanism studies in invertebrates and fish have covered a range of mechanisms. Pörtner et al. (2007) provided an integrated view across scales, from molecular to whole organism, of adaptation to cold extremes in Antarctic marine ectotherms. Since then additional novel methods of temperature compensation have been described. Morley et al. (2007b) showed the Antarctic clam, *L. elliptica* burrows at the same rate as species from warmer latitudes. This is only the second activity out of 8 studied previously in Antarctic species that shows such compensation (Peck et al. 2006). The compensation is achieved by an increase in the size of the burrowing organ, the foot, which is 2–3 times larger than that of warmer water species (Morley et al. 2007b). The only previous activity documented for Antarctic species, sustained swimming in fish is achieved from an increase in mitochondrial density in the red muscle (Johnston et al. 1998). Similar increases in mitochondrial density appear to occur in articulated brachiopod adductor muscles (Lurman et al. 2010), but not in the foot muscle of the Laternulid clams (Lurman et al. in review). On the limited evidence available it seems that increasing mitochondrial density with latitude may be a widespread, but not universal adaptation to low temperature.

9.4 Gene Level Adaptations

It has been known since the early 1990s that specific amino acid changes have occurred in the proteins of Antarctic marine species that enhance performance in the cold. For, example, hydrophobic remodelling and increased molecule flexibility were documented in tubulin dimers, A_4-lactate dehydrogenases, microtubule motors and the endoplasmic reticulum protein translocation channels of Antarctic fish (Detrich et al. 1989, 1992; Fields and Somero 1998; Römisch et al. 2003). There has been significant recent progress in this area of research in Antarctica, both for microbial groups and also for invertebrates and fish.

For microbial groups new molecular and proteomic approaches have allowed the identification of adaptations at levels not previously possible. Here Pucciarelli et al. (2009) used a comparative genomic approach to show that, similarly to the situation in the fish, there was cold adaptation of the mechanisms for producing tubulins in the protist *Euplotes focardii* from the Ross Sea. This system provided additional insights into protein-level structural changes that probably enhance microtubule nucleation and polymerization in a cold energy poor environment. They also demonstrated that the regulation of gene expression in the cold differed from that of temperate congeners. Ting et al. (2010) used a quantitative proteomic analysis to identify a low temperature adapted protein folding system involving the proteins GroESL, DnaK, DnaJ, GrpE, SecB, ClpB and PPIase, in the bacterium *Sphingopyxis alaskensis*. They also noted differences between warm and cold cultures of this bacterium in enzymes in fatty acid metabolism and energy generation, and also a de novo synthesis of polyunsaturated fatty acids in the membrane and cell wall. In terms of complexity, cold water adaptations are easier to describe and manipulate in micro-organisms. To date most gene level descriptions of adaptations in the macroinvertebrates and lower vertebrates (fish) are restricted to single genes or related gene family members.

9.5 Antifreeze Proteins

Perhaps the best known adaptation of Antarctic fish to their extreme environmental conditions is the production of antifreeze proteins to protect tissues from ice crystal damage. Antifreezes were first identified in Antarctic Notothenioids in the 1960s (DeVries and Wohlschlag 1969). Since then the concentrations and tissues secreting antifreeze proteins and glycoproteins (AFGP) have been studied and the evolution of AFGP from a trypsinogen-like precursor has also been elucidated (Chen et al. 1997). Multiple gene copies occur for all of the major types of antifreeze protein and glycoprotein so far identified (Zhang et al. 2009). The recent rapid improvement in genomic technologies has allowed significant progress in this area. Genome wide studies of adaptation to extreme low marine temperature in Antarctica have been begun (Chen et al. 2008). We also now know that some AFGPs evolved by neofunctionalization (Deng et al. 2010), that there are AFGP deficient early larval stages that are freeze resistant, indicating other resistance mechanisms in early stages, and that the liver is not the source for AFGP in notothenioid blood (Cheng and Detrich 2007; Bilyk and DeVries 2010). The new technologies have also allowed the identification of AFGPs in a wide range of other taxa, including marine diatoms (Gwak et al. 2010), bacteria (Wilson and Walker 2010) and calanoid copepods. In the latter AFGPs have been suggested to be acquired by horizontal gene transfer from other species (Kiko 2010).

9.6 Haemoglobin

Since the first description of clear blooded Antarctic fish (Ruud 1954), there has been a fascination with understanding oxygen delivery in Southern Ocean species. Initial discoveries of the radical deletion of haemoglobin genes (di Prisco et al. 2002) in the icefish (Channichthyidae) have now been supplemented by recent work showing altered oxygen carrying capacities in red-blooded Notothenioids. Here Verde et al. (2006) and di Prisco et al. (2007) identified that a decreased overall amount of haemoglobin in tissues and serum combined with an increased number of haemoglobin types is correlated with life in the cold Southern Ocean, and this appears adaptive in response to the low temperature and increased oxygen in the environment (see Chap. 11 Verde et al. 2012). The low temperature also increases oxygen solubility in the circulating haemolymph, and this further decreases the requirement for haemoglobin as an oxygen carrier. Further adaptations in the red-blooded fish include reduced rates of haemoglobin synthesis, the regulation of this synthesis by allosteric effectors and changes in oxygen affinity and pH sensitivity (Verde et al. 2008; Giordano et al. 2010). Very recently oxygen binding molecules, neuroglobins, have been identified in neurons of Antarctic fish (Cheng et al. 2009a, b). Neuroglobins are known to play a protective role in neurons during hypoxic stress, as both neuronal hypoxia and cerebral ischemia induce neuroglobin expression in Antarctic notothenioids. These genes were first found in mammals, and have been shown to be in various temperate and tropical fish. They have now been shown to be widely present in both red-blooded Antarctic fish and also in clear-blooded Channichthyid icefish (Cheng et al. 2009b). This is an interesting adaptation in the icefish, because they have lost other oxygen binding pigments and hence the intriguing possibility is presented for these molecules to contribute to increased oxygen capacity in these species.

9.7 Heat-Shock Proteins

Another area of ecophysiology that has progressed rapidly because of the use of novel genomic technologies has been the investigation of the heat shock response (HSR) in Antarctic marine species (reviewed in Clark and Peck 2009a). The initial observation of Hofmann et al. (2000) that Antarctic Notothenioid fish lacked the heat shock response, previously seen as ubiquitous in macroinvertebrates and vertebrates, stimulated further work on this subject. Later studies found that in Antarctic fish there was no up-regulation of gene expression for the family of heat shock proteins (HSP) either on exposure to the heavy metal cadmium, or to an inhibitor of proteasome activity (MG132). Both of these lead to the accumulation or production of misfolded proteins and commonly elicit the upregulation of HSP genes in temperate and tropical marine species (Buckley et al. 2004). This led to the development of the paradigm that Antarctic marine species are unusual in that they lack HSR. Recent molecular studies have demonstrated that the situation is more complex (Fig. 9.3).

The Antarctic amphipod *Paraceradocus gibber* and star fish *Odontaster validus* also lack an HSR (Clark et al. 2008a), but two molluscs, the clam *L. elliptica* and the limpet *Nacella concinna*, on the other hand do exhibit upregulation of HSP genes when heated (Clark et al. 2008b). Their HSR is, however, only elicited when warmed to above 10°C, temperatures not seen in the Southern Ocean for millions of years. It was thus thought the observed HSR was redundant, but further work on the limpet showed that intertidal specimens increase HSP gene expression in the field in response to the complex stresses of a tidal cycle (Clark et al. 2008c). More recently the stress response in the limpet has been shown to be even more complex (Clark and Peck 2009b). The duplicated HSP70 genes (HSPA and B) responded to acute heat shocks, a related HSP70 family member, glucose-regulated protein (GRP78) acts as a generalised stress response, and the heat shock protein cognate (HSC70) is the major chaperone invoked in response to long-term stresses of varying types. Clark and Peck (2009b) also demonstrated a decrease in expression of HSP family members in mid winter, when protein folding problems might be expected due to low temperatures. Interestingly this was followed by an upregulation of HSC70 when emerging from the winter period and increasing sea temperatures from -1.9 to -1.6°C. To date it seems that most Antarctic marine species studied lack the classical HSR. A few species do possess the mechanisms to produce an HSR, but the circumstances where it becomes important have only been identified in one species. There is now a need to characterise the stress response more widely in Antarctic marine species to identify any common features and trade-offs associated with the production of this energetically expensive protein (Sørensen and Loeschcke, 2007). The improvement in genomic technologies promises to be an important avenue in developing this approach, particularly with whole transcriptome level descriptions, as heat shock proteins are increasingly viewed as central hubs in complex transcription networks (Csermely 2004).

9.8 Antioxidants

The production of reactive oxygen species (ROS) in cells, primarily in mitochondria causes many problems including damage to proteins and DNA. These problems should be greatest in species living in environments that have high

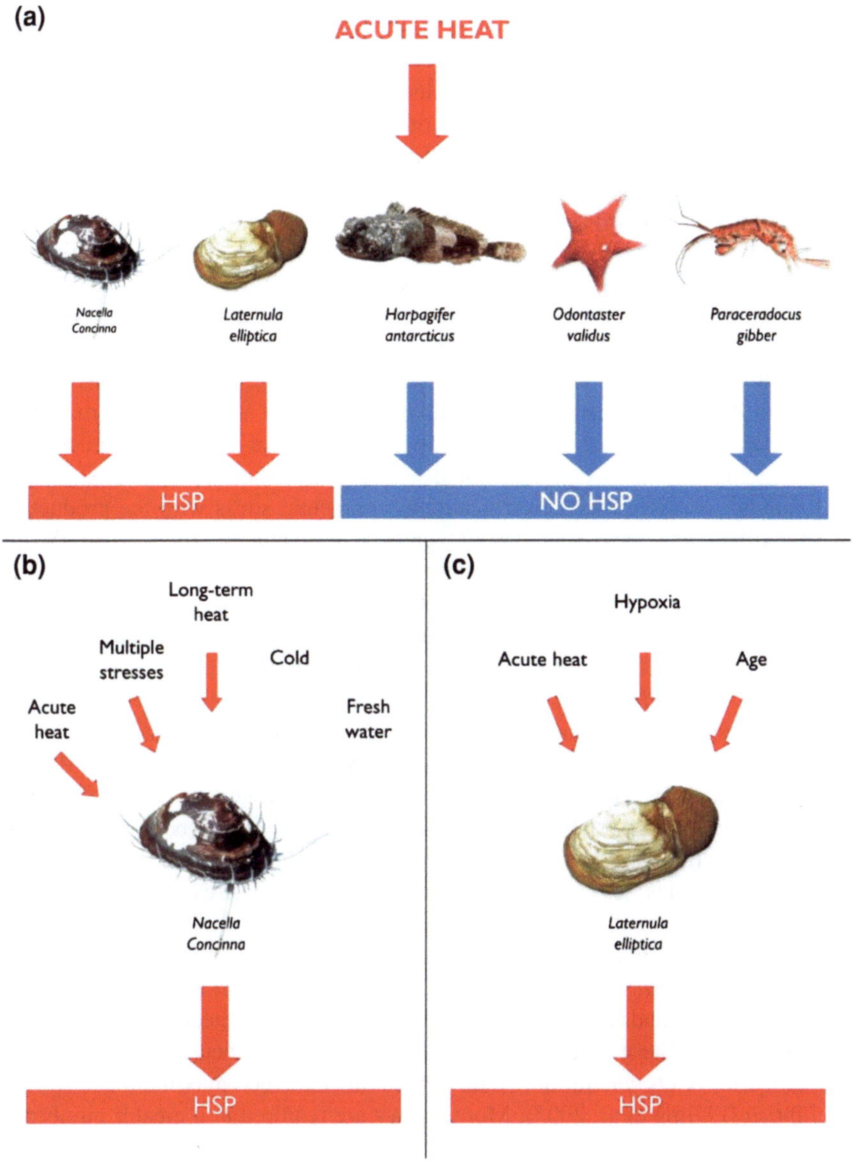

Fig. 9.3 **a** Antarctic Peninsula marine species tested for the heat shock response at both +10 and +15°C. *Red* arrows below the species indicate presence of an inducible HSP70 heat shock response, whilst *grey* arrows indicate that no production of inducible HSP70 s was recorded. **b** Different stress experiments carried out on the limpet *N. concinna*. *Red* arrows above the animal indicate that these stresses elicited an HSP70 response, whilst *grey* arrows indicate that no HSP70 response was produced. **c** Different stress experiments carried out on the clam *Laternula elliptica*. Significance of the colour of the arrows, above the animal. Modified from Clark and Peck (2009a)

oxygen availability and those with rapid oxygen use, or high metabolic rates (Buttemer et al. 2010). Thus the permanently cold waters of the Southern Ocean, with increased levels of dissolved oxygen, may be expected to present an additional ROS threat to the inhabiting species. However, the low metabolic rates of Antarctic animals mean that generation of ROS during energy production by the mitochondria is at much lower rates in these species. Hence, these two variables may be expected to cancel each other out, but this does raise the question of how these animals deal with increased ROS production, for example under conditions that further exacerbate this, such as environmental change. This is a continually productive field of research in two areas: stress and ageing. Initially Abele et al. (1998) showed that elevated temperature caused oxidative stress and an antioxidant response in intertidal specimens of the limpet *N. concinna,* but also subsequently that antioxidant defences play a significant role in the delayed ageing process in Antarctic clams (Philipp et al. 2005; Abele et al. 2009; Buttemer et al. 2010). Hence *L. elliptica* has been proposed as one of the model species for understanding cellular events associated with ageing (Abele et al. 2009).

A number of studies have correlated environmental stress with the production of antioxidant enzymes, such as UV defence in Antarctic microalgae (Wang et al. 2009; Janknegt et al. 2009) and aerial exposure during the tidal cycle of *N. concinna* (Weihe et al. 2010). The expression profiles of specific genes, such as peroxiredoxin have also been correlated with thermal stress in the infaunal bivalve *L. elliptica* (Park et al. 2008; Truebano et al. 2010), but curiously were not observed in the fish *Harpagifer antarcticus* (Thorne et al. 2010). The latter may be linked to the observation that Antarctic fish may have only a limited capacity for decomposing hydrogen peroxide via catalase (Benedetti et al. 2008, 2010), and alternative strategies based on low molecular mass scavengers has been suggested. It would appear that over time, certainly in the fish, there has been an evolution of reduced antioxidant defences (Benedetti et al. 2008, 2010), which would make these species less capable of surviving thermal stress, because of the reduced ability to defend against oxidative damage on warming. The situation with *L. elliptica* is different and emphasises the need to study this process in a variety of species. In this example, antioxidant defences are higher, but also the proton leak across the mitochondrial membrane is higher too, which theoretically mitigates ROS formation. These data, along with the low standard metabolic rate and low mitochondrial H_2O_2 generation, could explain the extended maximum life span in *L. elliptica* (Philipp et al. 2005). More studies are required in a greater number of species.

9.9 Novel Genes and Proteins

In addition to the previously well known gene systems (described above), the improvement in genomic technologies in the past decade has allowed a dramatic improvement in the ability to identify novel genes and proteins that may be specific either to cold adaptation in general or to individual Antarctic species. This field has significant interest because of the obvious potential for economic benefit, for example, from identifying and exploiting enzymes that work well at low temperatures or novel pharmacologically active compounds. Much of the recent work here has been in microbial targets, and cold active lipases have been identified in bacteria by Parra et al. (2008) and de Pascale et al. (2008). Other novel microbial enzymes identified include hydrolytic enzymes (Acevedo et al. 2008), an alkaline protease (Wang et al. 2010) and a β glucosidase (Makowski et al. 2007). Novel alkane monooxygenase (oil degrading) genes were also identified in Antarctic sediments by Kuhn et al. (2009). An interesting development in this area is the identification of pheromones and pheromone genes in Antarctic protozoa. Vallesi et al. (2009) and La Terza et al. (2009) both described pheromones from the Antarctic ciliate *Euplotes nobilii*. These pheromones play a key role in the mating of different strains of theses protozoa, and induce a switch between the growth stage and the sexual stage of the life cycle.

Secondary metabolites, and their role in chemical defence have been studied in Antarctica for over 35 years (Winston and Bernheimer 1986). During this time a wide range of Antarctic marine species have been demonstrated to possess chemical defences, including ascidians (e.g. Koplovitz et al. 2009), macroalgae (e.g. Amsler et al. 2009) and sponges (Ma 2009). In sponges the incidence of chemical defences along the Antarctic Peninsula is high, and has been measured at 78%, which is higher than in similar tropical surveys (McClintock et al. 2010). Secondary metabolites have also been shown to have antifouling properties in Antarctic sponges, deterring the settlement and establishment of diatoms (Peters et al. 2010). Recent technologies aimed at identifying the end productions of gene transcription events, such as metabolomics, have generally not been applied in this field in Antarctica, but promise much for the future. Additionally, reductions in the costs of whole genome sequence scans using Next Generation Sequencing approaches, could allow the rapid screening of species, and the detection of species possessing such chemical defences at times when they may not be being expressed.

9.10 Genome Level Changes

To date, molecular investigations in Antarctic species have generally been limited to a small number of genes. However, new Next Generation pyrosequencing technologies, have lead to a democratisation of sequencing access, such that it is now possible to generate gigabases (1,000,000,000 bp) of DNA relatively cheaply even in non-model species (Vera et al. 2008). Having said that, routine whole genome assemblies are probably still the domain of taxa with small genomes, such as bacteria. Allen et al. (2009) sequenced the genome of the psychrophilic archaeon *Methanococcoides burtonii*. This allowed them to ask questions of evolution and cold adaptation. The genome plasticity mechanisms nucleotide skew, horizontal gene transfer and transposase activity were identified as important in modifying the genome to facilitate adaptation to the cold, and allowing this species to adapt to lacustrine systems from marine origins.

To date, large-scale EST projects have only been achieved in one Antarctic species, the toothfish: *Dissosticus mawsoni.* Sequencing of 33,560 reads produced 3,114 non-redundant protein families, of which 177 were shown to be over-expressed compared with temperate relatives (Chen et al. 2008). It was suggested that constitutive up-regulation of these transcripts were required for life in the cold. Indeed the presence of genes involved in protein biosynthesis, protein folding and lipid metabolism in this up-regulated list, validated some of the proposed costs of living in permanent cold with the problems of decreased protein folding efficiency and membrane fluidity (Privalov 1990; Hazel 1995). The generation and sequencing of EST libraries has been a very effective and long-standing technique for identifying genes and expression profiles, particularly in species where there is no genome data. However, this technique has largely been superseded. By pyrosequencing technologies which are now capable of generating the long reads needed to produce transcriptomes or gene catalogues in non-model species. Recently the first two 454 transcriptomes were published for Antarctic species; the laternulid clam *Laternula elliptica* and the crustacean, krill, *Euphausia superba* (Clark et al. 2010; 2011). These generated assemblies containing 18,290 and 22,177 non-redundant transcripts respectively. Both were data-mined for different aspects; the clam transcriptome was generated from mantle tissue, the shell secreting organ of the animal with the aim of identifying genes involved in calcium regulation and shell secretion (Clark et al. 2010), whilst the krill was comprehensively analysed for chaperone proteins and anti-oxidants, biochemical pathways identified as important for life in the cold (Clark et al. 2011). As all these data are now publicly available, they represent a significant resource, not only for the researchers who generated the data, but the Antarctic community as a whole. Given the relatively small size of the Antarctic research community, it is vital that all molecular data is deposited in public databases. The new technologies produce far more data than can be analysed by a single group and making it freely available should stimulate a greater interest in Antarctic species from many different research fields, which can only be of benefit to Polar Science.

9.11 Genetic Relatedness and Geneflow

To expand the understanding of organism functioning to the population or ecosystem level, it is essential to evaluate genetic relatedness and gene flow between populations. This is an area of science that is benefitting markedly from, and promises to be enhanced in the near future by the use of modern genomic approaches. A number of recent studies have documented genetic restrictions in Antarctic species. For example, gene flow has been demonstrated to be limited in the crinoid *Promachocrinus kerguelensis* that was previously thought to be a single circum-Antarctic species, but clearly has at least 5 species level clades (Wilson et al. 2007). However, in contrast, the brooding brittle star *Astrotoma agassizii* was shown to have significant connectivity across the Drake Passage, which should be a strong barrier to gene flow (Hunter and Halanych 2008). In a study of nemertean larvae along the Antarctic Peninsula Mahon et al. (2010) identified 20 species. 19 of which have not previously been genetically described. In the fish *Gobionotothen gibberifrons*, strong connectivity over large distances was accomplished via passive larval dispersal; this was established using mitochondrial DNA sequence and eight microsatellite markers (Matschiner et al. 2010).

These studies, using a limited number of genes or neural markers are being expanded by more comprehensive genome level scans using techniques such as Amplified Fragment Length Polymorphisms (AFLPs). These have been used recently to quantify gene flow along the Antarctic Peninsula in the broadcast spawning gastropod *N. concinna* (Hoffman et al. 2010a) and the brooding *Margarella antarctica* (Hoffman et al. 2011). Patterns of population structure were profoundly different between the species indicating more restricted gene flow between populations in the brooding species. Linked to the *Nacella* Peninsula-level population study, Hoffman et al. (2010b) showed that the short range population structure of this mollusc was genetically homogeneous across a 25 m depth range. This species had previously been differentiated into two morphotypes, an intertidal "*polaris*" form and a subtidal "*concinna*" type that is found below 4 m depth (Strebel 1908; Powell 1951). Hence it documents the high level of phenotypic plasticity in the genome of this organism for shell shape. This study highlights the importance of understanding the extent of plastic characters in the genome, as it can help with the interpretation of physiological studies. Hence when Morley et al. (2010a) showed that limpets from different depths had different optimum and maximum temperatures for performing activity (righting), this could be directly correlated with a cline in shell shape and phenotypic plasticity, not population differentiation. To date, the *Nacella* work is one of the few examples correlating function with population diversity in the Antarctic. The numbers of such will naturally increase as we expand our knowledge of physiology, function and genetics across a wider range of species.

AFLPs and genome scans represent a step-wise change in our ability to quantify genetic variation between populations, however, these too, may soon be superseded by the new sequencing technologies with the development and utilisation of pyrosequencing and RAD tags for non-model species (Baird et al. 2008; Emerson et al. 2010). These have yet to be used on Antarctic species, but it is only a matter of time, and one which should not be that far away, given the speed these technologies are moving.

9.12 Environmental Change and Organismal Responses

9.12.1 Temperature Effects

Understanding how species have adapted to life in the cold is a major area of study in Antarctic biology. It is also an essential prerequisite in order to predict how these animals will respond in the face of climate change events, which are rapidly warming certain areas of Antarctica, mainly the Peninsula (Meredith and King 2005). The abilities of organisms to respond to a change in the environment are governed by their physiological flexibility and their capacity to alter the genetic makeup of populations (Somero 2010). Physiological flexibility and population genetic modification is a product of the evolutionary history of the fauna and the specific adaptations of the organisms inhabiting a given environment. Several of the adaptations in Antarctic marine species outlined above are likely to limit abilities to resist or respond the environmental change. This is especially the case for slow physiological rates, slow rates of protein synthesis, enzymes tailored to low temperature activity, a reduced or absent heat shock response, reduced haemoglobin levels in fish, the production of antifreeze and reduced antioxidant defences. From this it might be expected that capacities to resist or adapt to environmental change or insult would be poorer than species from lower latitudes.

There have been several studies of resistance to elevated temperatures by marine species in Antarctica in recent years. Clarke et al. (2007) summarised the understanding of potential and measured responses to change in Antarctic species, showing they have poorer abilities to acclimate to elevated temperature than species from lower latitudes, probably because of their reduced general physiological performance. Since then Peck et al. (2008) showed that when warmed by 3°C every 5–7 days, the starfish *Odontaster validus* ceases feeding at 8–10°C, loses the ability to right itself at around 10°C and survives to 15°C. On the same species Kidawa et al. (2010) showed responses to food odour was reduced at 4–5°C and there was also some loss of motor coordination. Janecki et al. (2010) found that warming from 0 to 5°C impaired the ability of the isopod *Serolis polita* to respond to food odours and also to right. Morley et al. (2009) showed that the population of *N. concinna* living at South Georgia and experiencing a higher temperature regime had lower critical and maximum temperature limits than those at Signy or Rothera on the Antarctic Peninsula. This superficially counter-intuitive finding suggests that in this species capacities are compromised when living at its range edge, indicating problems when adapting to warmer conditions.

A significant finding in this field was the demonstration that the rate of warming has a marked effect on the temperature limit obtained when testing the resilience of marine species (Peck et al. 2009a). When warmed at 1°C per day temperature limits for a range of Antarctic benthic species ranged from 8 to 17.5°C. However, when the rate of warming was slowed to 1°C per month these limits fell to between 1 and 4°C (Fig. 9.4). This makes 2 important points: that care must be taken when interpreting studies of the effects of warming to account for how rapidly the

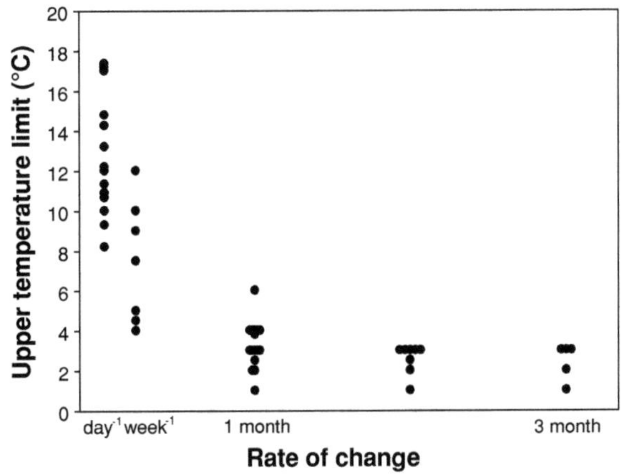

Fig. 9.4 Mean upper temperature tolerance limits vary exponentially with rate of temperature rise. Data shown are upper temperature limits for 25 species of Antarctic marine invertebrates. Modified from Peck et al. (2009a). Data points denote the temperature where 50% of the individuals studied became non-responsive to physical stimuli. Rates of warming ranged from 1 day^{-1} to 1°C every 3 months

warming was conducted, and also that markedly different temperature limits are obtained at different rates of warming. At the very slow rates of warming in the Peck et al. (2009a) study, animals kept at 2–3°C for several months appeared to be acclimating to the elevated temperatures. Acclimation is the mechanism whereby an organisms physiological processes are remodelled, or modulated to function long-term in a new set of conditions, primarily an altered temperature. It is usually observed in experimental systems and often distinguished from acclimatisation which is the seasonal change in physiological state commonly seen in temperate species. Acclimation has been argued to be the most important physiological mechanism deciding success and failure in changing environments (e.g. Stillman 2003; Somero 2010).

There have been several recent investigations of acclimation in Antarctic fish. These have generally shown that Antarctic fish can acclimate to a temperature of 4°C. Thus Robinson and Davison (2008a) showed how oxygen consumption and ventilation rate were compensated after 1 month at 4°C compared to controls at −1°C in *Pagothenia borchgrevinki*, whereas acute rises to 4°C caused increases in metabolic rate. Furthermore Robinson and Davison (2008b) held the same species at 4°C for 6 months with no mortality. However, individuals infected with xcell disease were incapable of acclimating to this temperature. In a study of the responses of blood characteristics to 4-month acclimation at 1.6 or 3.8°C Hudson et al. (2008) showed that haemolymph osmolarity was reduced, but cortisol and haematocrit levels were maintained. They interpreted this as evidence that warm acclimation did not induce a long-term stress response, and that these fish are capable of living long-term at these temperatures. In a comprehensive analysis of the acclimation capacities of species from McMurdo Sound and along the Antarctic Peninsula Bilyk and DeVries (2010) demonstrated small but consistent differences between fish from the two sites in both upper temperature limits and capacity to acclimate. Acute upper temperature limits (CT_{max}) were tested by warming specimens at 0.3°C min^{-1}, and limits identified when fish lost

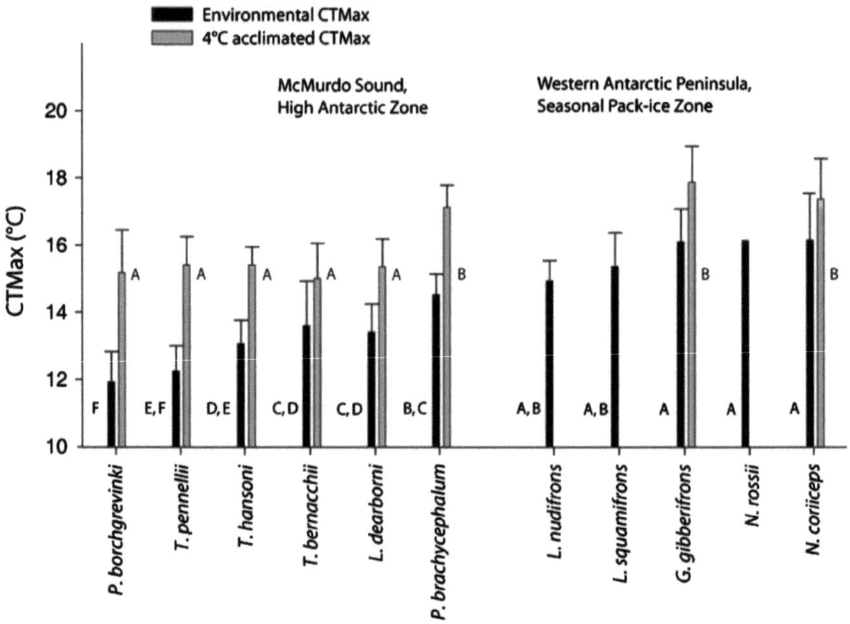

Fig. 9.5 CTMax of Antarctic fishes endemic to both the High Antarctic Zone (McMurdo Sound) and the Seasonal Pack-ice Zone (Antarctic Peninsula). Both environmental CTMax (*black bars*) and 4°C acclimated CTMax (*grey bars*) are shown as their mean with brackets to illustrate standard deviations. Standard deviation is not reported for *N. rossii* as only two specimens were available. Significant groupings were determined using the Student–Newman Keuls test for both environmental and acclimated CTMax and are displayed as the letters next to bars. Within acclimation temperatures species that do not share a letter significantly differ, note though that environmental and acclimated CTMax were tested independently and overlapping letters between these do not have any meaning. Figure reprinted from Comp Biochem Physiol A volume 158, Bilyk and De Vries, Heat tolerance and its plasticity in Antarctic fishes, Fig. 1, page 385 (2010) with permission from Elsevier

equilibrium. When fish with no experimental manipulation (normal field condition) were assessed species from the high Antarctic had CT_{max} values that ranged from 12.0 to 14.5°C, whereas values for those from the western Antarctic Peninsula ranged from 15.1 to 16.2°C (Fig. 9.5). These values probably reflect adaptation to the slightly higher and more variable temperature regime along the Antarctic Peninsula compared to McMurdo Sound over extended evolutionary periods. After acclimation to +4°C CT_{max} values rose by between 1.6 and 3.2°C for the high Antarctic species, but by only 1.2 and 1.8°C in the 2 Antarctic Peninsula species tested. It thus appears, although only based on data for 2 species from one of the sites, that acclimation to 4°C has a stronger effect on resistance to elevated temperature in high Antarctic fish species than those from the Peninsula.

Acclimation in Antarctic marine invertebrates is somewhat more variable. The species with the poorest capacity to acclimate reported to date, the brittle star

Ophionotus victoriae, cannot acclimate to +2°C, a temperature less than 0.5°C above currently experienced maximum temperatures at the site studied (Peck et al. 2009b).

In a further study of 6 invertebrate species, only the gastropod *Marseniopsis mollis* showed clear evidence of acclimation, by an alteration of its acute upper temperature limit after 60 days at +3°C (Peck et al. 2010b). This might suggest that invertebrates in general are less capable of acclimating to elevated temperatures than fish. However, the starfish *O. validus* has been shown to be able to survive over 2 months at 6°C, and to perform normal functions of feeding and activity at this temperature (Peck et al. 2008). In ongoing studies the urchin *S. neumayeri* appears able to complete acclimation after 3–6 months exposure to +3°C, but not after 2 months (Morley, pers. comm.), and the anemone *Urticinopsis antarcticum* and some individuals of the sea cucumber *Heterocucumis steineni* both appear able to survive in excess of 6 months at 6°C (LSP and MSCL pers obs). It thus appears that Antarctic marine invertebrates may be more variable than fish in their abilities to acclimate to elevated temperatures, and this may have interesting consequences for future responses to environmental change at the assemblage or ecosystem level.

In line with moves outside Antarctica to produce more widely encompassing methods for predicting future responses to change there have been recent moves to link studies of physiological flexibility and resistance to temperature elevation with more ecological, distribution, or biogeographical analyses. Thus Barnes and Peck (2008) and Barnes et al. (2010) looked at the use of physiological limits data in predicting responses to environmental change, or abilities to migrate in response to change. One outcome was that if species need to migrate past areas such as those along the Antarctic Peninsula and out towards the South Shetland Islands there are temperature elevations of 1–2°C that would be experienced in periods of days or weeks. These are not dissimilar to rates of warming in many of the studies of acclimation. There is thus direct immediate value in understanding such resistance mechanisms and capacities, as well as future potential. Further to this Morley et al. (2010b) identified changes in distribution across latitudes in Antarctic marine invertebrate species that reflect their temperature tolerances. Species living in shallow water along the Antarctic Peninsula inhabit deeper sites around South Georgia, where temperatures are lower, and conversely some species living deeper on the Peninsula in warmer mid water occur shallower at lower latitudes reflecting adaptation to a warmer regime. In a recent more temperate-based study Tomanek (2010) analysed the heat-shock response in species from different sea temperature regimes. He concluded that species from very stable systems that lack the heat-shock response, like many in Antarctica, are particularly vulnerable to warming. He also concluded that species from highly variable environments, such as the intertidal, are also vulnerable, because they appear unable to increase HSP production at temperatures above those experienced naturally. These animals are using their heat-shock response under current conditions at a such a level that acclimatory capacity in this trait is either largely or entirely absent. Species from moderately variable temperature systems appear least vulnerable on these criteria.

Overall many Antarctic marine species appear to be living in temperature regimes that are close to their limits, and some are very close. They also appear to have poor capacities to change their physiologies by acclimation. There is a clear need to understand the variability of different species in these capacities, but also the mechanisms limiting their survival at higher temperatures, which clearly will differ depending on the rate of warming used. This is where molecular techniques can bring a finer level of detail to physiological observations. Truebano et al. (2010) demonstrated some of the advantages of using transcriptomic approaches in this field when they identified the recruitment of anaerobic metabolic processes at around a degree lower than those found previously by physiological methods in the infaunal clam *L. elliptica*. Clark and Burns (2008) also used genomic technologies to identify a warm acclimated protein (wap65) that was expressed when the fish *H. antarcticus* was warmed to +6°C. Thorne et al. (2010) used transcriptome profiling to identify responses to temperature stress in *H. antarcticus*. They identified increased expression of genes associated with the classical vertebrate inflammatory response and a whole suite of different genes, which were all regulated by hypoxia inducible factor 1 (HIF1). It is interesting to note that in both cases of the gene chip work on the clam and the fish there was no up-regulation of expression of any heat shock protein genes. These transcriptome profiling studies used very short term warming experiments and custom-made gene chips, however, given the accumulated physiology data, it is expected that more studies will be under-taken at different rates of temperature warming and in particular long-term experiments in excess of several months. Also that gene chips will be replaced with cheaper short read pyrosequencing transcriptional profiling.

9.12.2 Ocean Acidification

Along with the threat of sea warming in the future, is that of altered pH, commonly known as Ocean Acidification. Since the observations that the ocean is becoming progressively more acidic because of human generation of CO_2 (Canadell et al. 2007), and the observation that some species, especially coccolithophores appear vulnerable to reduced pH because their skeletons dissolve (Muller et al. 2010), there has been a general drive from marine ecologists and physiologists to understand the likely outcomes for marine species in general of the predicted continued acidification of the oceans. It has been pointed out that because of the temperature effects on saturation states of calcium carbonate the polar oceans are likely to be the first to show marked effects of altered pH on organisms, especially those with external aragonitic skeletons (Fabry et al. 2009). McNeil and Matear (2008) have calculated that an atmospheric level of 450 ppm CO_2 is likely to be a tipping point in the Southern Ocean for the onset of widespread dissolution of shells and problems with larval development, because this is the point of significant undersaturation for the whole ocean. They add that there are likely to be larger problems in winter when many species have their larval phases. In respect of this both McClintock et al. (2009, 2010) and Cummings et al. (2011) have shown that the shells of the Antarctic marine benthos dissolve rapidly at lowered pH.

In recent years there have also been investigations of the effects of reduced pH on developmental stages and larval success in echinoderms, with Clark et al. (2009), showing that survival of pluteus larvae of the urchin *S. neumayeri* are markedly affected by being cultured in lowered pH, but only when the pH was reduced below 7.0 from the normal of 8.0. The size of larvae was also reduced, but calcification levels in larvae were unaffected. Following this Ericson et al. (2010) demonstrated that lowered pH had little effect on fertilisation in the same species or in the nemertean worm *Parborlasia corrugatus*. They also showed that early development in these species was unaffected by lowered pH until values of 7.3 or below were reached, and these levels are not predicted in oceanic waters for well over 100 years. To date there have been no significant molecular studies on the effects of altered pH on Antarctic organisms, however, 454 pyrosequencing has produced significant sequence resources of *L. elliptica*. Clark et al. (2010) used to study the mantle transcriptome of the large infaunal bivalve *L. elliptica* to identify genes associated with the deposition of calcium in the shell and these are now being used in further studies in this species.

9.13 Summary

Overall Antarctic marine species posses a range of adaptations, several of which are unique or unusual compared to species from lower latitudes, and that are allowed by the low temperature environment. These adaptations appear to provide extra problems when faced with environmental change such as warming or acidification of the oceans. This is especially so for the slow physiological rates, slow growth and very long developmental periods. This means that there is very little generational turnover over relatively long timescales (tens of years to centuries) in which genetic modifications can occur. Hence, acclimation and survival of an animal will be largely reliant upon phenotypic plasticity and excess acclimatory capacity (cf Tomanek, 2010). Certainly some of the unique adaptations found in Antarctic marine species (such as gigantism in invertebrates, giant muscle fibres and reduction in number in fish, deletion of haemoglobin genes, lack of an HSR), whilst not exactly evolutionary cul-de-sacs, cannot be accommodated by slightly modifying the odd gene or biochemical pathway when the animal is challenged with a changing environment. Indeed, the ageing work on the clam raised interesting questions about the extended life span of this species which could also apply to other species. It was suggested that the longer life span may either be a population management issue; they need longer life spans to ensure survival of the stock, or alternatively the different membrane lipid composition which enables the higher proton leak and hence longer life span was a by-effect of polar adaptation (Abele et al. 2009). This is just one of the many intriguing questions yet to be answered in Antarctic biology. The Southern Ocean may be one of the first to see large scale effects of climate change on the organisms living there and we now have more tools than ever before to understand how this unique fauna will cope.

References

Abele D, Brey T, Philipp E (2009) Bivalve models of aging and the determination of molluscan lifespans. Exp Gerontol 44:307–315

Abele D, Burlando B, Viarengo A (1998) Exposure to elevated temperatures and hydrogen peroxide elicits oxidative stress and antioxidant response in the Antarctic intertidal limpet *Nacella concinna*. Comp Biochem Physiol B–Biochem Molec Biol 120:425–435

Acevedo JP, Reyes F, Parra LP, Salazar O, Andrews BA, Asenjo JA (2008) Cloning of complete genes for novel hydrolytic enzymes from Antarctic sea water bacteria by use of an improved genome walking technique. J Biotechnol 133:277–286

Allen MA, Lauro FM, Williams TJ, Burg D, Siddiqui KS, De Francisci D, Chong KWY, Pilak O, Chew HH, De Maere MZ, Ting L, Katrib M, Ng C, Sowers KR, Galperin MY, Anderson IJ, Ivanova N, Dalin E, Martinez M, Lapidus A, Hauser L, Land M, Thomas T, Cavicchioli R (2009) The genome sequence of the psychrophilic archaeon, *Methanococcoides burtonii*: the role of genome evolution in cold adaptation. ISME J 3:1012–1035

Amsler CD, Iken K, McClintock JB, Baker BJ (2009) Defenses of polar macroalgae against herbivores and biofoulers. Bot Mar 52:535–545

Baird NA, Etter PD, Atwood TS, Currey MC, Shiver AL, Lewis ZA, Selker EU, Cresko WA, Johnson EA (2008) Rapid SNP discovery and genetic mapping using sequenced RAD markers. Plos One 3:7

Barnes DKA, Peck LS (2005) Extremes of metabolic strategy in Antarctic bryozoa. Mar Biol 147(4):979–988

Barnes DKA, Peck LS (2008) Examining vulnerability of Antarctic shelf biodiversity to predicted climate warming. Clim Res 37:149–163

Barnes DKA, Peck L, Morley S (2010) Ecological relevance of laboratory determined temperature limits: colonisation potential, biogeography and resilience of Antarctic invertebrates to environmental change. Global Change Biol 16:3164–3169

Barnes DKA, Webb KE, Linse K (2007) Growth rate and its variability in erect Antarctic bryozoans. Polar Biol 30:1069–1081

Benedetti M, Martuccio G, Nigro M, Regoli F (2008) Comparison of antioxidant efficiency in the Antarctic nototheniod species, *Trematomus bernacchii*, *Trematomus newnesi* and *Trematomus hansoni*. Mar Env Res 66:98–99

Benedetti M, Nigro M, Regoli F (2010) Characterisation of antioxidant defences in three Antarctic notothenioid species from Terra Nova Bay (Ross Sea). Chem Ecol 26:305–314

Billups K, Kelly C, Pierce E (2008) The late Miocene to early Pliocene climate transition in the Southern Ocean. Palaeogeogr Palaeoclimatol Palaeoecol 267:31–40

Bilyk KT, DeVries AL (2010) Delayed onset of adult antifreeze activity in juveniles of the Antarctic icefish *Chaenocephalus aceratus*. Pol Biol 33:1387–1397

Bosch I, Beauchamp KA, Steele ME, Pearse JS (1987) Development, metamorphosis and seasonal abundance of embryos and larvae of the Antarctic sea urchin *Sterechinus neumayeri*. Biol Bull 173:126–135

Bowgen A, Fraser KP, Peck LS, Clarke A (2007) The energetic cost of synthesizing proteins is not temperature dependent. Am J Physiol 292:R2266–R2274

Brockington S, Peck LS (2001) Seasonality of respiration and ammonia excretion in the Antarctic echinoid *Sterechinus neumayeri*. Mar Ecol Progr Ser 259:159–168

Buckley BA, Place SP, Hofmann GE (2004) Regulation of heat shock genes in isolated hepatocytes from an Antarctic fish, *Trematomus bernacchii*. J Fish Biol 207:3649–3656

Buttemer WA, Abele D, Costantini D (2010) From bivalves to birds: oxidative stress and longevity. Funct Ecol 24:971–983

Campbell H, Davison W, Fraser KPP, Peck LS, Egginton S (2009) Heart rate and ventilation in Antarctic fishes are largely determined by ecotype. J Fish Biol 74:535–552

Campbell HA, Fraser KPP, Bishop CM, Peck LS, Egginton S (2008) Hibernation in an Antarctic Fish: on ice for winter. Plos One 3:9

Canadell JG, Le Quere C, Raupach MR, Field CB, Buitenhuis ET, Ciais P, Conway TJ, Gillett NP, Houghton RA, Marland G (2007) Contributions to accelerating atmospheric CO_2 growth from economic activity, carbon intensity, and efficiency of natural sinks. Proc Natl Acad Sci USA 104:18866–18870

Chen LB, DeVries AL, Cheng C-HC (1997) Evolution of antifreeze glycoprotein gene from a trypsinogen gene in Antarctic notothenioid fish. Proc Natl Acad Sci USA 94:3811–3816

Chen ZZ, Cheng C-HC, Zhang JF, Cao LX, Chen L, Zhou LH, Jin YD, Ye H, Deng C, Dai ZH, Xu QH, Hu P, Sun SH, Shen Y, Chen LB (2008) Transcriptomic and genomic evolution under constant cold in Antarctic notothenioid fish. Proc Natl Acad Sci USA 103:10491–10496

Cheng CH-C, Detrich HW (2007) Molecular ecophysiology of Antarctic notothenioid fishes. Phil Trans R Soc B 362:2215–2232

Cheng CH-C, di Prisco G, Verde C (2009a) The "Icefish Paradox." Which is the task of neuroglobin in Antarctic hemoglobin-less icefish? IUBMB Life 61:184–188

Cheng CH-C, di Prisco G, Verde C (2009b) Cold-adapted Antarctic fish: the discovery of neuroglobin in the dominant suborder Notothenioidei. Gene 433:100–101

Clark D, Lamare M, Barker M (2009) Response of sea urchin *pluteus* larvae (Echinodermata: Echinoidea) to reduced seawater pH: a comparison among a tropical, temperate, and a polar species. Mar Biol 156:1125–1137

Clark MS, Burns G (2008) Characterisation of the warm acclimated protein gene (wap65) in the Antarctic plunderfish (*Harpagifer antarcticus*). DNA Seq 19:50–55

Clark MS, Peck LS (2009a) HSP70 Heat shock proteins and environmental stress in Antarctic marine organisms: a mini-review. Mar Gen 2:11–18

Clark MS, Peck LS (2009b) Triggers of the HSP70 stress response: environmental responses and laboratory manipulation in an Antarctic marine invertebrate (*Nacella concinna*). Cell Stress Chaperones 14:649–660

Clark MS, Fraser KPP, Peck LS (2008a) Lack of an HSP70 heat shock response in two Antarctic marine invertebrates. Polar Biol 31:1059–1065

Clark MS, Fraser KPP, Peck LS (2008b) Antarctic marine molluscs do have an HSP70 heat shock response. Cell Stress Chaperones 13:39–49

Clark MS, Geissler P, Waller C, Fraser KPP, Barnes DKA, Peck LS (2008c) Low heat shock thresholds in wild Antarctic inter-tidal limpets (*Nacella concinna*). Cell Stress Chaperones 13:51–58

Clark MS, Thorne MAS, Toullec T-Y, Meng Y, Guan L, Peck LS, Moore S (2011) Krill 454 pyrosequencing reveals chaperone and stress transcriptome. PLoS One 6:E15919

Clark MS, Thorne MAS, Vieira FA, Cardoso JCR, Power DM, Peck LS (2010) Insights into shell deposition in the Antarctic bivalve *Laternula elliptica*: gene discovery in the mantle transcriptome using 454 pyrosequencing. BMC Genomics 11:362

Clarke A (1988) Seasonality in the Antarctic Marine Environment. Comp Biochem Physiol B—Biochem Molec Biol 90:461–473

Clarke A, Gaston KJ (2006) Climate, energy and diversity. Proc Royal Soc B 273:2257–2266

Clarke A, Peck LS (1991) The physiology of polar marine zooplankton. In: Sakshaug E, Hopkins C, Oritsland N (eds) In: Proceedings of Pro Mare Symposium on Polar Marine Ecology, Polar Research, Trondheim, vol 10, pp 355–369

Clarke A, Murphy EJ, Meredith MP, King JC, Peck LS, Barnes DKA (2007) Climate change and the marine ecosystem of the western Antarctic Peninsula. Phil Trans R Soc 362:149–166

Csermely P (2004) Strong links are important–but weak links stabilise them. Trends Biochem Sci 29:331–334

Cummings V, Hewitt J, Van Rooyen A, Currie K, Beard S, Thrush S, Norkko J, Barr N, Heath P, Halliday NJ, Sedcole R, Gomez A, McGraw C, Metcalf V (2011) Ocean acidification at high latitudes: potential effects on functioning of the Antarctic bivalve *Laternula elliptica*. Plos One 6:11

Deng C, Cheng C-HC, Yea H, He X, Chen L (2010) Evolution of an antifreeze protein by neofunctionalization under escape from adaptive conflict. Proc Natl Acad Sci USA 107: 21593–21598

Detrich HW, Johnson KA, Marcheseragona SP (1989) Polymerization of Antarctic Fish tubulins at low temperatures: energetic aspects. Biochem 28:10085–10093

Detrich HW, Williams RC (1992) Dynamic instability of Antarctic fish microtubules. Mol Biol Cell 3:A167–A167

de Pascale D, Cusano AM, Autore F, Parrilli E, di Prisco G, Marino G, Tutino ML (2008) The cold active Lip1 lipase from the Antarctic bacterium *Pseudoalteromonas haloplanktis* TAC125 is a member of the new bacterial lipolytic enzyme family. Extremeophiles 12:311–323

DeVries AL, Wohlschlag DE (1969) Freezing resistance in some Antarctic fishes. Science 163:1073–1075

di Prisco G, Cocca E, Parker SK, Detrich HW (2002) Tracking the evolutionary loss of hemoglobin expression by the white-blooded Antarctic icefishes. Gene 295:185–191

di Prisco G, Eastman JT, Giordano D et al (2007) Biogeography and adaptation of Notothenioid fish: hemoglobin function and globin-gene evolution. Gene 398:143–155

Emerson KJ, Merz CR, Catchen JM, Hohenlohe PA, Cresko WA, Bradshaw WE, Holzapfel CM (2010) Resolving postglacial phylogeography using highthroughput sequencing. Proc Natl Acad Sci USA 107:16196–16200

Ericson JA, Lamare MD, Morley SA, Barker MF (2010) The response of two ecologically important Antarctic invertebrates (*Sterechinus neumayeri* and *Parborlasia corrugatus*) to reduced seawater pH: effects on fertilisation and embryonic development. Mar Biol 157:2689–2702

Fabry VJ, McClintock JB, Mathis JT, Grebmeier JM (2009) Ocean acidification at high Latitudes: the Bellweather. Oceanogr 22:160–171

Fields PA, Somero GN (1998) Hot spots in cold adaptation: localized increases in conformational flexibility in lactate dehydrogenase A(4) orthologs of Antarctic notothenioid fishes. Proc Natl Acad Sci USA 95:11476–11481

Forcada J, Trathan PN (2009) Penguin responses to climate change in the southern ocean. Glob Change Biol 15:1618–1630

Fraser KPP, Rogers AD (2007) Protein metabolism in marine animals: the underlying mechanism of growth. Adv Mar Biol 52:267–362

Giordano D, Russo R, Coppola D, di Prisco G, Verde C (2010) Molecular adaptations in haemoglobins of notothenioid fishes. J Fish Biol 76:301–318

Gwak IG, Jung WS, Kim HJ, Kang S-H, Jin ES (2010) Antifreeze protein in Antarctic marine diatom, *Chaetoceros neogracile*. Mar Biotech 12:630–639

Hazel JR (1995) Thermal adaptation in biological membranes: is homeoviscous adaptation the explanation? Ann Rev Physiol 57:19–42

Hoffman JI, Clarke A, Linse K, Peck LS (2010a) Strong population genetic structure in a broadcast-spawning Antarctic marine invertebrate. J Heredity 102:55–66

Hoffman JI, Clarke A, Linse K, Peck LS (2011) Effects of brooding and broadcasting reproductive modes on the population genetic structure of two Antarctic gastropod molluscs. Mar Biol 158:287–296

Hoffman JI, Peck LS, Hillyard G, Zieritz A, Clark MS (2010b) No evidence for genetic differentiation between Antarctic limpet *Nacella concinna* morphotypes. Mar Biol 157:765–778

Hofmann GE, Buckley BA, Airaksinen S, Keen JE, Somero GN (2000) Heat-shock protein expression is absent in the Antarctic fish *Trematomus bernacchii* family Nototheniidae. J Exp Biol 203:2331–2339

Hudson HA, Brauer PR, Scofield MA, Petzel DH (2008) Effects of warm acclimation on serum osmolality, cortisol and hematocrit levels in the Antarctic fish, *Trematomus bernacchii*. Pol Biol 31:991–997

Hunter RL, Halanych KM (2008) Evaluating connectivity in the brooding brittle star *Astrotoma agassizii* across the Drake Passage in the southern ocean. J Hered 99:137–148

Isely N, Lamare M, Marshall C, Barker M (2009) Expression of the DNA repair enzyme, photolyase, in developmental tissues and larvae, and in response to ambient UV-R in the Antarctic sea urchin *Sterechinus neumayeri*. Photochem Photobiol 85:1168–1176

Janecki T, Kidawa A, Potocka M (2010) The effects of temperature and salinity on vital biological functions of the Antarctic crustacean *Serolis polita*. Pol Biol 33:1013–1020

Janknegt PJ, de Graaff CM, van de Poll WH, Visser RJW, Helbling EW, Buma AGJ (2009) Antioxidative responses of two marine microalgae during acclimation to static and fluctuating natural UV nadiation. Photochem Photobiol 85:1336–1345

Johnston IA, Calvo J, Guderley H, Fernandez D, Palmer L (1998) Latitudinal variation in the abundance and oxidative capacities of muscle mitochondria in perciform fishes. J Exp Biol 201:1–12

Kidawa A, Potocka M, Janecki T (2010) The effects of temperature on the behaviour of the Antarctic sea star *Odontaster validus*. Polar Res 31:273–284

Kiko R (2010) Acquisition of freeze protection in a sea-ice crustacean through horizontal gene transfer? Polar Biol 33:543–556

Kock K-H, Everson I (1998) Age, growth and maximum size of Antarctic notothenioid fish revisited. In: di Prisco G, Pisano E, Clarke A (eds) Fishes of Antarctica: a Biological Overview. Springer, Milan, pp 29–40

Koplovitz G, McClintock JB, Amsler CD, Baker BJ (2009) Palatability and chemical anti-predatory defenses in common ascidians from the Antarctic Peninsula. Aquat Biol 7:81–92

Kuhn E, Bellicanta GS, Pellizari VH (2009) New alk genes detected in Antarctic marine sediments. Env Microbiol 11:669–673

Lamare MD, Barker MF, Lesser MP (2007) In situ rates of DNA damage and abnormal development in Antarctic and non-Antarctic sea urchin embryos. Aquat Biol 1:21–32

La Mesa M, Ashford J (2008) Age and growth of ocellated icefish, *Chionodraco rastrospinosus* DeWitt-Hureau 1976, from the South Shetland Islands. Polar Biol 31:1333–1342

La Mesa M, De Felice A, Jones CD, Kock KH (2009) Age and growth of spiny icefish (*Chaenodraco wilsoni* Regan 1914) off Joinville-D'Urville Islands (Antarctic Peninsula). CCAMLR Sci 16:115–130

La Terza A, Dobri N, Alimenti C, Vallesi A, Luporini P (2009) The water-borne protein signals (pheromones) of the Antarctic ciliated protozoan *Euplotes nobilii*: structure of the gene coding for the En-6 pheromone. Can J Microbiol 55:57–62

Lister KN, Lamare MD, Burritt DJ (2010) Sea ice protects the embryos of the Antarctic sea urchin *Sterechinus neumayeri* from oxidative damage due to naturally enhanced levels of UV-B radiation. J Exp Biol 213:1967–1975

Lurman G, Blaser T, Lamare M, Peck LS, Morley SA (2010) Mitochondrial plasticity in brachiopod (*Liothyrella* spp.) smooth adductor muscle as a result of season and latitude. Mar Biol 157:907–913

Ma WS, Mutka T, Vesley B et al (2009) Norselic acids A-E, highly oxidized anti-infective steroids that deter mesograzer predation, from the Antarctic sponge *Crella* sp. J Nat Prod 72:1842–1846

Mahon AR, Thornhill DJ, Norenburg JL, Halanych KM (2010) DNA uncovers Antarctic nemertean biodiversity and exposes a decades-old cold case of asymmetric inventory. Polar Biol 33:193–202

Makowski K, Bialkowska A, Olczak J, Kur J, Turkiewicz M (2007) Antarctic, cold-adapted beta-galactosidase of *Pseudoalteromonas* sp 22b as an effective tool for alkyl galactopyranosides synthesis. Enz Microb Technol 44:59–64

Matschiner M, Hanel R, Salzburger W (2010) Gene flow by larval dispersal in the Antarctic notothenioid fish *Gobionotothen gibberifrons*. Mol Ecol 18:2574–2587

McClintock JB, Amsler CD, Baker BJ (2010) Overview of the chemical ecology of benthic marine invertebrates along the Western Antarctic Peninsula. Integr Comp Biol 50:967–980

McClintock JB, Angus RA, Mcdonald MR, Amsler CD, Catledge SA, Vohra YK (2009) Rapid dissolution of shells of weakly calcified Antarctic benthic macroorganisms indicates high vulnerability to ocean acidification. Antarctic Sci 21:449–456

McNeil BI, Matear RJ (2008) Southern Ocean acidification: a tipping point at 450-ppmatmospheric CO_2. Proc Natl Acad Sci USA 105:18860–18864
Meredith MP, King JC (2005) Rapid climate change in the ocean west of the Antarctic Peninsula during the second half of the 20th century. Geophys Res Lett 32:L19604
Moore M, Manahan DT (2007) Variation among females in egg lipid content and developmental success of echinoderms from McMurdo Sound, Antarctica. Polar Biol 30:1245–1252
Morley SA, Peck LS, Miller A, Pörtner H-O (2007a) Hypoxia tolerance associated with activity reduction is a key adaptation for *Laternula elliptica* seasonal energetic. Oecologia 153:29–36
Morley SA, Peck LS, Tan KS, Martin SM, Pörtner H-O (2007b) Latitudinal insensitivity of burrowing capacity in the bivalve *Laternula*. Mar Biol 151:1823–1830
Morley SA, Hirse T, Portner HO, Peck LS (2009) Geographical variation in thermal tolerance within Southern Ocean marine ectotherms. Comp Biochem Physiol A153:154–161
Morley SA, Clark MS, Peck LS (2010a) Depth gradients in shell morphology correlate with thermal limits for activity and ice disturbance in Antarctic limpets. J Exp Mar Biol Ecol 390:1–5
Morley SA, Griffiths HJ, Barnes DKA, Peck LS (2010b) South Georgia: a key location for linking physiological capacity to distributional changes in response to climate change. Antarctic Sci 22:774–781
Muller MN, Schulz KG, Riebesell U (2010) Effects of long-term high CO_2 exposure on two species of coccolithophores. Biogeosci 7:1109–1116
Obermüller B, Peck LS, Barnes DKA, Morley SA (2010) Seasonal physiology of Antarctic marine benthic predators and scavengers. MEPS 415:109–126. doi:10.3354/meps0873
Pace DA, Manahan DT (2007) Cost of protein synthesis and energy allocation during development of Antarctic sea urchin embryos and larvae. Biol Bull 212:115–129
Pace DA, Maxson R, Manahan DT (2010) Ribosomal analysis of rapid rates of protein synthesis in the Antarctic sea urchin *Sterechinus neumayeri*. Biol Bull 218:48–60
Park H, Ahn IY, Kim H, Cheon J, Kim M (2008) Analysis of ESTs and expression of two peroxiredoxins in the thermally stressed Antarctic bivalve *Laternula elliptica*. Fish Shellfish Immunol 25:550–559
Parra LP, Reyes F, Acevedo JP, Salazar O, Andrews BA, Asenjo JA (2008) Cloning and fusion expression of a cold-active lipase from marine Antarctic origin. Enz Microbial Technol 42:371–377
Pearce I, Davidson AT, Bell EM, Wright S (2007) Seasonal changes in the concentration and metabolic activity of bacteria and viruses at an Antarctic coastal site. Aquat Microbiol Ecol 43:11–23
Peck LS (2002) Ecophysiology of Antarctic marine ectotherms: limits to life. Polar Biol 25:31–40
Peck LS, Clark MS, Morley SA, Massey A, Rosetti H (2009a) Animal temperature limits: effects of size, activity and rates of change. Funct Ecol 23:248–256
Peck LS, Convey P, Barnes DKA (2006) Environmental constraints on life histories in Antarctic ecosystems: tempos, timings and predictability. Biol Rev 81:75–109
Peck LS, Massey A, Thorne M, Clark MS (2009b) Lack of acclimation in *Ophionotus victoriae*: brittle stars are not fish. Polar Biol 32:399–402
Peck LS, Barnes DKA, Cook AJ, Fleming AH, Clarke A (2010a) Negative feedback in the cold: ice retreat produces new carbon sinks in Antarctica. Glob Change Biol 16:2614–2623
Peck LS, Morley SA, Clark MS (2010b) Poor acclimation capacities in Antarctic marine ectotherms. Mar Biol 157:2051–2059
Peck LS, Powell DK, Tyler PA (2007) Very slow development in two Antarctic bivalve molluscs, the infaunal clam, *Laternula elliptica* and the scallop *Adamussium colbecki*. Mar Biol 150:1191–1197
Peck LS, Webb KE, Clark MS, Miller A, Hill T (2008) Temperature limits to activity, feeding and metabolism in the Antarctic starfish *Odontaster validus*. Mar Ecol Prog Ser 381:181–189
Philipp E, Brey T, Portner HO, Abele D (2005) Chronological and physiological ageing in a polar and a temperate mud clam. Mech Age Dev 126:598–609

Place SP, Zippay ML, Hofmann GE (2004) Constitutive roles for inducible genes: evidence for the alteration in expression of the inducible hsp70 gene in Antarctic notothenioid fishes. Am J Physiol Reg Integr Comp Physiol 287:R429–R436

Pörtner H-O, Somero GA, Peck LS (2007) Thermal limits and adaptation in marine Antarctic ectotherms: an integrative view. In: Rogers A, Murphy E (eds) Antarctic ecology, from genes to ecosystems. Special Volume Phil Trans R Soc 362:2233–2258

Powell AWB (1951) Antarctic and subanctarctic mollusca: pelecypoda and gastropoda. Discovery Rep (USA) 26:49–196

Privalov PL (1990) Cold denaturation of proteins. Crit Rev Biochem Mol Biol 25:281–305

Pucciarelli S, La Terza A, Ballarini P et al (2009) Molecular cold-adaptation of protein function and gene regulation: the case for comparative genomic analyses in marine ciliated protozoa. Mar Gen 2:57–66

Robinson E, Davison W (2008a) The Antarctic notothenioid fish *Pagothenia borchgrevinki* is thermally flexible: acclimation changes oxygen consumption. Polar Biol 31:317–326

Robinson E, Davison W (2008b) Antarctic fish can survive prolonged exposure to elevated temperatures. J Fish Biol 73:1676–1689

Rodrigues E, Santos MRD, Rodrigues E, Gannabathula V, Lavrado HP (2009) Arginine metabolism of the Antarctic bivalve *Laternula elliptica* King-Broderip 1831: an ecophysiological approach. Polar Biol 32:691–702

Römisch K, Collie N, Soto N, Logue J, Lindsay M, Scheper W, Cheng CHC (2003) Protein translocation across the endoplasmic reticulum membrane in cold-adapted organisms. J Cell Sci 116:2875–2883

Ruud JT (1954) Vertebrates without erythrocytes and blood pigment. Nature 173:848–850

Schofield O, Ducklow HW, Martinson DG et al (2010) How do polar marine ecosystems respond to rapid climate change? Science 328:1520–1523

Somero GN (2010) The physiology of climate change: how potentials for acclimatization and genetic adaptation will determine 'winners' and 'losers'. J Exp Biol 213:912–920

Sørensen JG, Loeschcke V (2007) Studying stress responses in the post-genomic era: its ecological and evolutionary role. J Biosci 32:447–456

Stanwell-Smith DP, Peck LS (1998) Temperature and embryonic development in relation to spawning and field occurrence of larvae of 3 Antarctic echinoderms. Biol Bull Woods Hole 194:44–52

Stillman JH (2003) Acclimation capacity underlies susceptibility to climate change. Science 301:65

Strebel H (1908) Die Gastropoden. Wissenschaftliche Ergebn Schwedisch Sudpolar-Expedition 1901–1903 6:1–112

Sutton CP, Manning MJ, Stevens DW, Marriot PM (2008) Biological parameters for icefish (*Chionobathyscus dewitti*) in the Ross Sea, Antarctica. CCAMLR Sci 15:139–165

Thorne MAS, Burns G, Fraser KPP, Hillyard G, Clark MS (2010) Transcription profiling of acute temperature stress in the Antarctic plunderfish *Harpagifer antarcticus*. Mar Gen 3:35–44

Ting L, Williams TJ, Cowley MJ, Lauro FM, Guilhaus M, Raftery MJ, Cavicchioli R (2010) Cold adaptation in the marine bacterium, *Sphingopyxis alaskensis*, assessed using quantitative proteomics. Environ Microbiol 12:2658–2676

Tomanek L (2010) Variation in the heat shock response and its implication for predicting the effect of global climate change on species' biogeographical distribution ranges and metabolic costs. J Exp Biol 213:971–979

Truebano M, Burns G, Thorne MAS et al (2010) Transcriptional response to heat stress in the Antarctic bivalve *Laternula elliptica*. J Exp Mar Biol Ecol 391:65–72

Vallesi A, Alimenti C, La Terza A, Di Giuseppe G, Dini F, Luporini P (2009) Characterization of the pheromone gene family of an Antarctic and Arctic protozoan ciliate, *Euplotes nobilii*. Mar Gen 2:27–32

Vera JC, Wheat CW, Fescemyer HW et al (2008) Rapid transcriptome characterization for a nonmodel organism using 454 pyrosequencing. Mol Ecol 17:1636–1647

Verde C, Giordano D, Russo R, di Prisco G (2012) The adaptive evolution of polar fishes. Lessons from the function of hemoproteins. In: di Prisco G, Verde C (eds) Adaptation and evolution in marine environments—The impacts of global change on biodiversity, vol 1. Series "From Pole to Pole". Springer, Berlin, pp 197–213

Verde C, Parisi E, di Prisco G (2006) The evolution of thermal adaptation in polar fish. Gene 385:137–145

Verde C, Vergara A, Mazzarella L, di Prisco G (2008) The hemoglobins of fishes living at polar latitudes–current knowledge on structural adaptations in a changing environment. Curr Prot Pept Sci 9:578–590

Waller RG, Tyler PA, Smith CR (2010) Fecundity and embryo development of three Antarctic deep-water scleractinians: *Flabellum thouarsii*, *F-curvatum* and *F-impensum*. Deep-Sea Res Part II 55:2527–2534

Wang F, Hao JH, Yang CY, Sun M (2010) Cloning, expression, and identification of a novel extracellular cold-adapted alkaline protease gene of the marine bacterium strain YS-80-122. Appl Biochem Biothechnol 162:1497–1505

Wang QF, Hou YH, Miao JL, Li GY (2009) Effect of UV-B radiation on the growth and antioxidant enzymes of Antarctic sea ice microalgae *Chlamydomonas* sp ICE-L. Acta Physiol Plant 31:1097–1102

Weihe E, Kriews M, Abele D (2010) Differences in heavy metal concentrations and in the response of the antioxidant system to hypoxia and air exposure in the Antarctic limpet *Nacella concinna*. Mar Env Res 69:127–135

Wilson NG, Hunter RL, Lockhart SJ, Halanych KM (2007) Multiple lineages and absence of panmixia in the "circumpolar" crinoid *Promachocrinus kerguelensis* from the Atlantic sector of Antarctica. Mar Biol 152:895–904

Wilson SL, Walker VK (2010) Selection of low-temperature resistance in bacteria and potential applications. Env Technol 32:943–956

Winston JE, Bernheimer AW (1986) Hemolytic activity in an Antarctic bryozoan. J Nat Hist 20:369–374

Zachos J, Pagani M, Sloan L, Thomas E, Billups K (2001) Trends, rhythms, and aberrations in global climate 65 Ma to present. Science 292:686–693

Zhang JF, Deng C, Wang JS, Chen LB (2009) Identification of a two-domain antifreeze protein gene in Antarctic eelpout Lycodichthys dearborni. Polar Biol 32:35–40

Chapter 10
The Challenges of Low Temperature in the Evolution of Bacteria

Guido di Prisco, Daniela Giordano, Roberta Russo and Cinzia Verde

10.1 Introduction

It is currently recognised that extreme environments, by virtue of their extension and often unique features, are the most important part of the Earth's biosphere. Their study is still limited and is often hampered by logistic constraints; however, extreme environments are now becoming more and more accessible thanks to technological progress and research on adaptations to extreme conditions. Several Earth-based extreme environments, in particular the high-latitude regions, especially ice shelves and sheets and subglacial lakes, are considered by astrobiology research as model habitats for extraterrestrial life and thus apt to provide insights into the origins and evolution of life (Deming 2002).

The Antarctic marine habitats are unique natural laboratories for fundamental research on the evolutionary processes that shape biological diversity in extreme environments. The Antarctic biota evolved under the influence of a suite of

G. di Prisco · D. Giordano · R. Russo · C. Verde (✉)
Institute of Protein Biochemistry, National Research Council,
Via Pietro Castellino 111, 80131 Naples, Italy
e-mail: c.verde@ibp.cnr.it

G. di Prisco
e-mail: g.diprisco@ibp.cnr.it

D. Giordano
e-mail: d.giordano@ibp.cnr.it

R. Russo
e-mail: r.russo@ibp.cnr.it

G. di Prisco and C. Verde (eds.), *Adaptation and Evolution in Marine Environments, Volume 1*, From Pole to Pole, DOI: 10.1007/978-3-642-27352-0_10,

geological and climatic factors, including geographic isolation of the landmass and continental shelves, extreme low temperature and intense seasonality. Unlike deep oceans, polar marine environments are subject to large seasonal variations in sea-ice cover, greatly affecting the biology of organisms (Moline et al. 2008).

Psychrophilic bacteria growing in the range of 0–30°C cover a large proportion of the diversity in terms of biomass and distribution. Cold-loving extremophiles are a remarkable model to unravel the molecular basis of survival at low temperature (Cavicchioli et al. 2002; Deming 2002; Feller and Gerday 2003; Georlette et al. 2004; Rodrigues and Tiedje 2008). In the past, the diversity of microorganisms in cold environments was investigated in terms of distribution, paying little attention to their functional role in some important processes.

The "*omic*" methodologies currently offer suitable tools to investigate new biochemical pathways and to understand the evolutionary principles of adaptation and tolerance/resistance to extreme conditions. However, there are still many difficulties in interpreting amino-acid changes in the context of thermal adaptation without considering the general evolution and genetic drift. Recent evidence obtained by molecular approaches suggests that in some engineered bacteria the change of a limited number of genes is sufficient to vary the optimal temperature for an essential process such as growth (Duplantis et al. 2010).

In 2003 it was reported that the heterologous expression of chaperonin-encoding *cpn*60 and *cpn*10 genes from *Oleispira antarctica* RB8 was sufficient to allow *Escherichia coli* to grow at low temperatures (Ferrer et al. 2003). Thus inserting one or two genes isolated from a psychrophile made *E. coli* capable to thrive in the cold. However, different results were obtained with the expression of chaperonins isolated from the Antarctic marine bacterium *Pseudoalteromonas haloplanktis* TAC125 (*Ph*TAC125). Their expression in *E. coli* failed to promote growth of the bacterium at low temperature (Médigue et al. 2005).

More recently, Duplantis et al. (2010), in an attempt to develop new ideas for vaccines, engineered a variant of *Francisella tularensis* subsp. *novicida*, a typical mesophile, replacing an inventory of essential genes with the corresponding ones recruited by the cold-adapted bacterium *Colwellia psychrerythraea*, with the aim to shift the lifestyle of *Francisella* entirely towards cold dependence. The genes essential for inducing a psychrophile phenotype included the enzymes involved in DNA repair, e.g. DNA ligase, and in heme biosynthesis.

In this review we summarise how cold temperatures affect the physiology of microorganisms and we focus on the molecular mechanisms revealed by recent biochemical and genetic studies that shed light on microbial cold adaptations.

In the first section, we review current knowledge on molecular adaptations to cold and describes how 'omic' technologies (Casanueva et al. 2010) have contributed to our understanding. Lastly, a short overview on the biological function of oxygen-binding proteins in bacteria and their potential role in radical scavenging and cellular metabolism will be provided, together with the concept of the central role of oxygen and oxidative/nitrosative stress in regulating adaptive responses at cellular and molecular levels. Because of its features, the Antarctic bacterium *Ph*TAC125 was selected as a model for these studies.

10.2 Molecular Adaptations in Polar Microorganisms and Temperature-Dependent Features in Proteins

Cold-adapted microorganisms are generally classified as psychrophilic and psychrotolerant on the basis of their growth properties (Russell 1998). In this chapter the term psychrophile will be used in a generic sense to indicate all cold-adapted organisms.

Microorganisms grow successfully in the extreme conditions of cold habitats, through a variety of structural and physiological adjustments in their genomes. These strategies include synthesis of unique factors, such as cold-shock proteins (Cavicchioli et al. 2000), molecular chaperones (Motohashi et al. 1999; Watanabe and Yoshida 2004), compatible solutes (Carpenter and Crowe 1988; Pegg 2007) and structural modifications leading to the maintenance of membrane fluidity (Russell 1998, 2007; Chintalapati et al. 2004). In addition to adaptations at the cellular level, a key adaptive strategy is modification of enzyme kinetics, allowing maintenance of adequate reaction rates at thermal extremes. Enzyme catalysis is based on increased flexibility in certain regions of cold-active-enzyme architecture and high activity, with concomitant increase in thermolability (Georlette et al. 2004). An example of the adaptive relationships between stability and activity in psychrophilic enzymes is provided by mutational studies in cold-active α-amylase of *Ph*TAC125. Insertion of five stabilising weak interactions, absent in the psychrophilic enzyme but present in the mesophilic homologue, conferred stabilisation to the psychrophilic mutant, displaying activation and kinetic parameters identical to those of the mesophilic enzyme (Feller and Gerday 2003).

However, the adaptations to protein architecture essential to cold-active enzymes are still not well understood, and this study is an active area of investigation (Marx et al. 2004). The take-home message from the more recent studies is that only minor structural modifications are needed to change thermostability in psychrophilic enzymes, and that local rather than global flexibility may play a role. In fact, the overall fold is generally conserved in homologous thermophilic, mesophilic and psychrophilic proteins.

A successful method to evaluate the flexibility of cold-adapted proteins was applied by Tehei et al. (2004) looking at macromolecular dynamics in bacteria from different environments by neutron scattering. The analysis of the structural differences between homologous high- and low-temperature- adapted proteins showed that thermal adaptation is linked to the concept of resilience considered as an index of rigidity. These studies strongly support the concept that cold adaptation in proteins may be achieved by decreasing the macromolecular resilience and that the difference in the free energy of stabilisation is dominated by enthalpy terms rather than entropic terms. The resilience values increase with temperature, allowing to maintain the balance between macromolecular stability and flexibility within the range requested for biological activity.

Modulation of electrostatics is important in temperature adaptation. Psychrophilic proteins use electrostatic effects to ensure protein solvation, and flexibility is achieved by destabilisation of bonds made by charged residues. Kumar and Nussinov (2004) studied the structural properties of citrate synthase from hyperthermophilic, mesophilic, and psychrophilic organisms. It was evident that there are many differences between thermophilic and psychrophilic citrate synthases. In the latter, the salt-bridge networks were distributed all over the protein structure, whereas in thermophilic proteins the networks are mainly localised in the active-site regions. In general, the real challenge for psychrophilic proteins is to overcome their own expected lower solubility as well as those of the substrates in water, coupled with low reaction rates at cold temperature.

Nevertheless, the biochemical properties of cold-active enzymes make them attractive for exploitation in biochemical, bioremediation, and industrial processes (Feller and Gerday 2003).

10.3 Membrane Adaptation

The ability of microorganisms to adapt to low temperature is strongly linked to their capacity to sense changing temperature conditions. This feature has been attributed to DNA, RNA and cell membranes (Eriksson et al. 2002). In the process of adaptations to temperature, phenotypic and genotypic changes in lipid composition of membranes are important and there are many studies on these topics (Russell 1998, 2007). Conversely, much less is known on the changes in membrane proteins in response to low temperature. To correctly function in the cell, the lipid bilayer must be in the liquid-crystalline phase, thus allowing integral membrane proteins to change their conformation and diffuse laterally. When temperature decreases, lipids lose fluidity and consequently the molecules are packed much more tightly with reduced motions (Russell 1998). Cells produce unsaturated, polyunsaturated and methyl-branched fatty acids, and/or fatty acids having acyl-chains with shorter length, which allow membranes not to become less fluid at cold temperatures by introducing steric constraints that change the packing order or reduce the number of interactions in the membrane (Russell 1998; Chintalapati et al. 2004). Mutants of two cyanobacteria, *Anabaena variabilis* and *Synechocystis* sp. strain PCC 6803 defective in desaturation of fatty acids have a lower growth rate at low temperatures (Wada and Murata 1989). Although increased unsaturation and decreased chain length of fatty acids are the major modifications of cell membranes, other components associated to membrane may well play important roles in adaptation to low temperatures (Jagannadham et al. 1991; Chauhan and Shivaji 1994; Ray et al. 1998). For example, in vitro studies of Antarctic psychrotrophic bacteria suggested that carotenoids may have a role in buffering membrane fluidity (Jagannadham et al. 1991).

Being the membrane the first interface between the organism and the environment, changes in membrane fluidity in response to cold may well be the first signal of adaptation. The membrane generally possesses both saturated and unsaturated fatty acids, and therefore changing the proportion of these molecules may effectively alter the fluidity of the membrane, but two questions still need to be addressed: (1) what is the nature of the sensor and (2) what are the downstream events following sensing temperature change? The proteomic and genetic approaches have revealed that the key genes and proteins induced after downshifts in temperature include: (1) genes for fatty-acid desaturases; (2) genes involved in replication, transcription and translation. The transduction pathway of the signal occurs through two components: a membrane-associated sensor and a cytoplasmic response: as a consequence, cold-regulated genes are activated. Changes in DNA topology due to change of temperature may also induce mechanisms of cold response. Inducible proteins repair the damage caused by cold stress. Changes in RNA secondary structure, changes in translation and alteration in protein conformation may also act as temperature sensors (Shivaji et al. 2010 and references therein).

However, the mechanisms of activation of the cascade of genes are still unexplored. We foresee that the regulatory networks involved in cold sensing and response will be a challenge for future research.

10.4 Genomics, Metagenomics and Proteomics in Cold-Adapted Bacteria

The current knowledge of polar marine microorganisms, based on ecological and genomic perspectives, is in the early phase of an exponential growth.

To date, approximately 30 psychrophilic microbial genomes have been fully sequenced (Table 10.1, adapted from Murray et al. 2007; Russo et al. 2010). Genome and metagenome sequencing of cold-adapted bacteria has allowed to identify the structural determinants of psychrophily in a somewhat global evolutionary context. When analysing the individual protein sequences, it is often difficult to establish whether the observed residue changes are due to thermal adaptation or to general evolution.

The sequences of many microorganisms are accompanied by publications, e.g. euryarchaeota *Methanogenium frigidum* and *Methanococcoides burtonii* (Saunders et al. 2003) from the Ace Lake in the Antarctic region of the Vestfold Hills, the γ-proteobacterium *C. psychrerythraea* 34H (Methé et al. 2005) and *P. haloplanktis* TAC125 *Ph*TAC125 (Médigue et al. 2005) and γ-proteobacterium *Desulfotalea psychrophila* (Rabus et al. 2004). Many others are in various stages of completion. Recently, the genome of the *Exiguobacterium sibiricum* strain isolated from 3-million-year old permafrost was sequenced and annotated (Rodrigues et al. 2008).

The comparative analysis of the genome of *M. frigidum* and *M. burtonii* was the first study encompassing psychrophile-hyperthermophyle lifestyles (Saunders et al. 2003). Preliminary studies on proteins have revealed the presence in their genome of cold-shock-domain folds and the typical properties of cold-adapted proteins, namely increased glutamyl and threonyl usage. In order to improve molecular flexibility, cold-adapted proteins and enzymes display a reduced number of Pro in their primary structure to mitigate the negative effect of Pro isomerisation upon folding (Feller and Gerday 2003; Feller 2010).

Table 10.1 Polar bacterial and archaeal genomes. The status of genome sequencing without accession number is still in progress or available by URL (adapted from Murray et al. 2007; Russo et al. 2010)

Species	Strain origin	Status of genome sequencing/accession number or URL	Reference
Methanogenium frigidum	Ace Lake, Antarctica	draft/http://psychro.bioinformatics. unsw.edu. au/blast/mf_blast.php.	Saunders et al. (2003)
Methanococcoides burtonii DSM6242	Ace Lake, Antarctica	completed/CP000300	Saunders et al. (2003)
Colwellia psychrerythraea 34H	Arctic marine sediments	completed/CP000083	Methé et al. (2005)
Shewanella frigidimarina NCMB400	Sea ice, water, Antarctica	completed/CP000447	Not published
Shewanella baltica OS155	Baltic Sea	NC_009052	Not published
Shewanella baltica OS185		NC_009665	
Shewanella baltica OS195		NC_009997	
Shewanella baltica OS223		NC_011663	
Psychrobacter arcticus 273-4	Siberian permafrost	complete/CP000082	Ayala-Del-Río et al. (2010)
Psychrobacter cryohalolentis K5	Siberian permafrost	completed/CP000323, CP000324	Not published
Oleispira antarctica RB-8	Rod Bay, Ross Sea, Antarctica	in progress	
Pseudoalteromonas haloplanktis TAC 125	Coastal Antarctic water, Terre Adélie, Antarctica	completed/CR954246, CR954247	Medigue et al. (2005)
Desulfotalea psychrophila LSv54	Arctic marine sediments, Svalbard	completed/CR522870, CR522871, CR522872	Rabus et al. (2004)
Exiguobacterium sibiricum 255-15	Siberian permafrost	completed/AADW00000000	Rodrigues et al. (2008)
Psychroflexus torquis ATCC 700755	Sea ice algal assemblage, Prydz Bay, Antarctica	draft/AAPR00000000	
Polaribacter filamentous 215	Surface water, north of Deadhorse, Alaska	in progress	
Polaribacter irgensii 23-P	Nearshore waters off Peninsula	draft/AAOG00000000	
Psychromonas ingrahamii 37	Sea ice, off Point Barrow, northern Alaska	completed/CP000510	Riley et al. (2008)
Marine Actinobacterium PHSC20C1	Nearshore waters, Peninsula	draft/AAOB00000000	

In *Ph*TAC125, a significant bias towards Asn residues was found (Médigue et al. 2005). Comparative genome analyses suggest that the psychrophilic lifestyle is most likely conferred by a combination of changes in overall genome content and amino-acid composition. Significantly high levels of polar residues (particularly Ser), substitution of Asp for Glu, and a general decrease in charged residues on the surface of proteins from the Arctic marine bacterium *C. psychrerythraea* 34H were identified as structural features that enable proteins to retain flexibility at low temperature (Methé et al. 2005). A reduction in Pro and Arg was observed in the genome of *Psychrobacter arcticus*, especially in genes responsible for cell growth and reproduction (Ayala-Del-Río et al. 2010).

Proteomic analyses have shown the presence of a large number of genes induced at low temperatures. The proteomes expressed at 4 and 18°C by *Ph*TAC125 have demonstrated that translation, protein folding, membrane integrity and antioxidant activities are up-regulated at 4°C (Piette et al. 2010). In *Ph*TAC125 the major cold-repressed proteins, undetectable at 4°C, were heat-shock proteins involved in folding processes (Piette et al. 2011).

E. sibiricum is constitutively adapted to cold with differential gene expression between 4 and 28°C (Rodrigues et al. 2008). Similarly, the proteome of *P. arcticus* expressed at 4 and 22°C shows the presence of 33 proteins potentially involved in adaptation to low temperature (Zheng et al. 2007).

To preserve their function, proteins must reach a balance of structural rigidity and flexibility in their environments. Generally, enzymes isolated from psychrophiles living in perennially cold habitats are endowed with high catalytic efficiency at low temperature and low stability due to enhanced flexibility (Feller and Gerday 2003).

Among cold-adapted bacteria, the genus *Colwellia*, within γ-proteobacteria, is unusual, i.e. all characterised members are strictly psychrophilic (requiring temperatures of −20°C to grow on solid media) and live in stably cold environments, including deep sea and Arctic and Antarctic sea ice (Deming and Junge 2005). Many species produce extracellular polymeric substances relevant to biofilm formation and cryoprotection (Krembs et al. 2002) and enzymes capable of degrading high-molecular-mass organic compounds. The genome sequence of *C. psychrerythraea*, an obligatory psychrophilic Arctic bacterium, has provided an important opportunity to better understand its potential functions in the marine environment and to gain insight into adaptation (Methé et al. 2005). Environments in which *Colwellia* has been found include ice formations currently under study as models of past ice ages on Earth (Deming 2002).

10.5 The Role of the High Oxygen Concentration in Cold Environments

Gases and radicals are more soluble and stable at low temperature, with consequences which are already evident from genome annotations of cold-adapted bacteria. Cold-adapted bacteria have developed responses to strong oxidative stress. Indeed marine organisms have been exposed to permanent excess of oxygen, due to its high solubility at cold temperature, leading to oxygen reserves larger than those available in warmer waters. The apparent benefits of easier oxygen supply are contrasted by the constraints on kinetic effects at low temperature, which impair the functional capacities of molecules, and by increased production of Reactive Oxygen Species (ROS). Therefore, augmented capacities in antioxidant defence are likely to be important components of evolutionary adaptations in a cold and oxygen-rich environment. While stability of ROS is increased at low temperature, their reactivity increases at higher temperature. These features may in part explain why the oxidative stress-related proteins are repressed at 4°C and strongly induced at 18°C as a result of a stimulated metabolic activity with the consequence of enhanced production of ROS (Piette et al. 2011).

C. psychrerythraea (Methé et al. 2005) seems to have faced high oxygen concentration by developing enhanced antioxidant capacity owing to the presence of three copies of catalase genes as well as two different superoxide-dismutase (SOD) genes, one of which is a nickel-containing SOD, never before reported in proteobacteria. In contrast, the genome sequence of *Ph*TAC125 reveals that the bacterium copes with increased oxygen solubility by enhancing production of oxygen-scavenging enzymes and deleting entire metabolic pathways, such as those which generate ROS as side products. The deletion of the ubiquitous molybdopterin-dependent metabolism in the *Ph*TAC125 genome (Médigue et al. 2005) and the number of proteins such as hemoglobins involved in scavenging chemical groups (*see below*) can be seen in this perspective. Dioxygen-consuming lipid desaturases achieve both protection against oxygen and synthesis of lipids, making the membrane fluid. The cold environment of *Ph*TAC125 poses the problem of how this microorganism copes with ROS. High levels of ROS are potentially toxic for the cell, being involved in a large number of pathological mechanisms (Finkel 2003). ROS may act as signalling molecules during cell differentiation, cell-cycle progression and in response to extracellular stimuli (Sauer et al. 2001). Low temperatures should favour oxygen solubility and increase the stability of the oxygen-derived toxic compounds.

10.6 The 2/2 Hemoglobins in PhTAC125

The bacterial-hemoglobin superfamily is composed of three phylogenetically distinct lineages (Vinogradov et al. 2005). Two of these lineages include the proteins characterised by the 3/3 α-helical myoglobin-like structure, flavohemoglobins and sensor hemoglobins, respectively involved in nitrosative stress and in adaptive responses to variations of gaseous physiological messengers. The third lineage comprises hemoglobins displaying the 2/2 topology and is widely distributed in bacteria, microbial eukaryotes and plants (Pesce et al. 2000). Although the function of 2/2 hemoglobins is not well understood, it has been proposed that they may be involved in intracellular oxygen storage or transfer, binding and detoxification of ROS, enzymatic function(s), oxygen sensing (Wittenberg et al. 2002) and sulfide binding (Nicoletti et al. 2010).

The genome of cold-adapted *Ph*TAC125 contains multiple genes encoding three distinct monomeric hemoglobins exhibiting a 2/2 α-helical fold, and one flavohemoglobin (Giordano et al. 2007), a protein with two domains, (1) the heme-containing oxygen-binding domain, and (2) a FAD-containing reductase domain, involved in NO detoxification (Poole et al. 1996). In contrast, the *C. psychrerythraea* genome does not possess genes encoding 2/2 hemoglobins. The unusually high number of 2/2 hemoglobins in *Ph*TAC125 strongly suggests that these proteins fulfil important physiological roles, perhaps related to the peculiar features of the Antarctic habitat.

Recent in vivo results demonstrate that inactivation of 2/2 hemoglobin encoded by the *PSHAa0030* gene (hereafter named *Ph*-2/2HbO) makes the mutant bacterial strain sensitive to high oxygen levels, hydrogen peroxide, and nitrosating agents (Parrilli et al. 2010). A transcriptional analysis of the genes encoding the three 2/2 hemoglobins and flavohemoglobin, carried out on the *Ph*TAC125 wild type and the *Ph*TAC125-*PSHAa0030* mutant, showed that the transcription of the flavohemoglobin-encoding gene is observed only in the mutant grown in micro-aerobiosis at 4°C, suggesting that the occurrence of the NO-induced stress is related to the absence of *Ph*-2/2HbO (Parrilli et al. 2010). However, the function of this protein still needs to be ascertained.

In addition, *Ph*-2/2HbO has an unusual extension of 15 residues at the N terminus (pre-A helix). A similar situation has also been found in *Mycobacterium tuberculosis* hemoglobin and appears to occur in many slow-growing *Mycobacterium* species. The X-ray structure of *M. tuberculosis* hemoglobin (1IDR) showed that the pre-A motif does not significantly contribute to the structural integrity, protruding out of the compact globin fold, but rather confers a basic contribution in regulating nitrogen-monoxide-dioxygenase activity (Lama et al. 2009).

Over-expressed *Ph*-2/2HbO was recently characterised by spectroscopy, kinetic measurements and computer simulation approaches (Howes et al. 2011; Giordano et al. 2011). The ensemble of results indicates unique adaptive structural properties conferring higher flexibility to the protein that may facilitate the functioning in the cold by providing greater freedom for the correct positioning of ligand(s), even at low temperatures. The recombinant protein is 6-coordinated in the ferric and ferrous forms, in strong dependence on pH and temperature (Giordano et al. 2007, 2011; Verde et al. 2009; Howes et al. 2011).

The 6-coordinated hemoglobins are generally observed in bacteria, unicellular eukaryotes, plants, invertebrates and some tissues of higher vertebrates (Vinogradov and Moens 2008), but only a few bacterial 2/2 hemoglobins have been examined and reported in the literature. The occurrence of distinct oxidation/coordination states in members of the hemoglobin superfamily is not uniform, suggesting that the functional roles of these oxidation states are multiple, possibly being a tool for modulating ligand-binding or redox properties. The fact that 6-coordination is widespread in Antarctic marine organisms, fish and bacteria, suggests that this conformation may be useful in presence of high oxygen concentration.

10.7 Conclusions

Psychrophilic organisms have successfully coped with the two main physical challenges they had to face, namely low thermal energy and high viscosity, both of which slow down the metabolic flux. The apparent benefits of easier oxygen supply are contrasted by the constraints on kinetic effects at low temperature (D'Amico et al. 2006), which impair the functional capacities of molecules, and by increased production of ROS. The adaptive modifications appear to rely on higher flexibility of key parts of the molecule and/or decreased stability, partially compensating the effects of low temperature. At all levels analysed, the functional adaptation to permanently low temperature seems to require the maintenance of molecule flexibility for supporting the cellular functioning. Proteins are the main factors of the ensuing mechanisms of adaptation.

Acknowledgments This study is financially supported by the Italian National Programme for Antarctic Research (PNRA). It is in the framework of the SCAR programme Evolution and Biodiversity in the Antarctic (EBA), the project CAREX (Coordination Action for Research Activities on Life in Extreme Environments), European Commission FP7 call ENV.2007.2.2.1.6. DG and RR acknowledge CNR (for Short-Term Mobility fellowships) and CAREX (for Transfer of Knowledge grants).

References

Ayala-Del-Río HL, Chain PS, Grzymski JJ, Ponder MA, Ivanova N, Bergholz PW, Di Bartolo G, Hauser L, Land M, Bakermans C, Rodrigues D, Klappenbach J, Zarka D, Larimer F, Richardson P, Murray A, Thomashow M, Tiedje JM (2010) The genome sequence of psychrobacter arcticus 273–4, a psychroactive siberian permafrost bacterium, reveals mechanisms for adaptation to low temperature growth. Appl Environ Microbiol 76: 2304–2312

Carpenter JF, Crowe JH (1988) The mechanism of cryoprotection of proteins by solutes. Cryobiol 25:244–255

Casanueva A, Tuffin M, Craig C, Cowan DA (2010) Molecular adaptations to psychrophily: the impact of "omic" technologies. Trends Microbiol 18:374–381

Cavicchioli R, Siddiqui KS, Andrews D, Sowers KR (2002) Low-temperature extremophiles and their applications. Curr Opin Biotech 13:253–261

Cavicchioli R, Thomas T, Curmi PM (2000) Cold stress response in archaea. Extremophiles 4:321–331

Chauhan S, Shivaji S (1994) Growth and pigmentation in *Sphingobacterium antarcticus*, a psychrothrophic bacterium from Antarctica. Polar Biol 15:215–219

Chintalapati S, Kiran MD, Shivaji S (2004) Role of membrane lipid fatty acids in cold adaptation. Cell Mol Biol (Noisy-le-grand) 50:631–642

D'Amico S, Collins T, Marx JC, Feller G, Gerday C (2006) Psychrophilic microorganisms: challenges for life. EMBO Rep 7:385–389

Deming JW (2002) Psychrophiles and polar regions. Curr Opin Microbiol 5:301–309

Deming JW, Junge K (2005) Colwellia. In: Staley GT, Benner DJ, Krieg NR, Garrity GM (eds) The proteobacteria, part B, bergey's manual of systematic bacteriology, vol 2, 2nd edn. Springer, New York, pp 447–454

Duplantis BN, Osusky M, Schmerk CL, Ross DR, Bosio CM, Nano FE (2010) Essential genes from Arctic bacteria used to construct stable, temperature-sensitive bacterial vaccines. Proc Natl Acad Sci USA 107:3456–13460

Eriksson S, Hurme R, Rhen M (2002) Low temperature sensors in bacteria. Phil Trans R Soc Lond B 357:887–893

Feller G (2010) Protein stability and enzyme activity at extreme biological temperatures. J Phys Condens Matter 22:323101

Feller G, Gerday C (2003) Psychrophilic enzymes: hot topics in cold adaptation. Nat Rev Microbiol 1:200–208

Ferrer M, Chernikova TN, Yakimov M, Timmis KN, Golyshin PN (2003) Chaperonins govern growth of *Escherichia coli* at low temperatures. Nat Biotechnol 21:1266–1267

Finkel T (2003) Oxidant signals and oxidative stress. Curr Opin Cell Biol 15:247–254

Georlette D, Blaise V, Collins T, D'Amico S, Gratia E, Hoyoux A, Marx JC, Sonan G, Feller G, Gerday C (2004) Some like it cold: biocatalysis at low temperatures. FEMS Microbiol Rev 28:25–42

Giordano D, Parrilli E, Dettaï A, Russo R, Barbiero G, Marino G, Lecointre G, di Prisco G, Tutino ML, Verde C (2007) The truncated hemoglobins in the Antarctic psychrophilic bacterium *Pseudoalteromonas haloplanktis* TAC125. Gene 398:69–77

Giordano D, Russo R, Ciaccio C, Howes BD, di Prisco G, Smulevich G, Marden MC, Hui Bon Hoa G-H, Coletta M, Verde C (2011) Ligand- and proton-linked conformational changes of the ferrous 2/2 hemoglobin of *Pseudoalteromonas haloplanktis* TAC125. IUBMB Life 63:566–573

Howes BD, Giordano D, Boechi L, Russo R, Mucciacciaro S, Ciaccio C, Sinibaldi F, Fittipaldi M, Martì MA, Estrin DA, di Prisco G, Coletta M, Verde C, Smulevich G (2011) The peculiar heme pocket of the 2/2 hemoglobin of cold adapted *Pseudoalteromonas haloplanktis* TAC125. J Biol Inorg Chem 16:299–311

Jagannadham MV, Rao VJ, Shivaji S (1991) The major carotenoid pigment of a psychrotrophic *Micrococcus roseus* strain: purification, structure, and interaction with synthetic membranes. J Bacteriol 173:7911–7917

Krembs C, Eicken H, Junge K, Deming JW (2002) High concentrations of exopolymeric substances in Arctic winter sea ice: implications for the polar ocean carbon cycle and cryoprotection of diatoms. Deep-Sea Res A 49:2163–2181

Kumar S, Nussinov R (2004) Different roles of electrostatic in heat and in cold: adaptation by citrate synthase. Chem-BioChem 5:280–290

Lama A, Pawaria S, Bidon-Chanal A, Anand A, Gelpí JL, Arya S, Martí M, Estrin DA, Luque FJ, Dikshit KL (2009) Role of Pre-A motif in nitric oxide scavenging by truncated hemoglobin, HbN, of *Mycobacterium tuberculosis*. J Biol Chem 284:14457–14468

Marx JC, Blaise V, Collins T, D'Amico S, Delille D, Gratia E, Hoyoux A, Huston AL, Sonan G, Feller G, Gerday C (2004) A perspective on cold enzymes: current knowledge and frequently asked questions. Cell Mol Biol Noisy-le-grand 50:643–655

Médigue C, Krin E, Pascal G, Barbe V, Bernsel A, Bertin PN, Cheung F, Cruveiller S, D'Amico S, Duilio A, Fang G, Feller G, Ho C, Mangenot S, Marino G, Nilsson J, Parrilli E, Rocha EP, Rouy Z, Sekowska A, Tutino ML, Vallenet D, von Heijne G, Danchin A (2005) Coping with cold: the genome of the versatile marine Antarctic bacterium *Pseudoalteromonas haloplanktis* TAC125. Genome Res 15:1325–1335

Methé BA, Nelson KE, Deming JW, Momen B, Melamud E, Zhang X, Moult J, Madupu R, Nelson WC, Dodson RJ, Brinkac LM, Daugherty SC, Durkin AS, DeBoy RT, Kolonay JF, Sullivan SA, Zhou L, Davidsen TM, Wu M, Huston AL, Lewis M, Weaver B, Weidman JF, Khouri H, Utterback TR, Feldblyum TV, Fraser CM (2005) The psychrophilic lifestyle as revealed by the genome sequence of *Colwellia psychrerythraea* 34H through genomic and proteomic analyses. Proc Natl Acad Sci USA 102:10913–10918

Moline MA, Karnovsky NJ, Brown Z, Divoky GJ, Frazer TK, Jacoby CA, Torres JJ, Fraser WR (2008) High latitude changes in ice dynamics and their impact on polar marine ecosystems. Annu NY Acad Sci 1134:267–319

Motohashi K, Watanabe Y, Yohda M, Yoshida M (1999) Heat-inactivated proteins are rescued by the DnaK.J-GrpE set and ClpB chaperones. Proc Natl Acad Sci USA 96:7184–7189

Murray AE, Grzymski JJ (2007) Diversity and genomics of Antarctic marine microorganisms. Phil Trans R Soc B 362:2259–2271

Nicoletti FP, Comandini A, Bonamore A, Boechi L, Boubeta F, Feis A, Smulevich G, Boffi A (2010) Sulfide binding properties of truncated hemoglobins. Biochemistry 49:2269–2278

Parrilli E, Giuliani M, Giordano D, Russo R, Marino G, Verde C, Tutino ML (2010) The role of a 2-on-2 haemoglobin in oxidative and nitrosative stress resistance of Antarctic *Pseudoalteromonas haloplanktis* TAC125. Biochimie 92:1003–1009

Pegg DE (2007) Principles of cryopreservation. Meth Mol Biol 368:39–57

Pesce A, Couture M, Dewilde S, Guertin M, Yamauchi K, Ascenzi P, Moens L, Bolognesi M (2000) A novel two-over-two alpha-helical sandwich fold is characteristic of the truncated hemoglobin family. EMBO J 19:2424–2434

Piette F, D'Amico S, Mazzuchelli G, Danchin A, Leprince P, Feller G (2011) Life in the cold: a proteomic study of cold-repressed proteins in the Antarctic bacterium *Pseudoalteromonas haloplanktis* TAC125. Appl Environ Microbiol 77:3881–3883

Piette F, D'Amico S, Struvay C, Mazzuchelli G, Renaut J, Tutino ML, Danchin A, Leprince P, Feller G (2010) Proteomics of life at low temperatures: trigger factor is the primary chaperone in the Antarctic bacterium *Pseudoalteromonas haloplanktis* TAC125. Mol Microbiol 76:120–132

Poole RK, Anjum MF, Membrillo-Hernàndez J, Kim SO, Hughes MN, Stewart V (1996) Nitric oxide, nitrite, and Fnr regulation of hmp (flavohemoglobin) gene expression in *Escherichia coli* K-12. J Bacteriol 178:5487–5492

Rabus R, Ruepp A, Frickey T, Rattei T, Fartmann B, Stark M, Bauer M, Zibat A, Lombardot T, Becker I, Amann J, Gellner K, Teeling H, Leuschner WD, Glöckner FO, Lupas AN, Amann R, Klenk HP (2004) The genome of *Desulfotalea psychrophila*, a sulfate-reducing bacterium from permanently cold Arctic sediments. Environ Microbiol 6:887–902

Ray MK, Kumar GS, Janiyani K, Kannan K, Jagtap P, Basu MK, Shivaji S (1998) Adaptation to low temperature and regulation of gene expression in Antarctic psychrotrophic bacteria. J Biosci 23:423–435

Riley M, Staley JT, Danchin A, Wang TZ, Brettin TS, Hauser LJ, Land ML, Thompson LS (2008) Genomics of an extreme psychrophile, psychromonas ingrahamii. BMC Genomics 9:210

Rodrigues D, Tiedje M (2008) Coping with our cold planet. Appl Environ Microbiol 74:1677–1686

Rodrigues D, Ivanova N, He Z, Huebner M, Zhou J, Tiedje M (2008) Architecture of thermal adaptation in an *Exiguobacterium sibiricum* strain isolated from 3 million year old permafrost: a genome and transcriptome approach. BMC Genomics 9:547

Russell NJ (1998) Molecular adaptations in psychrophilic bacteria: potential for biotechnological applications. Adv Biochem Eng Biotechnol 61:1–21

Russell NJ (2007) Psychrophiles: membrane adaptations. In physiology and biochemistry of extremophiles. In: Gerday C, Glansdorff N (eds) ASM Press, Washington, pp 155–164

Russo R, Giordano D, Riccio A, di Prisco G, Verde C (2010) Cold-adapted bacteria and the globin case study in the Antarctic bacterium *Pseudoalteromonas haloplanktis* TAC125. Mar Gen 3:125–131

Sauer H, Wartenberg M, Hescheler J (2001) Reactive oxygen species as intracellular messengers during cell growth and differentiation. Cell Physiol Biochem 11:173–186

Saunders NF, Thomas T, Curmi PM, Mattick JS, Kuczek E, Slade R, Davis J, Franzmann PD, Boone D, Rusterholtz K, Feldman R, Gates C, Bench S, Sowers K, Kadner K, Aerts A, Dehal P, Detter C, Glavina T, Lucas S, Richardson P, Larimer F, Hauser L, Land M, Cavicchioli R (2003) Mechanisms of thermal adaptation revealed from the genomes of the Antarctic Archaea *Methanogenium frigidum* and *Methanococcoides burtonii*. Gen Res 13:1580–1588

Shivaji S, Prakash Jogadhenu SS (2010) How do bacteria sense and respond to low temperature? Arch Microbiol 192:85–95

Tehei M, Franzetti B, Madern D, Ginzburg m, Ginzburg BZ, Giudici-Orticoneìi MT, Bruschi M, Zaccai G (2004) Adaptation to extreme environments: macromolecular dynamics in bacteria compared in vivo by neutron scattering. EMBO Reports 5:66–70

Verde C, Giordano D, Russo R, Riccio A, Vergara A, Mazzarella L, di Prisco G (2009) Hemoproteins in the cold. Mar Gen 2:67–73

Vinogradov SN, Hoogewijs D, Bailly X, Arredondo-Peter R, Guertin M, Gough J, Dewilde S, Moens L, Vanfleteren JR (2005) Three globin lineages belonging to two structural classes in genomes from the three kingdoms of life. Proc Natl Acad Sci USA 102:11385–11389

Vinogradov S, Moens L (2008) Diversity of globin function: enzymatic, transport, storage, and sensing. J Biol Chem 283:8773–8777

Wada H, Murata N (1989) *Synechocystis* PCC6803 mutants defective in desaturation of fatty acids. Plant Cell Physiol 30:971–978

Watanabe YH, Yoshida M (2004) Trigonal DnaK-DnaJ complex versus free DnaK and DnaJ; heat stress converts the former to the latter and only the latter can do disaggregation in cooperation with ClpB. J Biol Chem 279:15723–15727

Wittenberg JB, Bolognesi M, Wittenberg BA, Guertin M (2002) Truncated hemoglobins: a new family of hemoglobins widely distributed in bacteria, unicellular eukaryotes, and plants. J Biol Chem 227:871–874

Zheng S, Ponder MA, Shih JY, Tiedje JM, Thomashow MF, Lubman DM (2007) A proteomic analysis of psychrobacter arcticus 273–4 adaptation to low temperature and salinity using a 2-D liquid mapping approach. Electrophoresis 28:467–488

Chapter 11
The Adaptive Evolution of Polar Fishes: Lessons From the Function of Hemoproteins

Cinzia Verde, Daniela Giordano, Roberta Russo and Guido di Prisco

11.1 Biogeography of Polar Fish

11.1.1 Antarctic Notothenioidei

The perciform suborder Notothenioidei, mostly confined within Antarctic and sub-Antarctic waters, dominates the modern Southern Ocean ichthyofauna. Notothenioids probably appeared in the early Tertiary and began to diversify on the Antarctic shelf in the middle Tertiary, adapting to progressive cooling (Eastman 1993). Notothenioids are morphologically and ecologically diverse, and account for 77% of the shelf fish diversity, 92% of abundance and 91% of biomass (Eastman 2005). They are monophyletic (Balushkin 2000; Chen et al. 2003; Near et al. 2004).

Over the past 40 million years the Antarctic shelf has undergone tectonic and oceanic events that began to alter the composition of the fish fauna and to initiate replacement. A key event was the opening of the Drake Passage, 23.5–41 million years ago (mya), between the tip of South America and the Antarctic Peninsula (Thomson 2004; Scher and Martin 2006). The Drake Passage generated the Antarctic Circumpolar Current (ACC), partially responsible for the cooling of waters. The Antarctic Polar Front (APF), the northern boundary of the ACC, is a

C. Verde · D. Giordano · R. Russo · G. di Prisco (✉)
Institute of Protein Biochemistry, National Research Council,
Via Castellino 111, 80131 Naples, Italy
e-mail: g.diprisco@ibp.cnr.it

C. Verde
e-mail: c.verde@ibp.cnr.it

D. Giordano
e-mail: d.giordano@ibp.cnr.it

R. Russo
e-mail: r.russo@ibp.cnr.it

G. di Prisco and C. Verde (eds.), *Adaptation and Evolution in Marine Environments, Volume 1*, From Pole to Pole, DOI: 10.1007/978-3-642-27352-0_11,

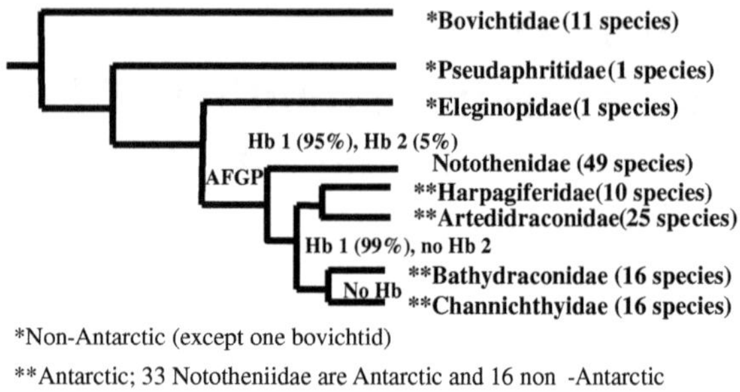

Fig. 11.1 Cladogram of the families of Notothenioidei. The cladogram is a strict consensus of four trees resulting from maximum parsimony analysis of the complete gene 16S rRNA dataset (Near et al. 2004). AFGP in the Antarctic clade and lack of Hb are mapped

roughly circular oceanic feature between 50 and 60°S. Just north of the APF, the surface water temperature is ca. 3°C warmer. Although the APF is somewhat "leaky", allowing northward transport of planktonic invertebrates (Clarke et al. 2005), it promoted initial isolation and diversification of cold-adapted notothenioids. The water column south of the APF is close to oxygen saturation at all depths. Oxygen solubility increases with temperature decrease, thus the cold seas are an oxygen-rich habitat. Antarctic notothenioids are stenothermal, and their ability to cope with the ongoing increases in temperature might be reduced. Thus, the question to what extent Antarctic fish may adapt to environmental change is a very important issue.

Eight families encompass 129 species (Fig. 11.1). Seven families have hemoglobin-containing erythrocytes. Channichthyidae are devoid of hemoglobin (Hb) (Ruud 1954). Three families (Bovichtidae, only one out of ten species is Antarctic; monotypic Pseudaphritidae and Eleginopidae) became established in waters areas around New Zealand, Australia and high-latitude South America.

As water temperatures decreased and ice appeared, Antarctic notothenioids acquired antifreeze glycoproteins (AFGPs), a key innovation and physiological adaptation that allows them to survive and diversify in ice-laden seawater that reaches temperatures of nearly −2°C (DeVries and Cheng 2005). Bovichtids, pseudaphritids and eleginopids have no AFGP genes, indicating that they diverged before isolation and cooling (Cheng et al. 2003). The split between eleginopids and the five families having AFGPs (100 species distributed south of the APF) is variously estimated to have occurred 5–14 (Chen et al. 1997), 27 (Bargelloni et al. 2000) and 40 mya (Near 2004). Near (2004) analyses the reasons for these discrepancies.

11.1.2 Non-Antarctic Notothenioidei (Sub-Antarctic and Temperate)

North of the APF, non-Antarctic notothenioids (sub-Antarctic and some temperate species) account for 22% (28 of 129 species) of notothenioid biodiversity (Eastman 2005). Bovichtidae (11 species), with the exception of *Bovichtus elongatus* (northern Antarctic Peninsula) are non-Antarctic. *B. diacanthus* from Tristan da

Cunha (37°S) lives near the northern limit for notothenioids (Eastman 1993) at temperatures of 13–19°C, which reach 27.4°C in summer tide pools. Morphological and karyological data (Voskoboinikova 2004; Eastman 2006; Mazzei et al. 2006) confirm that bovichtids are the phyletically basal family.

The single species of the monotypic family Pseudaphritidae, *Pseudaphritis urvillii*, a freshwater fish, is found in coastal waters, estuaries and rivers of Tasmania, Victoria, New South Wales and South Australia. Morphological analyses described *P. urvillii* as phyletically basal (Balushkin 1992, 2000), but this hypothesis was not supported by molecular data (Lecointre et al. 1997; Bargelloni et al. 2000; Near et al. 2004).

The status of the other monotypic family Eleginopidae as the sister group of Antarctic notothenioids is supported by phylogenetic analyses based on morphological and molecular data (Balushkin 2000; Bargelloni et al. 2000; Near et al. 2004). *Eleginops maclovinus* is non-Antarctic, and became established in cool temperate inshore marine and freshwater habitats. It is one of the two euryhaline notothenioids. It inhabits coastal waters, sounds and tidal creeks in the Falkland Islands and coastal waters and estuaries in southern South America, where *E. maclovinus* experiences annual temperatures of $\approx$0–15°C.

11.1.3 The Arctic Ichthyofauna

Biological comparison of species inhabiting the polar regions identifies the differences in evolutionary pressures between the two ecosystems. The main differences between the regions are the greater age and isolation of the Antarctic. The Arctic lies between North America, Greenland, Europe and Asia. Unlike the Southern Ocean, the sea north of the Arctic Circle is almost completely enclosed and influenced by extensively populated terrestrial areas and industrial activities. Exchange of Atlantic and Arctic waters occurs through the passage between Greenland and the Svalbard Islands. About 10–15 mya, Arctic land masses reached their present positions and temperatures dropped below freezing. Glaciation events may have begun 6–10 mya, much later than in Antarctica; however recent evidence suggests a "bipolar symmetry" in climate cooling (Moran et al. 2006; Krylov et al. 2008), namely simultaneous evolution of ice with a bipolar transition from "greenhouse" to "icehouse", highlighting the role of greenhouse warming. The earliest Arctic cooling events would then be dated much earlier, ca. 45 mya.

Due to human activities and unrelated to natural variability, there is a current loss of polar sea ice, most notably in the Arctic. The ocean may become ice-free during summer by the end of the century. This loss may have dramatic effects on species associated with sea ice for feeding and reproduction. The sub-Arctic is highly ecologically sensitive, and warming will affect habitat and species resilience and induce dramatic changes in community dynamics and structure. In the Antarctic Peninsula region, loss of sea and shelf ice is occurring at the same speed as in the Arctic. In polar environments changes occur much faster than in other regions.

Arctic fish have higher biodiversity. Being exposed to seasonal temperature variations, they exhibit higher physiological plasticity than the stenothermal Antarctic counterparts (experiencing stable water temperatures). Some cold-water features, e.g. identical AFGPs (Chen et al. 1997), are in common, but many physiological features do not show cold adaptation.

11.2 Adaptive Evolution of Fishes of Southern and Northern Oceans

The current changes make comparative work at the poles a source of invaluable indications on the evolution of adaptations. High latitudes and cold climates are common to the Antarctic and the Arctic, but in many respects the two regions are more dissimilar than similar. Faunal composition and diversity are strongly linked to geological history; hence there are historical, physical and biological differences between the two modern habitats. The modern polar faunas differ in age, endemism and physiological tolerance to various environmental parameters.

There are advantages in using organisms from both poles in evolutionary studies. Fish offer many opportunities for comparative approaches to understand cold adaptation and how to counteract ongoing climate changes. Zoarcidae and Liparidae are represented in both oceans. On the other hand, hypotheses of adaptation to a common environmental parameter have greater certainty in phyletically unrelated taxa. The two polar faunas include a number of such taxa, and the comparative approach permits molecule-to-organism analysis of convergent and parallel evolutionary trends to similar habitats.

11.3 The Oxygen Transport: Structure, Function, Adaptations of Hemoproteins

Oxygen carriers are a useful system for studying physiological adaptations. Few proteins have been studied in such a wide array of organisms as *hemoglobin* (Hb). Hbs, found in bacteria, fungi, plants and animals, are ancient proteins that probably evolved from enzymes that protected against toxic oxygen levels. During 2 billion years, the oxygen levels in the atmosphere had been very low, then reached the current levels about 540 mya. At those times, the Hb ancestor was likely to have adapted to scavenge excessive oxygen and/or to have been involved in detoxification of nitrogen monoxide.

Hbs are highly sensitive to temperature, and the comparison between cold- and non-cold-adapted Hbs may be informative for many biological questions. Psycrophilic bacteria, will yield useful indications. The genome of Antarctic marine

Pseudoalteromonas haloplanktis TAC125 has been sequenced and annotated, shedding light on several molecular features selectively developed in cold environments (see Chap. 10: di Prisco et al. 2012). These studies are likely to afford unequivocal conclusions more easily than in vertebrates, whose physiology and metabolism is much more complex. Comparison between phylogenetically related bacteria living in freezing and non-freezing habitats may provide additional insights into globin evolution. Such an approach in fish is hampered by lack of genomic sequences.

Most animals carry oxygen to the tissues by one or more respiratory proteins. In fish, sensitive sentinels of environmental challenge and responses to temperature adaptation, Hbs evolved structural and functional diversity to adapt to all types of selective pressure; however, both the helical structure and a large number of amino-acid residues are well conserved.

Higher-vertebrate Hbs have developed a common molecular mechanism based on ligand-linked conformational change in a multi-subunit structure; generally, the molecule exhibits positive cooperativity between oxygen-binding sites (homotropic interactions). The pioneering studies of Max Perutz demonstrated that Hb can exist in alternate quaternary molecular conformations (Perutz et al. 1987), with transition between low- and high-affinity states, tense (T) and relaxed (R) (Monod et al. 1965), involved in the modulation of oxygen affinity by physiological effectors (heterotropic interactions), which may preferentially bind to the T or R state, thereby lowering or enhancing the overall oxygen affinity.

Elucidating molecular mechanisms of adaptation is one of the main goals in evolutionary biology. Evolutionary processes have allowed marine organisms to colonise habitats, conserving the essential feature of Hb *vis-à-vis* of widely different environmental conditions. The replacement of few key amino-acid residues may lead to a different function (Perutz 1983). In polar Hbs, however, it has not been possible to ascribe thermal adaptation to specific substitutions. The situation is much more complex, and is probably linked to the combination and interplay of a number of factors. One or more residues do not necessarily produce a specific adaptation and/or function. Cold adaptation of oxygen transport in high-Antarctic notothenioids seems based on evolutionary changes involving levels of biological organisation higher than the Hb structure. These include changes in the rate of Hb synthesis or in regulation by allosteric effectors, which affect the amount of transported oxygen. These factors are very important for short-term response to environmental challenges. In the same time span, sequence changes altering oxygen affinity would occur to a much lower extent, and are therefore considered long-term adaptations. Fish may also modulate oxygen delivery by changing the concentrations of effectors, and by expression of multiple, functionally different Hbs.

Novel hemoproteins have recently been found in vertebrates, e.g. *neuroglobin and cytoglobin*. Neuroglobin, mainly expressed in retinal neurons and fibroblast-like cells, plays a neuroprotective role during hypoxic stress (Brunori and Vallone 2007; Halder et al. 2007). Cytoglobin is ubiquitously expressed in vertebrate tissues.

The physiological role of these hemoproteins is unclear. A role in oxygen transport cannot be excluded, but seems unlikely. Neuroglobin (1) may scavenge oxygen under hypoxic conditions and supply it for aerobic respiration; (2) may function as terminal oxidase by oxidising NADH under hypoxic conditions, enhancing ATP production; (3) may be an oxygen-sensor, activating other proteins with regulatory function; (4) may be involved in nitrogen-monoxide metabolism (reviewed in di Prisco et al. 2011). The discovery of the neuroglobin gene in the brain of red-blooded notothenioids and in at least 13 of the 16 icefish species suggests a crucial biological function (Cheng et al. 2009). The finding that icefish retain the gene despite having lost Hb, and myoglobin in most species, may have important implications in the physiology and pathology of the brain. Polar fish will be a suitable model to learn more about the function of these proteins.

11.3.1 Hemoglobins of Fishes of the Southern-Ocean–Notothenioidei

The hematological features of *high-Antarctic Notothenioidei* have been extensively investigated in the past few decades. Notothenioids have evolved reduced erythrocyte number and Hb concentration/multiplicity (Everson and Ralph 1968; Wells et al. 1980). This may be advantageous in coping with increased viscosity of body fluids at low temperature (Wells et al. 1990), and finds partial compensation in the increased blood volume and higher cardiac output. One family has abolished Hb as oxygen carrier (Ruud 1954): the blood of Channichthyidae (icefishes) lacks Hb and erythrocytes and is thus colourless. These are the only known adult vertebrates showing such an astonishing adaptation (see below).

Most notothenioids are bottom dwellers. The red-blooded families generally have a single Hb (Hb 1) accounting for 95–99% of the total (except in two Bovichtidae, see below), accompanied by the minor or "embryonic" components Hb C (in traces) and Hb 2 (ca. 5% of the total), having one of the globins in common with Hb 1 (di Prisco et al. 1991; Verde et al. 2006b). In three Nototheniidae, i.e. *Trematomus newnesi* (D'Avino et al. 1994), *Pagothenia borchgrevinki* (Riccio et al. 2000) and *Pleuragramma antarcticum* (Tamburrini et al. 1996), minor Hbs are expressed at higher levels (ca. 25% of the total). Species of modern Artedidraconidae and Bathydraconidae lack minor Hbs. Multiplicity of Hbs in fish is linked to the need to respond to variable environmental conditions associated with a given habitat. Furthermore, the availability of several genes for α and β globins may provide protection against deleterious mutations in an individual gene. Lack of multiple globin genes is not detrimental in thermostable environments, as shown in the modern families (see below). The oxygen affinity of Hbs of many high-Antarctic species is quite low (di Prisco et al. 1988), as indicated by the values of p_{50} (partial pressure to achieve half-saturation). The Hbs of active species appear to have the lowest affinity (di Prisco et al. 1991).

The red-blooded modern families Artedidraconidae and Bathydraconidae show some intriguing features. Adult fish only have Hb 1. Artedidraconid Hbs display weak pH and effector regulation, and very low cooperativity of oxygen binding (Tamburrini et al. 1998), thus having–although modern–functional properties typical of ancestral organisms. An additional physiological role of Hb might be that of an "oxygen store" under hypoxic or anoxic conditions.

Comparison of adaptations of cold-adapted Antarctic notothenioids with *sub-Antarctic and temperate notothenioids*, and with Arctic fish, has been a powerful tool to understand whether (and to what extent) an extreme environment has required specific adaptations (Verde et al. 2004, 2006b; Giordano et al. 2006; di Prisco et al. 2007; Coppola et al. 2010).

Non-Antarctic Bovichtidae, the basal notothenioid family (Eastman 1993), do not have any of the physiological and biochemical adaptations to the extreme environmental conditions shared by most notothenioids. Unlike Antarctic notothenioids but similar to many other fish, adult *Cottoperca gobio*, thriving in sub-Antarctic waters north of the Polar Front, has two major Hbs sharing the α chain. Higher multiplicity has also been observed in *B. diacanthus*, one of the most northern notothenioids. The more complex oxygen-transport system in *C. gobio* and *B. diacanthus* may have been maintained by positive selection to deal with the large temperature changes north of the APF.

Similar to most high-Antarctic notothenioids, *P. urvillii* (Pseudapritidae) and *E. maclovinus* (Eleginopidae) have Hb 1 and Hb 2; the latter, similar to *T. newnesi*, also has Hb C. The low amount of Hb 2 can be a synapomorphy linking *P. urvillii* and *E. maclovinus* to the other notothenioids. *E. maclovinus* is the sister taxon of non-bovichtid and non-pseudaphritid notothenioids, consistent with the sub-Antarctic status of AFGP-less *E. maclovinus*. A switch to exclusive expression of the embryonic (minor) β-globin gene has occurred in adult *C. gobio*. Such switch may have occurred exclusively along the lineage leading to *C. gobio*, because *P. urvillii* conforms to the Hb pattern of most notothenioids. An embryonic character in the *C. gobio* ancestor can be interpreted as neoteny, possibly consequential to the maintenance of high expression of the embryonic adult genes followed by impaired expression of the major "adult" β-globin gene (Giordano et al. 2006).

Oxygen affinity is very high in Hbs of non-Antarctic *C. gobio, B. diacanthus, P. urvillii* and *E. maclovinus*; p_{50} values indicate that a decrease in affinity occurred along the lineage of the high-Antarctic notothenioids, with the exceptions of *P. antarcticum*, *Trematomus bernacchii* and *Artedidraco orianae*, whose Hbs have higher oxygen affinity than other Antarctic notothenioids. The relationship between higher oxygen affinity in non-Antarctic notothenioid Hbs and habitat features remains an open question. It is worth noting that spectroscopic and modelling studies on *P. urvillii* Hb 1 have shown that all the non-conservative replacements in the primary structure of α and β chains leave the conformation and electrostatic field surrounding the heme pocket essentially unmodified with respect to high-Antarctic Hb 1 (Verde et al. 2004).

The relevance of studying Hbs of non-Antarctic species in the evolution of Notothenioidei is shown by the analysis of the features of the Hb system of another temperate notothenioid, the more recently evolved *Notothenia angustata* (family Nototheniidae), common near the coast of southern New Zealand. The sequence identity between Hbs of non-cold adapted *N. angustata* and cold-adapted *Notothenia coriiceps* is the highest ever found among notothenioids (Fago et al. 1992). It is devoid of AFGP but has the AFGP genes (Cheng et al. 2003), which may become functional upon cold acclimation. These findings support the hypothesis that, unlike *P. urvillii*, this species had developed cold adaptation before migration from the Antarctic continental shelf to temperate latitudes in a relatively recent geological time, much later than *P. urvillii*. In comparison with Antarctic cold-adapted notothenioids, similar to *P. urvillii*, *N. angustata* displays high oxygen affinity.

11.3.1.1 The Family Channichthyidae and the Reduced Role of Hb

Icefishes (Channichthyidae) are the notothenioid crown group. Radiation of species within the icefish clade appears to have been confined to the last one million years (Bargelloni et al. 1994). Lack of Hb is balanced by high oxygen solubility and low metabolic rates in the cold. Icefish developed compensatory adaptations that reduce oxygen demand and enhance oxygen transport (e.g. decreased metabolic rates, enhanced gas exchange by large, well-perfused gills and through scaleless skin, large increases in cardiac output and blood volume). Delivery occurs by transport of oxygen dissolved in the plasma. These compensatory adaptations argue that lack of Hb and erythrocytes ("disaptation", see Chap. 7 Tota et al. 2012) is maladaptive under physiological stress.

The icefish evolution to the Hb-less phenotype arose from large-scale deletional events, which removed all globin genes, including embryonic/juvenile, with the exception of the 3′ transcriptionally inactive end of adult α-globin gene (Cocca et al. 1995; Zhao et al. 1998; di Prisco et al. 2002). *Neopagetopsis ionah* has a complete, but non-functional, adult $\alpha\beta$-globin complex (Near et al. 2006). This pseudogene complex may be an intermediate "genomic fossil" revealing key mechanisms on the pathway to loss of expression by all icefish.

Only icefish have taken such a radical course, whereas the other Antarctic families have only partial Hb reduction. The benefits include reduced costs for protein synthesis, simplified metabolic pathways and lower amounts of oxygen radicals. However, the shift to oxygen transport based on diffusion may cause higher vulnerability to warmer temperatures.

The question arises: does Hb remain absolutely vital for adequate oxygen transport in red-blooded notothenioids, or is it a vestigial relict which may be redundant under stress-free living conditions? Despite gradual reduction of the hematocrit in cannulated red-blooded specimens, and reversibly "poisoning" Hb by carbon monoxide (lethal for organisms whose life depends on oxygen), survival occurred with no obvious ill effects. The answer is: routine oxygen transport is possible in the absence of functional Hb, even during bouts of enforced exercise (di Prisco et al. 1992; di Prisco 2000). Similar to icefish, red-blooded Antarctic fish can carry oxygen in the plasma, suggesting that, in the cold and stable Antarctic sea, Hb is not essential for this function; a single Hb in limited amount may be a consequence of its reduced role as oxygen carrier (Verde et al. 2006b). In temperate and tropical fish the essential role of Hb in oxygen transport is undisputed.

11.3.2 Hemoglobins of Arctic Fish

Many Arctic species display Hb multiplicity. For instance, the spotted wolffish *Anarhichas minor*, a benthic, sedentary fish of the family Anarhichadidae (Zoarcoidei), has three functionally distinct major components, whose amino-acid sequences and oxygen-binding properties have been described (Verde et al. 2002). High multiplicity and functional differences have also been observed in three Gadidae, namely the Arctic cod *Arctogadus glacialis*, the polar cod *Boreogadus saida* and the Atlantic cod *Gadus morhua* (Verde et al. 2006a); they also have three major Hbs. *A. glacialis* is sedentary, living in the Greenland fjords. *B. saida* and *G. morhua* are pelagic and migratory, the former thriving in polar seas, whereas *G. morhua* is also found in northern temperate waters. Multiple Hbs provide a strategy for finely tuned regulation of oxygen transport in response to environmental variability and/or variations in metabolic demands (e.g. in physiological hypoxia), illustrating how in some instances, adaptive modifications of physiological pathways may arise from opportunistic retention of plesiomorphic characters.

11.3.3 The Root Effect

The decreased oxygen affinity of Hb at lower pH values in the physiological range is known as the alkaline *Bohr effect*. In many fish Hbs, when pH is lowered, the oxygen affinity decreases to such an extent that Hbs cannot be fully saturated even at very high oxygen pressure. At low pH, cooperativity is lost and the oxygen capacity of blood undergoes reduction of 50% or more of the value measured at alkaline pH. This feature is known as the *Root effect*. Root-effect Hbs are so strongly pH dependent that they can unload a large amount of bound oxygen at low pH and against a pressure gradient. The Root effect dictates to what extent the oxygen tension can be raised in acid-producing tissues.

The physiological significance of Root-effect Hbs has been linked to the presence of at least one of two anatomical structures requiring high oxygen pressure: the *rete mirabile*, supplying the gas gland that inflates fish swim bladders with oxygen (regulating buoyancy), and the choroid *rete*, a vascular structure which supplies oxygen to the poorly vascularised retina (Wittenberg and Wittenberg 1974). The Root effect evolved 100 million years before the appearance of the choroid *rete* (Berenbrink et al. 2005), whereas the swim bladder evolved independently at least four times. Arctic fish have the swim bladder, whereas all notothenioids do not. The eye choroid *rete* probably represents the most ancient anatomical structure associated with Root-effect Hbs (Wittenberg and Haedrich 1974).

The investigated Arctic species possess the choroid *rete*. Notothenioids display a more variable scenario (Eastman 1988, 2006): all non-Antarctic species have a well-developed *rete*, but many high-Antarctic notothenioids have lost it, although several have a small/vestigial form. Icefish and Bathydraconidae lack the *rete*, but *Racovitzia glacialis* retains a vestigial form. When analysing the evolution of the Root effect (Verde et al. 2006b), the lineages leading to Arctic fish are characterised by a strong effect, that increases even further in non-Antarctic notothenioids. A weakening is seen in the Antarctic notothenioid lineage but, similar to the choroid *rete*, there are a few exceptions, i.e. some species display a strong Root effect.

The question whether the Root effect in high-Antarctic notothenioid Hbs is related or not to environmental conditions still remains to be answered. The fact that these fish still have Root-effect Hbs also when the choroid *rete* is absent suggests that this function is subjected to neutral selection, not representing a disadvantage for the species. It is yet impossible to ascribe the presence or absence of the Root effect to substitutions of a few amino-acid residues, or to a single explanation. On the other hand, its weakening in many high-Antarctic notothenioids indicates that it is not an all-or-nothing phenomenon, suggesting that it may be generated by combination and interplay of several factors.

11.4 The Molecular Evolution of Hemoglobin in Polar Fish

To address some of the questions regarding the physiological and biochemical adaptations in polar fish, part of this review has been focussed on Hb phylogeny (see also Chap. 6 Lecointre 2012).

Investigating the evolutionary processes leading to adaptations in the oxygen transport requires a well-resolved phylogenetic hypothesis. Published molecular and morphological hypotheses of notothenioid phylogeny are strongly congruent. Figure 11.2 exhibits the interrelationships among the main three taxonomic components, namely notothenioids, zoarcoids and gadids. In the molecular phylogenetic analysis the globins of major and minor Antarctic fish Hbs cluster in two separate, strongly supported groups, with the globins of temperate fish Hbs forming the first divergence lineage.

Figures 11.3 and 11.4 report the Neighbour Joining trees for α and β globins. As a result of the isolation of Antarctica, the genotype of Notothenioidei diverged with respect to other fish groups in a way interpreted as typical of a species flock (Eastman and McCune 2000).

Fig. 11.2 Fish interrelationship, also supported by classification from anatomical evidence. Others: non-acanthomorph sequences of the current data set. Adapted from Verde et al. 2006a

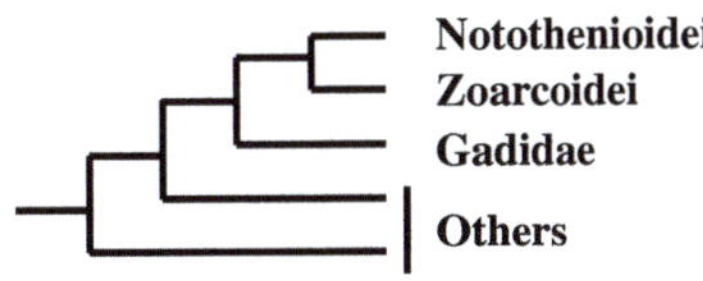

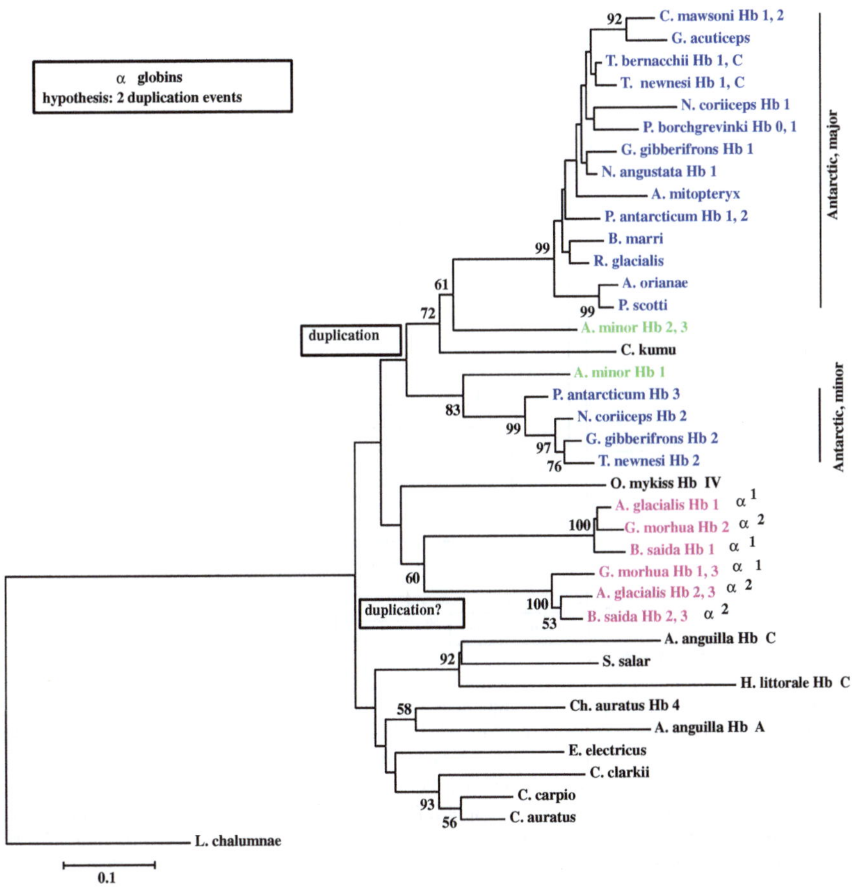

Fig. 11.3 Neighbour Joining tree of amino-acid sequences of α chains of Arctic, Antarctic and temperate fish Hbs. Bootstrap proportions (BP, percentage of 10,000 replicates) are given at the nodes. Globin sequences of notothenioids are in blue, of zoarcoids in green, of gadids in pink, of other fishes in black. Adapted from Dettaï et al. (2008)

When evolutionary pressures and rates of change are the same across taxa, similarity is proportional to phylogeny, and in that case the gene (or protein) tree reflects the species tree. The topologies suggest different evolutionary histories for the α and β chains. Globin paralogs (gene copies originated by duplication in a given genome)

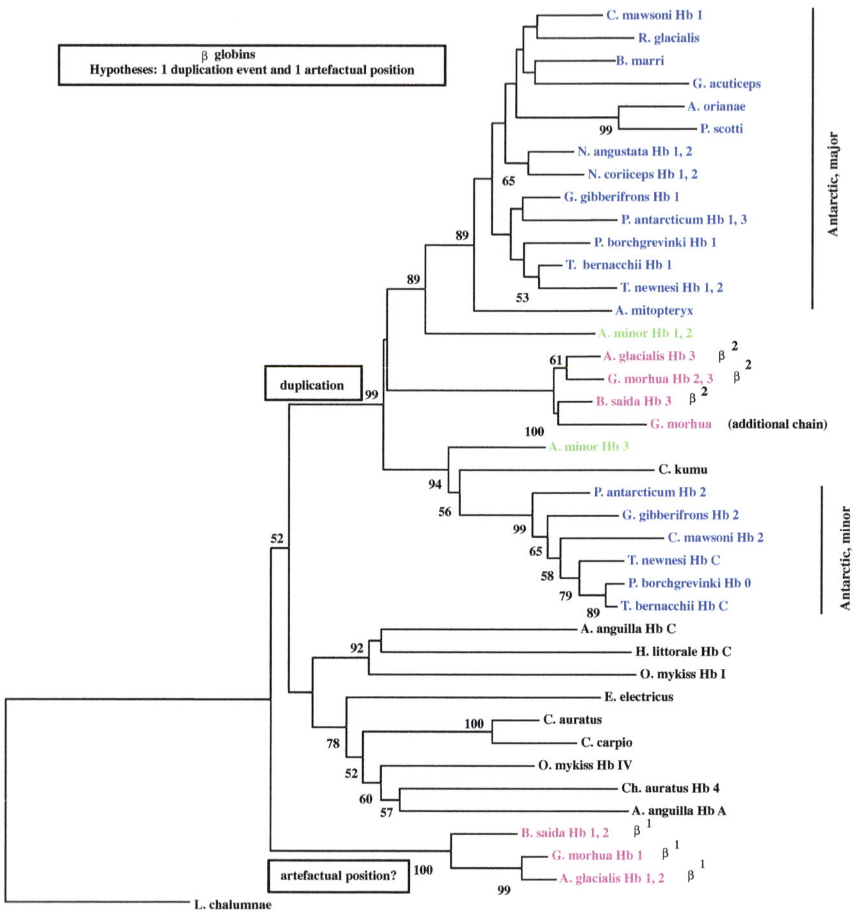

Fig. 11.4 Neighbour Joining tree of sequences of β chains of Arctic, Antarctic and temperate fish Hbs. For details and colours, see Fig. 11.3. Adapted from Dettaï et al. (2008). In both Figures: *C. mawsoni*, *Cygnodraco mawsoni*; *G. acuticeps*, *Gymnodraco acuticeps*; *G gibberifrons*, *Gobionotothen gibberifrons*; *A. mitopterix*, *Aethotaxis mitopterix*; *B. marri*, *Bathydraco marri*; *A. orianae*, *Artedidraco orianae*; *P. scotti*, *Pogonophryne scotti*; *C. kumu*, *Chelidonichthys kumu*; *O. mykiss*, *Oncorhynchus mykiss*; *S. salar*, *Salmo salar*; *A. anguilla*, *Anguilla anguilla*; *H. littorale*, *Hoplosternum littorale*; *Ch. auratus*, *Chrysophrys auratus*; *E. electricus*, *Electrophorus electricus*; *C. clarkii*, *Catostomus clarkii*; *C. carpio*, *Cyprinus carpio*; *C. auratus*, *Carassius auratus*; *L. chalumnae*, *Latimeria chalumnae*

currently found in Antarctic fish diverged 250 mya; hence, unlike AFGPs, whose appearance coincided with cooling, Hb diversification appears less stringently correlated to changes in the environment. Presumably, the two clusters of Antarctic major (adult) and minor (embryonic) Hbs were generated by gene-duplication events which occurred independently for the α- and β-globin genes. In the phylogenetic trees, the basal position of *P. urvillii* and *E. maclovinus* Hbs is congruent with the postulated divergence before the appearance of AFGPs (data not shown).

The globin sequences of the Arctic zoarcoid *A. minor* follow the track of species history, as *A. minor* appears close to the notothenioid clades as predicted by teleost phylogenies. By contrast, Arctic gadid sequences occupy different positions in the two trees with regard to temperate and Antarctic sequences. On one hand, gadid α chains appear related to the notothenioid-zoarcoid group. On the other, the β^1 chains of *A. glacialis*, *B. saida* and *G. morhua* are excluded from the β^2 chains of the same species and from major and minor Antarctic globins with very good bootstrap proportion (99%).

As clearly shown by phylogenetic analysis, under the constant conditions of marine habitats we can recover phylogeny (zoarcoids with notothenioids, gadids as sister-group to both) in Antarctic globins, whereas the variability typical of the Arctic Ocean corresponds to high sequence variation in gadid β globins (Verde et al. 2006a).

The well known position of gadids with regard to zoarcoids and notothenioids allows hypothesising that the basal position of the β^1 sequences of Arctic gadids in the β-globin tree (Fig. 11.3) is probably artefactual, whereas the α-globin tree mostly recovers the species tree plus a few duplications (Fig. 11.4). Such position may be interpreted as an effect of the extreme perturbation of the available mutational space in gadid β^1-globin sequences, possibly due to the variability of thermal conditions experienced by these migratory Arctic fish in comparison with the thermal stability in the life style of zoarcoids and notothenioids, two groups that display unperturbed phylogenetic signal in β sequences (Dettaï et al. 2008).

11.5 Concluding Remarks

Polar organisms are exposed to strong environmental constraints, and it is important to understand how they have adapted in the past to cope with these challenges, and to what extent current climate changes will impact on adaptations in the future. The recognition of the important role of the polar habitats in global climate change has awakened great interest in the evolutionary biology of the organisms that live there, leading to many important SCAR programmes, e.g. EASIZ, EVOLANTA, CAML (www.caml.aq), ICEFISH (www.icefish.neu.org). All of these have merged into EBA and will influence development of future initiatives (see Chap. 1 di Prisco and Convey 2012). EBA, a very large umbrella, was also an IPY core programme, and adaptations of Notothenioidei have been studied by several teams. Comparison with Arctic fish was developed in the framework of the IPY programme TUNU-MAFIG (see Chap. 3 Christiansen 2012).

The Southern Ocean offers a uniquely stable thermal environment where fish cold adaptation did not need the functional plasticity required with high variability (Somero 1995). Antarctic fish have evolved in an environment with the lowest temperature variations. They may have limited abilities to withstand an increase in temperature and be thus vulnerable to warming.

There is already compelling evidence for widespread changes in polar ecosystems due to climate change. The study of fish adapted to extreme polar conditions will allow to look at the impact and consequences of anthropogenic challenges and the role played by temperature in establishing species distribution (di Prisco and Verde 2006). In recent years, the urge to extend these studies to the North Pole has become stronger. A future challenge pertains to analysing the ability of polar fish to develop repair mechanisms to changes, as well as the ways in which responses feed back to influence these processes. In this perspective, the oxygen-transport system is a source of precious indications.

The urgent and challenging agenda for the next decade will be to incorporate thinking along the physiological/biochemical viewpoint into evolutionary biology. Such an integrative approach can provide answers to the question of how Antarctic and Arctic fish will respond, and whether they will be able to adjust, to ongoing Global Warming, already in full action in the polar regions.

Acknowledgments Part of the reviewed work is financially supported by the Italian National Programme for Antarctic Research (PNRA), is in the framework of the SCAR programme Evolution and Biodiversity in the Antarctic (EBA), of the Coordination Action for Research Activities on Life in Extreme Environments (CAREX), and is in the framework of the IPY core programmes EBA and Marine Fishes of North East Greenland (TUNU-MAFIG). DG and RR acknowledge CNR (for Short-Term Mobility fellowships) and CAREX (for Transfer of Knowledge grants).

References

Balushkin AV (1992) Classification, phylogenetic relationships, and origins of the families of the suborder Notothenioidei (Perciformes). J Ichthyol 32:90–110

Balushkin AV (2000) Morphology, classification, and evolution of notothenioid fishes of the Southern Ocean (Notothenioidei, Perciformes). J Ichthyol 40(Suppl 1):S74–S109

Bargelloni L, Marcato S, Zane L, Patarnello T (2000) Mitochondrial phylogeny of notothenioids: a molecular approach to Antarctic fish evolution and biogeography. Syst Biol 49:114–129

Bargelloni L, Ritchie PA, Patarnello T, Battaglia B, Lambert DM, Meyer A (1994) Molecular evolution at subzero temperatures: mitochondrial and nuclear phylogenies of fishes from Antarctica (suborder Notothenioidei), and the evolution of antifreeze glycopeptides. Mol Biol Evol 11:854–863

Berenbrink M, Koldkjær P, Kepp O, Cossins AR (2005) Evolution of oxygen secretion in fishes and the emergence of a complex physiological system. Science 307:1752–1757

Brunori M, Vallone B (2007) Neuroglobin, seven years after. Cell Mol Life Sci 64:1259–1268

Chen L, DeVries AL, Cheng C-HC (1997) Evolution of antifreeze glycoprotein gene from a trypsinogen gene in Antarctic notothenioid fish. Proc Natl Acad Sci USA 94:3811–3816

Chen W-J, Bonillo C, Lecointre G (2003) Repeatability of clades as a criterion of reliability: a case study for molecular phylogeny of Acanthomorpha (Teleostei) with larger number of taxa. Mol Phylogenet Evol 26:262–288

Cheng C-HC, Chen L, Near TJ, Jin Y (2003) Functional antifreeze glycoprotein genes in temperate-water New Zealand nototheniid fish infer an Antarctic evolutionary origin. Mol Biol Evol 20:1897–1908

Cheng C-HC, di Prisco G, Verde C (2009) The "icefish paradox". Which is the task of neuroglobin in Antarctic hemoglobin-less icefish? IUBMB Life 61:184–188

Clarke A, Barnes DKA, Hodgson DA (2005) How isolated is Antarctica? Trends Ecol Evol 20:1–3

Cocca E, Ratnayake-Lecamwasam M, Parker SK, Camardella L, Ciaramella M, di Prisco G, Detrich III HWIII (1995) Genomic remnants of α-globin genes in the hemoglobinless Antarctic icefishes. Proc Natl Acad Sci USA 92:1817–1821

Coppola D, Giordano D, Vergara A, Mazzarella L, di Prisco G, Verde C, Russo R (2010) The hemoglobins of sub-Antarctic fishes of the suborder Notothenioidei. Polar Sci 4:295–308

Christiansen JS (2012) The TUNU-programme: Euro-Arctic marine fishes–diversity and adaptation. In: di Prisco G, Verde C (eds) Adaptation and evolution in marine environments—The impacts of global change on biodiversity, vol 1. Series "From Pole to Pole". Springer, Berlin, pp 35–50

D'Avino R, Caruso C, Tamburrini M, Romano M, Rutigliano B, Polverino de Laureto P, Camardella L, Carratore V, di Prisco G (1994) Molecular characterization of the functionally distinct hemoglobins of the Antarctic fish *Trematomus newnesi*. J Biol Chem 269:9675–9681

Detta A, di Prisco G, Lecointre G, Parisi E, Verde C (2008) Inferring evolution of fish proteins: the globin case study. Meth Enzymol 436:535–566

DeVries AL, Cheng C-HC (2005) Antifreeze proteins and organismal freezing avoidance in polar fishes. In: Farrell AP, Steffensen JF (eds) The Physiology of Polar Fishes, Vol. 22 Fish Physiology, pp 155–201. Elsevier Academic Press, San Diego

di Prisco G (2000) Life style and biochemical adaptation in Antarctic fishes. J Mar Syst 27:253–265

di Prisco G, Convey P (2012) The origin of the SCAR programme "Evolution and Biodiversity in the Antarctic". In: di Prisco G, Verde C (eds) Adaptation and evolution in marine environments—The impacts of global change on biodiversity, vol 1. Series "From Pole to Pole". Springer, Berlin, pp 3–18

di Prisco G, Verde C (2006) Predicting the impacts of climate change on the evolutionary adaptations of polar fish. Rev Environ Sci Biotechnol 5:309–321

di Prisco G, Cocca E, Parker SK, Detrich III HW (2002) Tracking the evolutionary loss of hemoglobin expression by the white-blooded Antarctic icefishes. Gene 295:185–191

di Prisco G, D'Avino R, Caruso C, Tamburrini M, Camardella L, Rutigliano B, Carratore V, Romano M (1991) The biochemistry of oxygen transport in red-blooded Antarctic fish. In: di Prisco G, Maresca B, Tota B (eds) Biology of Antarctic Fish. Springer, Berlin, pp 263–281

di Prisco G, Eastman JT, Giordano D, Parisi E, Verde C (2007) The evolutionary adaptations in Antarctic marine organisms. Gene 398:143–155

di Prisco G, Giardina B, D'Avino R, Condò SG, Bellelli A, Brunori M (1988) Antarctic fish hemoglobin: an outline of the molecular structure and oxygen binding properties–II. Oxygen binding properties. Comp Biochem Physiol 90B:585–591

di Prisco G, Giordano D, Russo D, Verde C (2011) Haemoproteins in cold environments—An evolutionary view. In: Nagai M (ed) Hemoglobin: Recent developments and topics. Research Signpost, Trivandrum, pp 211–229

di Prisco G, Giordano D, Russo R, Verde C (2012) The challenges of low temperature in the evolution of bacteria. In: di Prisco G, Verde C (eds) Adaptation and evolution in marine environments—The impacts of global change on biodiversity, vol 1. Series "From Pole to Pole". Springer, Berlin, pp 183–195

di Prisco G, Macdonald JA, Brunori M (1992) Antarctic fishes survive exposure to carbon monoxide. Experientia 48:473–475

Eastman JT (1988) Ocular morphology in Antarctic notothenioid fishes. J Morphol 196:927–934

Eastman JT (1993) Antarctic fish biology: evolution in a unique environment. Academic, San Diego

Eastman JT (2005) The nature of the diversity of Antarctic fishes. Polar Biol 28:93–107

Eastman JT (2006) Aspects of the morphology of phyletically basal bovichtid fishes of the Antarctic suborder Notothenioidei (Perciformes). Polar Biol 29:754–763

Eastman JT, McCune AR (2000) Fishes on the Antarctic continental shelf: evolution of a marine species flock? J Fish Biol Suppl A 57((Suppl. A)):84–102

Everson I, Ralph R (1968) Blood analyses of some Antarctic fish. Br Antarctic Surv Bull 15:59–62

Fago A, D'Avino R, di Prisco G (1992) The hemoglobins of *Notothenia angustata*, a temperate fish belonging to a family largely endemic to the Antarctic Ocean. Eur J Biochem 210:963–970

Giordano D, Grassi -globin in L, Parisi E, Bargelloni L, di Prisco G, Verde C (2006) Embryonic the non-Antarctic notothenioid fish Cottoperca gobio (Bovichtidae). Polar Biol 30:75–82

Halder P, Trent J, Hargrove M (2007) Influence of the protein matrix on intramolecular histidine ligation in ferric and ferrous hexacoordinate hemoglobins. Proteins Struct Funct Bioinf 66:172–182

Krylov AA, Andreeva IA, Vogt C, Backman J, Krupskaya VV, Grikurov GE, Moran K, Shoji H (2008) A shift in heavy and clay mineral provenance indicates a middle Miocene onset of a perennial sea ice cover in the Arctic Ocean. Paleoceanography 23, PA1S06, doi:10.1029/2007PA001497

Lecointre G (2012) Phylogeny and systematics of Antarctic teleosts: methodological and evolutionary issues In: di Prisco G, Verde C (eds) Adaptation and evolution in marine environments—The impacts of global change on biodiversity, vol 1. Series "From Pole to Pole". Springer, Berlin, pp 97–117

Lecointre G, Bonillo C, Ozouf-Costaz C, Hureau J-C (1997) Molecular evidence for the origins of Antarctic fishes: paraphyly of the Bovichtidae and no indication for the monophyly of the Notothenioidei (Teleostei). Polar Biol 18:193–208

Mazzei F, Ghigliotti L, Lecointre G, Ozouf-Costaz C, Coutanceau JP, Detrich III HW, Pisano E (2006) Karyotypes of basal lineages in notothenioid fishes: the genus Bovichtus. Polar Biol 29:1071–1076

Monod J, Wyman J, Changeux JP (1965) On the nature of allosteric transitions: a plausible model. J Mol Biol 12:88–118

Moran K, Backman J, Brinkhuis H et al (2006) The Cenozoic palaeoenvironment of the Arctic Ocean. Nature 441:601–605

Near TJ (2004) Estimating divergence times of notothenioid fishes using a fossil-calibrated molecular clock. Antarctic Sci 16:37–44

Near TJ, Parker SW, Detrich III HW (2006) A Genomic fossil reveals key steps in hemoglobin loss by the Antarctic icefishes. Mol Biol Evol 23:2008–2016

Near TJ, Pesavento JJ, Cheng C-HC (2004) Phylogenetic investigations of Antarctic notothenioid fishes (Perciformes: Notothenioidei) using complete gene sequences of the mitochondrial encoded 16S rRNA. Mol Phylogenet Evol 32:881–891

Perutz MF (1983) Species adaptation in a protein molecule. Mol Biol Evol 1:1–28

Perutz MF, Fermi G, Luisi B, Shanan B, Liddington RC (1987) Stereochemistry of cooperative mechanisms in hemoglobin. Acc Chem Res 20:309–321

Riccio A, Tamburrini M, Carratore V, di Prisco G (2000) Functionally distinct hemoglobins of the cryopelagic Antarctic teleost *Pagothenia borchgrevinki*. J Fish Biol 57:20–32

Ruud JT (1954) Vertebrates without erythrocytes and blood pigment. Nature 173:848–850

Scher HD, Martin EE (2006) Timing and climatic consequences of the opening of Drake Passage. Science 312:428–430

Somero GN (1995) Proteins and temperature. Annu Rev Physiol 57:43–68

Tamburrini M, D'Avino R, Fago A, Carratore V, Kunzmann A, di Prisco G (1996) The unique hemoglobin system of *Pleuragramma antarcticum*, an Antarctic migratory teleost. Structure and function of the three components. J Biol Chem 271:23780–23785

Tamburrini M, Romano M, Carratore V, Kunzmann A, Coletta M, di Prisco G (1998) The hemoglobins of Antarctic fishes *Artedidraco orianae and Pogonophryne scotti*. Amino acid sequence, lack of cooperativity, and ligand binding properties. J Biol Chem 273:32452–32459

Thomson MRA (2004) Geological and palaeoenvironmental history of the Scotia Sea region as a basis for biological interpretation. Deep-Sea Res Part II 51:1467–1487

Tota B, Amelio D, Garofalo F, Pellegrino D (2012) Evolutionary adaptation and disaptation in the cold: the icefish paradigm. In: di Prisco G, Verde C (eds) Adaptation and evolution in marine environments—The impacts of global change on biodiversity, vol 1. Series "From Pole to Pole". Springer, Berlin, pp 121–141

Verde C, Balestrieri M, de Pascale D, Pagnozzi D, Lecointre G, di Prisco G (2006a) The oxygen-transport system in three species of the boreal fish family Gadidae. Molecular phylogeny of hemoglobin. J Biol Chem 281:22073–22084

Verde C, Carratore V, Riccio A, Tamburrini M, Parisi E, di Prisco G (2002) The functionally distinct hemoglobins of the Arctic spotted wolffish *Anarhichas minor*. J Biol Chem 277:36312–36320

Verde C, Howes BD, De Rosa MC, Raiola L, Smulevich G, Williams R, Giardina B, Parisi E, di Prisco G (2004) Structure and function of the Gondwanan hemoglobin of *Pseudaphritis urvillii*, a primitive notothenioid fish of temperate latitudes. Prot Sci 13:2766–2781

Verde C, Parisi E, di Prisco G (2006b) The evolution of thermal adaptation in polar fish. Gene 385:137–145

Voskoboinikova OS (2004) Ontogenetic bases of the origin and of relationships of the fishes from the suborder Notothenioidei (Perciformes). J Ichthyol 44:418–432

Wells RMG, Ashby MD, Duncan SJ, MacDonald JA (1980) Comparative studies of the erythrocytes and hemoglobins in nototheniid fishes from Antarctica. J Fish Biol 17:517–527

Wells RMG, MacDonald JA, di Prisco G (1990) Thin-blooded antarctic fishes: a rheological comparison of the hemoglobin-free icefishes, *Chionodraco kathleenae* and *Cryodraco antarcticus*, with a red-blooded nototheniid, *Pagothenia bernacchii*. J Fish Biol 36:595–609

Wittenberg JB, Haedrich RL (1974) The choroid rete mirabile of the fish eye. II. Distribution and relation to the pseudobranch and to the swimbladder rete mirabile. Biol Bull 146:137–156

Wittenberg JB, Wittenberg BA (1974) The choroid *rete mirabile*. I. Oxygen secretion and structure: comparison with the swimbladder rete mirabile. Biol Bull 146:116–136

Zhao Y, Ratnayake-Lecamwasam M, Parker SK, Cocca E, Camardella L, di Prisco G, Detrich III HW (1998) The major adult α-globin gene of Antarctic teleosts and its remnants in the hemoglobinless icefishes. Calibration of the mutational clock for nuclear genes. J Biol Chem 273:14745–14752

Conclusions

The contribution of Marine Biology to IPY has been outstanding. The high number of international collaborations and the remarkable boost observed in searching multidisciplinarity clearly appears from Volume 1 of "Adaptation and Evolution in Marine Environments—The Impacts of Global Change on Biodiversity". At first glance, the titles and abstracts of the chapters that will be assembled in the oncoming Volume 2 provide expectation that the same values will similarly characterise the next volume. Thus, "Adaptation and Evolution in Marine Environments—The Impacts of Global Change on Biodiversity" will be an invaluable component of the general scientific outcome of IPY.

The responses of cold-adapted polar organisms provide information to analyse the effect of changes in general and foresee their impact at lower latitudes. In this scenario, EBA's relevance to Global Change underscores the importance of this programme in environmental research during IPY 2007–2009. EBA addresses the impacts of change on Antarctic biodiversity, evolutionary adaptations and community dynamics, with the ambitious aim to seek implications and forecasts concerning the whole planet.

The programme envisages links with the Arctic. The latter is undergoing rapid climate change, with progressive and fast decrease of sea and land ice, only matched—for the time being—by what is happening in the Antarctic Peninsula.

The similarities and, conversely, the differences that the northern and southern polar environments are revealing as current consequences of global warming suggest that it is wise, in the context of future polar research, to as much as possible enhance the promotion of initiatives that envisage bipolar activities. The official polar international institutions, namely SCAR and IASC (International Arctic Science Committee), are well aware of the importance of this need, and have been customarily holding joint meetings since several years. An increasing number of national institutions is following this trend. The research projects reflect this factor, in that they are more and more hosting collaborative bipolar activities. This aspect, highlighted in several chapters in this volume, is also a hallmark of IPY.

G. di Prisco and C. Verde (eds.), *Adaptation and Evolution in Marine Environments, Volume 1*, From Pole to Pole, DOI: 10.1007/978-3-642-27352-0,

The EBA wide umbrella has fulfilled its important role for many years, providing fertile ground for organisation and coordination efforts, but now needs to be replaced by different research structures. The cooperative and cross-disciplinary research of EBA, in particular for evolutionary and biodiversity information, is a long-term legacy and must be retained. The Antarctic Peninsula and the sub-Antarctic islands are very important areas for collaborative investigations. It is necessary to keep in mind that Antarctic research is very expensive: it must be excellent, relevant, multi-national and well planned.

Since 2009, the EBA community has been actively engaged in planning the course that the programme will take in the future of polar science, because EBA is approaching the end of its planned lifespan in 2013. The future developments of EBA will inherit the message and values of the mother programme, and keep feeding inputs into IPY in the future.

Climate change and its effects on biological systems, evolution and biodiversity in a changing environment—two major multi-disciplinary themes—essentially are the core of EBA. These themes will be developed into new SCAR programmes. In the XXXI SCAR Meeting (2010), the proposal of two distinct but complementary Science Research Programmes was outlined and approved by the Delegates, and the programmes are currently being shaped to undergo final approval in the next SCAR Meeting in 2012, as outlined in Chap. 1.

Following approval and implementation, we are confident that the new research initiatives stemming from EBA will continue to successfully help IPY 2007–2009, just as EBA did, to conquer its deserved place in the history of international polar science.

Guido di Prisco
Cinzia Verde

Perspectives and Implications

The description of the SCAR (www.scar.org) international, multidisciplinary biology programme "Evolution and Biodiversity in the Antarctic—The Response of Life to Change" (EBA) appears the most appropriate introduction to the volume. EBA was launched in 2004 and will come to its end in 2013. The EBA and IPY activities were conceived in parallel, and EBA was chosen by the IPY Science Plan as a Lead IPY Project. Assembling almost one hundred teams and covering most of Antarctic biological research in the marine, terrestrial and freshwater realms, EBA is now encouraging its community to elaborate proposals for future, more focussed developments.

EBA will keep contributing to the Antarctic Treaty System, informing non-biological disciplines of the need to contribute to understand the impact of climate change on Antarctic ecosystems, providing links with the northern polar regions and liaising with physics, climatology, earth sciences and history. In its mission of protecting biodiversity, studies in the field and laboratory will predict how organisms and communities will respond to current and future environmental change, taking advantages of new technologies and of the recent explosive development of molecular approaches. As it is currently functioning, EBA will continue being a hallmark of IPY.

The contributions, addressing the two themes of Volume 1 (1. Biodiversity and the Environment; 2. Response to Stress—Adaptations), are both in the framework of EBA. As outlined in the Editorial Introduction, they are all closely interconnected and complementary. In some instances, studies merge into wider programmes, and three cases (CAML, ICEFISH and ANDEEP-SYSTCO) have been considered.

CAML contributed to IPY, and essentially is the part of CoML that deals with the Southern Ocean. Despite the end of this 5-year-project, its sampling sites will be revisited for further time comparisons. The legacy of CAML, funded by the Sloan Foundation, includes inventories, biodiversity databases, extensive use of

G. di Prisco and C. Verde (eds.), *Adaptation and Evolution in Marine Environments, Volume 1*, From Pole to Pole, DOI: 10.1007/978-3-642-27352-0,

genomic techniques (Barcode of Life), comparisons between the Arctic Ocean and the SO.

The series of ICEFISH cruises is aimed at specifically exploring the sub-Antarctic ocean and studying sub-Antarctic fishes, in particular those of the dominant suborder Notothenioidei. Sub-Antarctic notothenioids are critical for understanding the evolution, population dynamics, ecophysiology and biochemistry of their Antarctic relatives. The ICEFISH programme was designed to fill critical gaps in our knowledge, and ICEFISH-2004 (www.icefish.neu.edu) was the first international survey of the sub-Antarctic marine habitat of the South-Atlantic sector, focussing on comparative systematics and evolutionary studies, life-history strategies and population dynamics, physiology, biochemistry and molecular biology of adaptations, genomics of sub-Antarctic notothenioids, to monitor the impact of climate change on loss of biodiversity and community dynamics in the SO. Knowledge of the sub-Antarctic will allow to evaluate future changes in species distribution and survival. Hopefully, in the near future it will be possible to overcome the slowing down of the programmed legs of ICEFISH due to recent funding difficulties, and scientists from over 10 countries will thus be enabled to continue their meritorious activity. In addition to research in the field and home laboratories, ICEFISH will continue depositing specimens in museum collections around the world, distributing genomic resources to polar marine biologists, training PhD students, generating media coverage, publications, conference proceedings, input to databases and CCAMLR, interacting with other SCAR programmes, investigating evolution and adaptation of fishes from the Antarctic to the Arctic via the description of a geographical system which is climatically intermediate between the polar extremes.

Following EBA's intention, ANDEEP-SYSTCO does not only involve biologists, but also scientists from many other disciplines to shed light on atmospheric-pelagic-benthic coupling processes. It uses innovative technology, and is training new polar scientists. It addresses measurements of atmospheric parameters, plankton and nanoplankton, biogeochemistry of surface water, primary productivity, morphology and physiology of abyssal species, recruitment of benthic animals, food sinking, sedimentology, biogeochemistry, palaeontology, bathymetric mapping. Deep-sea isopod biodiversity is investigated, as well as species richness, distribution, endemism and faunal characteristics, and coupling processes between the pelagic and benthic realms. In the framework of EBA-IPY, ANDEEP-SYSTCO is conducting the first survey of megafaunal, macrofaunal and meiofaunal deep-water communities in the Atlantic sector of the SO. A new cruise is scheduled in 2012. On this occasion, it will be continued to address biodiversity, ecology of dominant abyssal species, functioning of abyssal communities, thereby trying to understand atmospheric-pelagic-benthic coupling and to gain insights into the trophic structure of the deep SO. Based on current climate change, potential future scenarios are hypothesised to establish a benchmark against which future faunal changes can be measured.

The IPY TUNU Programme (currently TEAM-Fish, TUNU-Programme: Euro-Arctic Marine Fishes—Diversity and Adaptation) is an ongoing international initiative addressing Arctic fishes across the Euro-Arctic region. Although Arctic, it is in the framework of EBA. The successful research of this programme will continue, as future cruises onboard the R/V Jan Mayen are already scheduled. TEAM-Fish includes PhD students and scientists from the EU, USA, and Russia. The speed of warming trends in the Arctic makes studies of the fish response an unprecedented challenge. Genetic and demographic structuring, trophic relationships and physiological adaptations are viewed in the context of climate and human stressors. A growing collection in the Bergen Museum is serving as the basis for taxonomy and phylogeny. Studies will continue addressing zoogeographical mapping and quantification of fish at sites along the NE Greenland coast, selected during the previous cruises, gathering hydrographical data at the sites, repeating investigations to monitor inter-annual changes. TEAM-Fish is inter- and multi-disciplinary, linking fish distribution and diversity to molecular genetics, cytogenetics, physiology, metabolism, and hydrography. Many Arctic fish species are adapted to live within a narrow thermal range; slight increases in temperature and reductions in salinity are deemed to have profound effects on their diversity and distribution. The reduction of sea-ice cover during the last three decades makes NE Greenland and its pristine fjords an excellent site to study effects of a changing environment on the marine fauna. Other fish species are migratory, and cover large distances, sometimes spawning in ice-free areas such as the coasts of Norway. The variations along such latitudinal gradient, and the consequent range of evolved adaptations, are an additional research area challenging the consequences of climate change, especially in comparison with adaptations evolved by the much more isolated Antarctic ecosystem.

Sea ice has influenced the life of marine plants and animals to such an extent as to cause the evolution of antifreeze biological responses. In such a manner, polar fish escape freezing. *Pleuragramma antarcticum*, a fish playing a pivotal role in the coastal system, uses seasonal sea ice as nursery, and this stimulates research on its life cycle and relationships with sea ice. Research within IPY highlights that such relationship is a major feature in its early life history and reproduction, calling for future work on predictions about impacts that changes in the sea-ice dynamics may have on the life style of this species and, consequently, on the whole coastal Antarctic ecosystem. These investigations are very urgent, considering the rapid loss in sea-ice cover already taking place in the Antarctic Peninsula, although not yet around the rest of the continent.

Gene flow, genetic drift, selection, and other factors, affect the evolution of biodiversity. The Antarctic Circumpolar Current ought to homogenise the structure of the fish populations of the SO, and strong connectivity is indeed found for some species, with genotypes being shared across the full range. However, species-specific life-history traits influence the patterns of most taxa such that distinct populations are identified. Fishing and climate change impact the genetic structure in some cases. Management measures are urgently needed, including implementation of quotas and identification of areas that need protection.

Molecular phylogeny is an area which has greatly advanced in recent years. Thanks to appropriate changes in systematics, to date molecular trees are characterised by better taxonomy. Access to the Antarctic for extensive population samplings has occurred during IPY. There are increasing interactions between traditional morphology-based taxonomic skills and DNA sequence-based approaches through "integrative taxonomy". Gene amplification, sequencing and computing power is driving the uprise of phylogenetics, for instance providing criteria from independent data to know whether a monophyletic group of species is reliable. Although not the only criterion, rapidly increasing barcoding is modifying the flow chart of taxonomy. In fish, identifications of eggs and larvae are increasingly reliable. Notothenioids are the most studied Antarctic fish group, and their phylogeny is now clear at the genus and interspecies levels, but needs further work, in particular for Nototheniidae, the largest (endowed with the most complex phylogeny) of the eight families of the suborder. Other important evolutionary issues (species flocks, dating of notothenioids, their origin) will also be further investigated.

The SO and the Antarctic continent pose extreme survival challenges for organisms, from bacteria to vertebrates. Considering the strong environmental constraints that polar organisms had to face to successfully cope with progressive cooling during dozens of millions of years, evolutionary adaptation has been, and will continue to be, a major theme of research in IPY. Climate change is calling for more and more work on the consequences that even a slight modification in the current climatic parameters may entail on cold-adapted organisms, whose physiology has succeeded in finely adapting and escaping extinction. In other words, it will be essential to increase our understanding of how polar organisms have adapted to cope with past challenges, to what extent adaptations may be upset by current climate changes, and—most important—whether it will be possible to minimise future threats of extinction, also at our latitudes.

The final five chapters of Volume 1 tackle evolutionary cold adaptation, describing research carried out on aspects of ecology, physiology, biochemistry, molecular biology, phylogeny. The variety of organisms includes vertebrates and invertebrates, namely fish, molluscs, bacteria, etc. Will they have the genetic opportunity to adapt, and/or the physiological plasticity to tolerate newly developing climate conditions?

The dominant fish suborder Notothenioidei offers invaluable opportunities for investigating cold adaptation. The evolutionary adaptations of the cardiovascular system have been vital for their survival and are beginning to be understood at the molecular level. Special attention is given to endemic Channichthyidae (icefishes), a notothenioid family that permanently lives at or near freezing. The icefish is a paradigm of "disaptation" among adult vertebrates, because of loss of hemoglobin and, in some species, myoglobin. Icefishes are natural "knockouts", permitting to analyse epigenetic compensatory mechanisms and multilevel integration in the original phenotype. In the future, some of the authors plan to continue investigating physiological cardiocirculatory compensations to face hypoxia induced by the loss of respiratory pigments, as well as the phenotypic plasticity/ vulnerability of species of this fish family. This hemoglobin-less organism delivers oxygen to tissues by carrying it physically dissolved in the blood, and not via a classical, specific oxygen carrier. In order to survive, it needs to take advantage of the very high levels of oxygen in sea water, due to freezing temperatures, but may have to face lower oxygen concentrations as a consequence of increased temperatures.

The function of hemoproteins has taught important lessons. Cold adaptation has produced unique specialisations in notothenioids, including decreased amounts and multiplicity of hemoglobins, reaching their absence in Channichthyidae. Some authors have elucidated the genomic changes leading to the hemoglobin- and myoglobin-less phenotype, and plan to investigate the expression, molecular structure and physiological role of other hemoproteins they have recently discovered in icefish, natural knockouts for hemoglobin and myoglobin. These authors plan to expand research on molecular phylogeny of Antarctic and Arctic globins, establishing correlations between the positions in the evolutionary trees and life styles under different thermal conditions experienced in the two environments, i.e. variability/stability in the Arctic and stability in the Antarctic.

One key adaptation is the fish ability to avoid freezing in sea water, constantly below the freezing point of the body fluids, by means of antifreeze compounds. A large enough increase in temperature may have consequences in the extent of biosynthesis of antifreeze compounds, which is highly seasonal in the Arctic and instead occurs year-round in the Antarctic. Other physiological aspects involve microtubule formation, membrane fluidity, heat-shock response. There is experimental evidence that a mere increase in temperature is in itself survivable, but will be accompanied by changes in ecological niches, that may alter the habitat and affect interactions with other species. The consequences of these climatic and ecological changes will be the object of future studies.

Whilst terrestrial species are adapted to very variable conditions, marine species face more severe problems in response to climate change, because the large thermal capacity of water means that the most stable thermal environments are aquatic. As a consequence, although warming will occur in the atmosphere to a much larger extent than in the ocean, the impacts of even a small increase in temperature in a marine environment are likely to affect organisms and their biodiversity much more severely. Therefore, studies on marine organisms provide the most important insights into physiological and molecular adaptation to temperature. In contrast, investigations in terrestrial organisms help to understand how organisms respond to short-term thermal challenges and how these affect their ecology.

Taking Antarctic invertebrates into account, rapid advances in genomic technologies are providing the opportunity to advance the knowledge of Antarctic biology (e.g. the clam *Laternula elliptica*, the krill *Euphausia superba*, and many other organisms). The "omics" technology is an increasingly essential tool, and molecular biology is gradually becoming a standard must in understanding basic adaptations and adaptive responses to changes in sea temperature and acidity and dynamics of gene flow). Also in studies on invertebrates, these integrated approaches will support efforts to identify novel genes and proteins, essential for life in this extreme environment. Integration with ecophysiology will also help in investigating thermal windows for survival and aspects of life style.

Hemoproteins have a pivotal role also in the adaptive evolution of polar microorganisms, not only of fish. Again, the dramatic advances in DNA sequencing have a key role. In the past, research on psychrophiles and psychrotolerant microorganisms paid very little attention to their functional role in some important processes. Thanks to "omics", new biochemical pathways, evolutionary adaptation and tolerance/resistance to extreme conditions can now be investigated, thus gaining insight on how low temperatures may affect the physiology of microorganisms, thereby shedding light on microbial adaptations to cold. Oxygen-binding proteins in bacteria have many biological functions. Further studies are planned to ascertain their role in radical scavenging and cellular metabolism, together with the role of oxygen and oxidative/nitrosative stress in regulating adaptive responses at cellular and molecular levels.

The comparison between phylogenetically related bacteria, living in freezing and non-freezing habitats, may help to understand whether extreme environments require adaptations at species level or the action of few selected genes sufficient for defining the preference for a given environment. Most of the studies of protein/gene adaptation to temperature may take advantages of one of two classical approaches. The structural/mutational approach often produces a detailed portrait (although often controversial) of thermal adaptation. However this approach, due to much labour and high costs, restricts analyses to a limited number of genes and proteins, leading to a potentially biased view of thermal adaptation. The second approach is comparative; it is less expensive and yet more comprehensive, and uses sequence and three-dimensional-structure comparisons of homologous high- and low-temperature-adapted proteins. In microorganisms, genome sequencing is easy and quick; in addition, fast generation times and cultivability allow the comparative analysis of genetic adaptations to extreme environments between generations. In higher organisms, e.g. fish, such an approach is still hampered by limitations, due for instance to the lack of polar genomic sequences.

Polar regions experience greater rates of climate change than elsewhere. Ecosystems are adapted to extreme environments, and may become vulnerable to climate changes. Efforts will continue to increase our knowledge of the Antarctic marine fauna of the continental shelf, the slopes and the deep sea, including tiny organisms. The urgent and challenging agenda for the next decade will be to incorporate thinking along the physiological/biochemical viewpoint into evolutionary biology.

This message appears to have been received and accepted by the authors of this volume, in which the scientific relevance of the research performed and of the implication for future research and relevance for regulatory purposes, as well as global processes, appears very clearly.

Guido di Prisco
Cinzia Verde

MIX
Papier aus verantwortungsvollen Quellen
Paper from responsible sources
FSC® C105338

If you have any concerns about our products,
you can contact us on
ProductSafety@springernature.com

In case Publisher is established outside the EU,
the EU authorized representative is:
Springer Nature Customer Service Center GmbH
Europaplatz 3, 69115 Heidelberg, Germany

Printed by Libri Plureos GmbH
in Hamburg, Germany